AF389095

Bioactive Formulations in Agri-Food-Pharma: Source and Applications, Volume II

Bioactive Formulations in Agri-Food-Pharma: Source and Applications, Volume II

Editors

Minaxi Sharma
Kandi Sridhar
Zeba Usmani

MDPI • Basel • Beijing • Wuhan • Barcelona • Belgrade • Manchester • Tokyo • Cluj • Tianjin

Editors

Minaxi Sharma
Département AgroBioscience
et Chimie
Haute Ecole Provinciale
de Hainaut-Condorcet
Ath
Belgium

Kandi Sridhar
Department of Food
Technology
Karpagam Academy
of Higher Education
Coimbatore
India

Zeba Usmani
Department of Applied
Biology
University of Science
and Technology
Meghalaya
India

Editorial Office
MDPI
St. Alban-Anlage 66
4052 Basel, Switzerland

This is a reprint of articles from the Special Issue published online in the open access journal *Bioengineering* (ISSN 2306-5354) (available at: www.mdpi.com/journal/bioengineering/special_issues/bioactive_formulations_food_pharma).

For citation purposes, cite each article independently as indicated on the article page online and as indicated below:

LastName, A.A.; LastName, B.B.; LastName, C.C. Article Title. *Journal Name* **Year**, *Volume Number*, Page Range.

ISBN 978-3-0365-6721-1 (Hbk)
ISBN 978-3-0365-6720-4 (PDF)

Contents

Preface to "Bioactive Formulations in Agri-Food-Pharma: Source and Applications, Volume II"

Over the years, bioactive compounds have received a significant amount of interest in agri-food, pharma, and chemical industries due to their biological and functional activities, such as antioxidant, anti-inflammatory, antimutagenic, antimicrobial, biosurfactant, biostimulants, cosmeceuticals, etc. Generally, reactive oxygen and nitrogen species and other oxygen free radicals are generated in the human body by various physiological and biochemical processes. These free radicals adversely alter lipids, proteins, and DNA, thereby resulting in damage to the cell and leading to a number of human diseases. By using the external source of antioxidants, the generation of free radicals can be neutralized. In this context, there has been a global trend toward the use of natural bioactive compounds present in plant sources as antioxidants to scavenge free radical or active oxygen molecules inside the human body. This, in turn, raised the use of natural antioxidants as an alternative to synthetic antioxidants in agri-food, pharma, and chemical industries in respect of low-cost and no side effects.

Development of new and novel bioactive chemicals and their formulations from natural sources are essential for the advancement of the agri-food, pharmaceutical, and health industries. Hence, advances in phytochemistry led to the identification of different novel bioactive compounds from microbes, as well as plant sources, which were further engineered for wider applications in food, cosmetic, and therapeutic industries. However, extraction technologies in the recovery of the bioactive compounds is a key factor and, thus, efforts should be directed to enhance the biological activity and advancement in the scale-up process of bioactive compounds. The improvement in extraction, identification, and enhancement of biological activity of bioactive compounds makes the process accessible and successfully implementable across various industries, particularly the agri-food pharma sector. In this regard, to improve scientific knowledge and understanding of valuable natural products from native biological sources, we organized the below Special Issue.

It is our immense pleasure to present this Special Issue entitled, "Bioactive Formulations in Agri-Food-Pharma: Source and Applications, Volume II". The Special Issue collection contains a total of 13 articles with most relevant and state-of-the-art high-quality research including, but not limited to, the following:

- Recent developments in the production of natural compounds from various bioresources, and their formulations as bioactives;
- Engineered approaches for enhanced biological properties;
- Potential industrial applications of new bioactive formulations through novel strategies;
- Identification of novel compounds and bioprocess tools to improve the quantity and quality of such products;
- Increasing knowledge on the bioactive formulations on regulation at the genomic and molecular levels.

Minaxi Sharma, Kandi Sridhar, and Zeba Usmani

Editors

 *bioengineering*

Editorial

Bioactive Formulations in Agri-Food-Pharma: Source and Applications

Kandi Sridhar [1],* , Zeba Usmani [2] and Minaxi Sharma [3],*

[1] Department of Food Technology, Karpagam Academy of Higher Education (Deemed to be University), Coimbatore 641021, India
[2] Department of Applied Biology, University of Science and Technology, Meghalaya 793101, India
[3] Haute Ecole Provinciale de Hainaut-Condorcet, 7800 Ath, Belgium
* Correspondence: sridhar4647@gmail.com (K.S.); minaxi86sharma@gmail.com (M.S.)

Citation: Sridhar, K.; Usmani, Z.; Sharma, M. Bioactive Formulations in Agri-Food-Pharma: Source and Applications. *Bioengineering* **2023**, *10*, 191. https://doi.org/10.3390/bioengineering10020191

Received: 28 January 2023
Accepted: 1 February 2023
Published: 2 February 2023

Bioactive compounds are the secondary metabolites produced by the plant cell through numerous metabolic pathways. Many epidemiological studies suggest that a diet rich in bioactive compounds can reduce the risk of degenerative diseases, such as diabetes, cancer, obesity, and cardiovascular complications by scavenging the free radicals generated in the body. Thus, there has been an increasing trend in the use of the natural bioactive compounds present in plant sources as antioxidants to scavenge free radicals. Recently, many agri-food pharma industries are exclusively searching for natural bioactive compounds and their formulations and approaches for the treatment of many degenerative diseases due to their low-cost production with no side effects. Hence, recent innovations in phytochemistry led to the extraction and identification of different novel bioactive compounds from a wide variety of plant sources, thereby application in food, cosmetic, and therapeutic industries. Nevertheless, appropriate extraction and identification strategies must be needed in order to recover the bioactive compounds from natural sources. These improvements in bioactive compound formulations make the process accessible and successfully implementable across various industries, particularly the agri-food pharma sector. Therefore, a group of researchers (Dr. Minaxi Sharma, Dr. Kandi Sridhar, and Dr. Zeba Usmani) with expertise in the area of bioactive compounds, plant-based foods, agri-food utilization, and plant-based drug development, organized the Special Issue on "Bioactive Formulations in Agri-Food-Pharma: Source and Applications, Volume II" to be published in *Bioengineering* (ISSN 2306-5354) under the section "Biochemical Engineering". The Special Issue explored the most relevant and state-of-the-art high-quality research based on the following thematic areas:

- Recent developments in the production of natural compounds from various bioresources and their formulations as bioactives;
- Engineered approaches for enhanced biological properties;
- Potential industrial applications of new bioactive formulations through novel strategies;
- Identification of novel compounds and bioprocess tools to improve the quantity and quality of such products;
- Increasing knowledge on the bioactive formulations on regulation at the genomic and molecular levels.

This Special Issue provided a multidisciplinary view for the identification, extraction, and engineered application of bioactive compounds in agri-food pharma industries. Under this Special Issue topic, we invited renowned global researchers and received high-quality submissions. A total of 29 submissions were received for vigorous peer-review consideration, in which a total of 13 articles were accepted for final publication in this Special Issue. The total number of submissions indicated the importance of the research topic among researchers, policy makers, and other non-academic communities.

The Special Issue published a total of 13 scientific contributions comprising 9 research and 4 review articles which explored the bioactive compound sources and applications in the agri-food-pharma sector. These published articles are freely accessible from the moment of publication and part of a wider open movement (Copyright © 2022 or Copyright © 2023 by authors) under the terms and conditions of Creative Commons Attribution (CC BY) license to encourage the free exchange of knowledge and attract greater public engagement. The take-home message of the exciting articles published in this Special Issue are introduced below.

A study reported by Chuetor et al. [1] investigated the recyclability of ionic liquid, 1-ethyl-3-methylimidazolium acetate (EMIM-Ac) by optimization using rice straw. Under the optimized conditions, the authors studied the suitability of anti-solvent among water, acetone, methanol, and combinations of them. The study revealed that methanol was the best anti-solvent with the highest sugar yield. The combination of methanol and EMIM-Ac improved the recyclability of EMIM-Ac by up to five recycles. The recycled EMIM-Ac was used for the production of ethanol (89%), thereby confirming the potential of recycled ionic liquid in ethanol production at a low cost.

A study by Jo et al. [2] investigated various parts of *Prunus mume* for vasorelaxant effects, which could be related to the NO/sGC/cGMP vascular prostacyclin pathway. Similarly, another study by a Slovenian research group attempted to develop a plant-based drug against highly contagious severe acute respiratory syndrome coronavirus 2 (SARS-CoV-2) [3]. For this, authors used 70% aqueous ethanolic extract of the rhizome bark of Japanese knotweed and tested it on Vero-E6 cells (i.e., mimics the mechanism of the real virus–host interaction). A dosage of 50.8 µg/mL of plant extract significantly displayed the antiviral effect against SARS-CoV-2. This study concluded the use of Japanese knotweed rhizome bark extract in the formulation and development of food supplements for the complementary treatment of COVID-19 patients.

The results of an international collaboration among researchers from Egypt and Saudi Arabia provided insight regarding the production of cytokine (i.e., human interferon-α), of which regulates the immune system's response to viral and bacterial infections from transgenic *Raphanus sativus* L. plants [4], indicating the *R. sativus* as a source for a biologically active compound that acts as an anticancer and antiviral drug. Similarly, another study by Wang et al. [5] investigated the neuroprotective effect of the Abelmoschus manihot flower extract obtained using ethanol, methanol, and supercritical fluid extraction against the cytotoxicity, oxidative stress, and inflammation in PC12 cells. A concentration of ethanolic and water extracts showed toxic effects on PC12 cells, while the supercritical extract caused 10% cell death with a more protective effect on the apoptosis of PC12 cells. The supercritical extracts up-regulated the expression of antioxidant enzymes, promoted the production of an intracellular antioxidant with reduced glutathione and reduced the reactive oxygen species generation in PC12 cells; this indicates the efficacy of supercritical extracts to repair the damage caused by oxidative stress and further delays the oxidative stress-induced neurodegenerative diseases. Likewise, Guo et al. [6] concluded the regulatory role of AdpA_1075 as a global regulator of morphological differentiation and secondary metabolism, promoting the biosynthesis of ansamitocin in *Actinosynnema pretiosum*.

Another research group from Portugal revealed the storage stability and in vitro bioaccessibility of the encapsulation of tomato extract and developed the yogurt enriched with the encapsulated tomato extract [7]. In vitro release studies showed a 63 and 13% release of encapsulated lycopene from the arabic gum and inulin particles in simulated gastric fluid, respectively. The microencapsulated lycopene added to yogurt showed the increased bioaccessibility during simulated gastrointestinal digestion compared to the microencapsulated lycopene alone, indicating the potential of encapsulation that can be used in the development of functional foods. By using the nanoencapsulation approach, Chen et al. [8] developed facile and novel lipid-based liposomes loaded with vitamins C and D3. This study indicated the high encapsulation efficiency for encapsulated vitamins with enhanced vitamin bioavailability. Another study by Alsubhi et al. [9] developed the polyphenol-enriched strawberry yogurt smoothie. In this study, authors extracted the polyphenols from pomegranate pomace and then added them to the strawberry yogurt smoothie. The addition of polyphenols enhanced the quality attributes and antioxidant activity of the strawberry yogurt smoothie, showing the potential of pomegranate pomace in the development of functional beverages.

A review by Das et al. [10] provided the overview on the cannabis (*Cannabis sativa* L.) bioactive compounds (i.e., cannabinoids), biosynthesis, post-harvest operations on the cannabinoid profile, drying treatments, and the effect of the different post-harvest operations on the cannabinoid yield. This review suggested the optimization of drying conditions, pre-treatment operations, and curing conditions in order to improve the extraction and identification of cannabinoids from cannabis. Since the bioactive compounds are sensitive to external and/or internal environments, many studies recommended the use of micro/nano encapsulation strategies, which were exemplified in a review by Put-tasiddaiah et al. [11]. This review concluded the need of nanofabrication technologies for the encapsulation of bioactive compounds, thereby controlling the release of bioactive compounds and retaining the shelf-life of bioactive compounds to develop functional foods. Similarly, Rachitha et al. [12] reviewed the antihyperlipidemic activity of *Emblica officinalis* synergized with nanotechnological approaches, indicating the need of nanotechnological approaches in enhancing the antihyperlipidemic activity of *E. officinalis*. Another review by Bala et al. [13] provided novel approaches for the valorization of agro-waste into value-added bioproducts and bioactive compounds and explored the potential applications in the agri-food-pharma sector.

In conclusion, the advanced extraction, identification, and application of bioactive compounds reported in aforementioned studies can further advance the field and generate scientific knowledge. The novel strategies in the application of bioactive compounds are of interest to all stakeholders, encouraging them to look for new strategies to obtain bioactive compounds from natural sources and their application in agri-food, pharma, and chemical industries. More importantly, many of the studies featured in this article collection revealed the possibility in future research to identify novel natural bioactive sources from plants, which will be documented in future Special Issues.

Author Contributions: Conceptualization, K.S. and M.S.; methodology, K.S. and M.S.; software, Z.U.; validation, K.S. and M.S.; formal analysis, K.S. and M.S.; investigation, K.S. and M.S.; resources, M.S.; data curation, K.S. and M.S.; writing—original draft preparation, K.S. and M.S.; writing—review and editing, K.S., Z.U. and M.S.; visualization, K.S. and M.S.; supervision, M.S.; project administration, K.S. and M.S.; funding acquisition, M.S. All authors have read and agreed to the published version of the manuscript.

Funding: This research received no external funding.

Acknowledgments: We are deeply indebted to Vijai Kumar Gupta, who always strongly supported, nurtured, and believed in our scientific abilities to be successful in the academic arena.

Conflicts of Interest: The authors declare no conflict of interest.

References

1. Chuetor, S.; Panakkal, E.J.; Ruensodsai, T.; Cheenkachorn, K.; Kirdponpattara, S.; Cheng, Y.-S.; Sriariyanun, M. Improvement of enzymatic saccharification and ethanol production from rice straw using recycled ionic liquid: The effect of anti-solvent mixture. *Bioengineering* **2022**, *9*, 115. [CrossRef] [PubMed]
2. Jo, C.; Kim, B.; Lee, K.; Choi, H.-Y. Vascular relaxation and blood pressure lowering effects of *Prunus mume* in rats. *Bioengineering* **2023**, *10*, 74. [CrossRef] [PubMed]
3. Jug, U.; Naumoska, K.; Malovrh, T. Japanese knotweed rhizome bark extract inhibits live SARS-CoV-2 in vitro. *Bioengineering* **2022**, *9*, 429. [CrossRef] [PubMed]
4. Kebeish, R.; Hamdy, E.; Al-Zoubi, O.; Habeeb, T.; Osailan, R.; El-Ayouty, Y. A biotechnological approach for the production of pharmaceutically active human interferon-α from *Raphanus sativus* L. plants. *Bioengineering* **2022**, *9*, 381. [CrossRef] [PubMed]
5. Wang, S.-W.; Chang, C.-C.; Hsuan, C.-F.; Chang, T.-H.; Chen, Y.-L.; Wang, Y.-Y.; Yu, T.-H.; Wu, C.-C.; Houng, J.-Y. Neuroprotective effect of *Abelmoschus manihot* flower extracts against the H_2O_2-induced cytotoxicity, oxidative stress and inflammation in PC12 cells. *Bioengineering* **2022**, *9*, 596. [CrossRef] [PubMed]

6. Guo, S.; Leng, T.; Sun, X.; Zheng, J.; Li, R.; Chen, J.; Hu, F.; Liu, F.; Hua, Q. Global regulator AdpA_1075 regulates morphological differentiation and ansamitocin production in *Actinosynnema pretiosum* subsp. auranticum. *Bioengineering* **2022**, *9*, 719. [CrossRef] [PubMed]

7. Corrêa-Filho, L.C.; Santos, D.I.; Brito, L.; Moldão-Martins, M.; Alves, V.D. Storage stability and in vitro bioaccessibility of microencapsulated tomato (*Solanum lycopersicum* L.) pomace extract. *Bioengineering* **2022**, *9*, 311. [CrossRef] [PubMed]

8. Chen, J.; Dehabadi, L.; Ma, Y.-C.; Wilson, L.D. Development of novel lipid-based formulations for water-soluble vitamin C *versus* fat-soluble vitamin D3. *Bioengineering* **2022**, *9*, 819. [CrossRef] [PubMed]

9. Alsubhi, N.H.; Al-Quwaie, D.A.; Alrefaei, G.I.; Alharbi, M.; Binothman, N.; Aljadani, M.; Qahl, S.H.; Jaber, F.A.; Huwaikem, M.; Sheikh, H.M.; et al. Pomegranate pomace extract with antioxidant, anticancer, antimicrobial, and antiviral activity enhances the quality of strawberry-yogurt smoothie. *Bioengineering* **2022**, *9*, 735. [CrossRef] [PubMed]

10. Das, P.C.; Vista, A.R.; Tabil, L.G.; Baik, O.-D. Postharvest operations of cannabis and their effect on cannabinoid content: A review. *Bioengineering* **2022**, *9*, 364. [CrossRef] [PubMed]

11. Puttasiddaiah, R.; Lakshminarayana, R.; Somashekar, N.L.; Gupta, V.K.; Inbaraj, B.S.; Usmani, Z.; Raghavendra, V.B.; Sridhar, K.; Sharma, M. Advances in nanofabrication technology for nutraceuticals: New insights and future trends. *Bioengineering* **2022**, *9*, 478. [CrossRef] [PubMed]

12. Rachitha, P.; Krishnaswamy, K.; Lazar, R.A.; Gupta, V.K.; Inbaraj, B.S.; Raghavendra, V.B.; Sharma, M.; Sridhar, K. Attenuation of hyperlipidemia by medicinal formulations of *Emblica officinalis* synergized with nanotechnological approaches. *Bioengineering* **2023**, *10*, 64. [CrossRef] [PubMed]

13. Bala, S.; Garg, D.; Sridhar, K.; Inbaraj, B.S.; Singh, R.; Kamma, S.; Tripathi, M.; Sharma, M. Transformation of agro-waste into value-added bioproducts and bioactive compounds: Micro/nano formulations and application in the agri-food-pharma sector. *Bioengineering* **2023**, *10*, 152. [CrossRef]

bioengineering

Article

Improvement of Enzymatic Saccharification and Ethanol Production from Rice Straw Using Recycled Ionic Liquid: The Effect of Anti-Solvent Mixture

Santi Chuetor [1], **Elizabeth Jayex Panakkal** [2], **Thanagorn Ruensodsai** [2], **Kraipat Cheenkachorn** [1], **Suchata Kirdponpattara** [1], **Yu-Shen Cheng** [3] and **Malinee Sriariyanun** [2,*]

[1] Department of Chemical Engineering, Faculty of Engineering, King Mongkut's University of Technology North Bangkok, Bangkok 10800, Thailand; santi.c@eng.kmutnb.ac.th (S.C.); kraipat.c@eng.kmutnb.ac.th (K.C.); suchata.k@eng.kmutnb.ac.th (S.K.)

[2] Department of Chemical and Process Engineering, The Sirindhorn International Thai-German Graduate School of Engineering, King Mongkut's University of Technology North Bangkok, Bangkok 10800,Thailand; elizabeth.jayex@gmail.com (E.J.P.); oventhanagorn@gmail.com (T.R.)

[3] Department of Chemical and Materials Engineering, National Yunlin University of Science and Technology, Yunlin 640, Taiwan; yscheng@gemail.yuntech.edu.tw

* Correspondence: macintous@gmail.com

Abstract: One of the major concerns for utilizing ionic liquid on an industrial scale is the cost involved in the production. Despite its proven pretreatment efficiency, expenses involved in its usage hinder its utilization. A better way to tackle this limitation could be overcome by studying the recyclability of ionic liquid. The current study has applied the Box–Behnken design (BBD) to optimize the pretreatment condition of rice straw through the usage of 1-ethyl-3-methylimidazolium acetate (EMIM-Ac) as an ionic liquid. The model predicted the operation condition with 5% solid loading at 128.4 °C for 71.83 min as an optimum pretreatment condition. Under the optimized pretreatment condition, the necessity of the best anti-solvent was evaluated among water, acetone methanol, and their combinations. The study revealed that pure methanol is the suitable choice of anti-solvent, enhancing the highest sugar yield. Recyclability of EMIM-Ac coupled with anti-solvent was conducted up to five recycles following the predicted pretreatment condition. Fermentation studies evaluated the efficacy of recycled EMIM-Ac for ethanol production with 89% more ethanol production than the untreated rice straw even after five recycles. This study demonstrates the potential of recycled ionic liquid in ethanol production, thereby reducing the production cost at the industrial level.

Keywords: anti-solvent; enzymatic saccharification; EMIM-Ac; ethanol; ionic liquid; lignocellulosic biomass; recycling

Citation: Chuetor, S.; Panakkal, E.J.; Ruensodsai, T.; Cheenkachorn, K.; Kirdponpattara, S.; Cheng, Y.-S.; Sriariyanun, M. Improvement of Enzymatic Saccharification and Ethanol Production from Rice Straw Using Recycled Ionic Liquid: The Effect of Anti-Solvent Mixture. *Bioengineering* 2022, 9, 115. https://doi.org/10.3390/bioengineering9030115

Academic Editors: Minaxi Sharma, Kandi Sridhar and Zeba Usmani

Received: 2 February 2022
Accepted: 10 March 2022
Published: 11 March 2022

1. Introduction

The growing urbanization and the tremendous growth of industries have led to a rise in energy requirements. However, the production of energy from non-renewable origins as the main source is debated worldwide regarding its impact on environmental sustainability. Therefore, renewable energy in various forms has gained popularity and importance in policies of governmental and private sectors. Lignocellulosic biomass, being more abundant in nature, is considered the major biological feedstock for energy production. Globally, it is estimated that nearly 180 billion tons of lignocellulosic biomass are produced annually, having the potential to be converted into various high value-added goods, including biofuels, biochemicals, and biomaterials via the biorefinery process [1]. Lignocellulosic [2,3] material comprises cellulose, hemicellulose, and lignin as biopolymeric constituents that can be broken down into reactive biomolecules, which are transformed into valuable products and fuels [4].

However, the utilization of lignocellulose biomass as raw material has several limitations, including its physical and chemical barriers, the crystallinity of cellulose, and complex and recalcitrant structures [5,6]. An inevitable process is a pretreatment that basically enhances the lignocellulosic valorization. The efficient pretreatment technology must render the lignocellulosic material more reactive to subsequent processes [7]. Moreover, the hard barrier of using lignocellulosic as feedstocks are the chemical and physical properties that hinder the hydrolysis reaction.

To overcome these limitations, the selection of efficient pretreatment should be considered as a crucial factor for lignocellulosic valorization. The deficient pretreatment technique results in lower conversion yields that consequently determine economic infeasibility on a high production scale. Aforementioned argumentation, the selection of lignocellulosic biomass pretreatment is, therefore, necessary to reduce the recalcitrance of lignocellulosic biomass structure and to eliminate some inhibitors including; 5-hydroxyethyl furfural, furfural, and acetic acid that are produced by sugar degradation during pretreatment. An efficient pretreatment should provide reactive material for further processes, including enzymatic accessibility and fermentation. To reach a high conversion yield, the presence of lignin components in the lignocellulosic structure is unnecessary due to its specific characteristic for hindering the enzymatic accessibility [8].

Pretreatment processes are generally grouped into four categories, namely chemical, physical, biological, and physicochemical pretreatment [9,10]. Each pretreatment technique affects different properties, including chemical and physical properties. As previously mentioned, the effects of these pretreatments are demonstrated in various lignocellulosic biomasses, which could be helpful for their application in the biorefining process [11]. Among the various pretreatment process, the most widely used process is chemical pretreatment because of its ability to alter the biopolymeric conformation of the biomass and its uncomplicated application compared to other pretreatment methods [11]. Chemical pretreatment helps in removing the chemical linkages associated with three biomolecules together that it facilitates further processes [12]. Various existing chemical pretreatment methods, including alkaline, acid, deep eutectic solvent extraction, and ionic liquid, were previously investigated to enhance sugar production and bioethanol production [13]. However, their application on an industrial scale is limited due to the investment cost. Pretreatment and hydrolysis processes are considered a high cost and time investment from an industrial point of view [14,15]. Hence, it is preferable to select a potential pretreatment method, which can reduce the total cost for its industrial application. Among the various types of chemical pretreatment techniques, a widely advanced technique is ionic liquid pretreatment, which has the potential to be recycled and reused.

Ionic liquids (ILs) were recently recommended for lignocellulosic pretreatment because of their properties such as high heat resistance, high chemical stability, low liquefaction point, noninflammable, high polarity, and less risky process [10,16]. The ILs mainly emphasize lignin removal and cellulose structural swelling [17]. They also have other applications in various types of lignocellulosic materials for enhancing conversion [17]. One important property of ILs that makes them attractive for pretreatment purposes is their recyclability [18,19]. The recycling capacity of the IL can help in reducing the cost of the pretreatment. The low volatility of IL makes it feasible for recycling studies [20]. The integration of IL to the biorefining process was also previously demonstrated to be compatible with IL-tolerant cellulases [21,22]. Additionally, cellulose present in the biomass pretreated with IL can be hydrolyzed easily when various anti-solvents such as water, acetone, ethanol, methanol, and dichloromethane are used in the washing process [23]. In fact, the efficacy of ionic liquid pretreatment does not only depend on the type of ionic liquid but also on the anti-solvent and pretreatment condition [24]. After the pretreatment process, anti-solvent can be removed easily through the process of evaporation, and the ionic liquid is retrieved for the next round of pretreatment. A previous study on pretreatment of cotton with EMIM-Ac and its recyclability showed that the ionic liquid maintained its efficiency even after five recycles in comparison with its initial usage [25]. Another study on the pretreatment of spruce and oak dust with EMIM-Ac revealed an enhancement in the sugar yield post pretreatment and the ionic liquid could be reused up to seven times [20]. However, another recycling study with ionic liquid, 1-ethyl-3-methylimidazolium acetate and 1-ethyl-3-methylimidazolium acetate/ethanolamine, showed that reused IL was less efficient after 5–7th recycling [26]. The study also revealed that the lignin and moisture content in the recycled ionic liquid could influence pretreatment. The study carried out on the pretreatment of wood meals with 1-ethyl-3-methylimidazolium acetate and its recyclability showed that the pretreatment with recycled IL was effective until three recycles even without the removal of the accumulated lignin after each recycle [27]. Even though researchers studied the recyclability of IL, there are few reports of the recyclability effect on bioethanol production from recycled IL.

Hence, this study focused on evaluating the performance of the IL, 1-ethyl-3-methylimidazolium acetate (EMIM-Ac) pretreatment for bioethanol production with recycled ionic liquid-solvent mixture under optimal pretreatment conditions. Optimization studies were conducted with a mathematical model (BBD) that could predict the optimum pretreatment condition based on the interaction between variables and factors. The study also focused on recycling EMIM-Ac and its effect on rice straw pretreatment to improve the reducing sugar and ethanol yield.

2. Materials and Methods

2.1. Biomass Preparation

Rice straw was procured from a local paddy field in the central part of Thailand. The moisture content from the collected biomass was removed by drying it in a hot-air oven (WOF-50, Daihan Scientific, Gangwon-do, Korea) at 80 °C until a constant weight was obtained. Then, the dried rice straw sample was reduced in size using a household blender and sieved through a 20-mesh-sized aluminum sieve to obtain a uniform particle size. The biomass composition of the rice straw samples was analyzed by following the Van Soest protocol [28].

The 1-ethyl-3-methylimidazolium acetate (EMIM-Ac) and commercial cellulase enzyme, CelluClast 1.5 L, produced by *Trichoderma reesei* used in this study was bought from Sigma-Aldrich (St. Louis, MO, USA). The enzyme β-glucosidase from *Aspergillus niger* was obtained from Megazyme (Wicklow, Ireland). The 3,5-Dinitrosalicyclic acid used in reducing sugar determination was purchased from Alfa Aesar (Heysham, UK). The other solvents used in this study were obtained from RCI Labscan (Bangkok, Thailand).

2.2. Optimization of Rice Straw Pretreatment for Reducing Sugar Production

A mathematical model from response surface methodology (RSM), namely, the Box–Behnken design (BBD), was employed for the optimization of pretreatment conditions using EMIM-Ac. The model considered three pretreatment factors, namely, solid loading ratio (X_1: 5–15 wt%), pretreatment temperature (X_2: 100–140 °C), and pretreatment time (X_3: 30–60 min). Each of these pretreatment factors was tested at 3 levels, high (+1), mid (0), and low (−1) (Table 1). The model predicted a total of 17 runs, from among which the optimal pretreatment condition was selected based on the highest sugar yield (Y). Design-Expert software version 7.0.0 (STAT-EASE Inc., Minneapolis, MN, USA) was used to analyze the optimization result [29]. The effects of each pretreatment factor and interacting effects between two factors on the highest sugar yield were analyzed by ANOVA with the significance level of p-value less than 0.05.

2.3. EMIM-Ac Pretreatment Procedure

Pretreatment was performed following the suggested pretreatment conditions from BBD. Rice straw pretreatment was conducted by mixing 1 g of dried rice straw with 19, 9, and 5.67 g of EMIM-Ac to obtain a solid loading ratio at 5, 10, and 15 wt%, respectively, in a screw-capped tube. Then, every reaction was proceeded to a targeted pretreatment temperature (100–140 °C) in a controlled-temperature hot air oven (±2.0 °C, Model: WOF-50,

Daihan Scientific, Gangwon-do, Korea) with the retention time of the targeted pretreatment time (30–90 min). After the pretreatment, the sample was taken out from the hot air oven and quickly cooled down to room temperature by placing the tube in a room-temperature water bath (Model: WB-22, Daihan Scientific, Gangwon-do, Korea) for 5 min to stop the progress of pretreatment effect. The solids in the pretreated biomass were regenerated by the addition of water as an anti-solvent in a 1:1 (w/w) ratio. Then, the pretreated biomass was separated into a solid and liquid fraction by centrifugation (Combi 514R, Hanil Scientific Inc., Gimpo, Korea) at 8000× *g* for 10 min. The separated solids after pretreatment were washed thoroughly with 50 mL of deionized water three times to remove any EMIM-Ac residues. Then, the pretreated solids were oven-dried in a controlled-temperature hot air oven (±2.0 °C, Model: WOF-50, Daihan Scientific, Gangwon-do, Korea) at 80 °C until moisture was removed completely and a constant weight was achieved. The solid biomass void of moisture was subsequently hydrolyzed enzymatically, and the amounts of reducing sugars released by hydrolysis of biomass were analyzed using the dinitrosalicylic method (DNS) [30]. The liquid fractions containing EMIM-Ac were collected after centrifugation and were proceeded to the recycling process for the next round of pretreatment (Figure 1).

Table 1. Value of pretreatment parameters with corresponding coded level.

Pretreatment Factor	Level of Factor		
	Low	Med	High
Abbreviation	−1	0	1
Loading ratio (wt%) (X_1)	5	10	15
Temperature (°C) (X_2)	100	120	140
Time (min) (X_3)	30	60	90

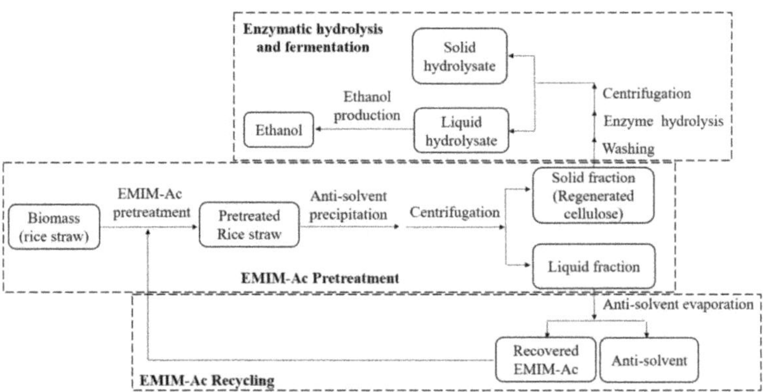

Figure 1. Process flow of EMIM-Ac pretreatment and enzymatic saccharification of biomass and process of EMIM-Ac recycling.

2.4. Recyclability of EMIM-Ac with Different Anti-Solvents

The EMIM-Ac recyclability studies were conducted by using 3 kinds of anti-solvents, comprising deionized water, acetone, and methanol. After the pretreatment of the rice straw sample, each type of anti-solvent was added to the mixture with the ratio of 1:1 (*w*/*w*). Solid and liquid fractionation was carried out by centrifugation as described in Section 2.3 (Figure 1). The liquid fraction containing EMIM-Ac was then filled in a rotary evaporator (Model: RV 10DS93, IKA, Tokyo, Japan) to recover EMIM-Ac and remove anti-solvents by setting the temperature of the water bath at 100, 56, and 65 °C for water, acetone, and methanol, respectively. The weight of the remaining liquid in the rotary evaporator was measured every 10 min, and the recycled EMIM-Ac was collected when the weight was consistent. The recycled EMIM-Ac was re-used further for the pretreating process of the new rice straw sample at the optimum pretreatment condition obtained from the RSM experiment for 5 rounds of recycling.

2.5. Enzymatic Hydrolysis and Measurement of a Sugar Yield

The efficacy of the EMIM-Ac pretreatment was estimated by enzymatic hydrolysis of the pretreated sample and measurement of reducing sugar concentration released from the hydrolyzed biomass based on the protocol published in our previous works [31,32]. An enzymatic hydrolysis reaction was set up in a screw-capped tube with 0.1 g biomass (2.5% w/v) in 4 mL of 50 mM citrate buffer (pH 4.7). A mixture of commercially available cellulase enzymes, CelluClast 1.5 L, was added into the mixture at the concentration of 20 FPU/g-biomass. To avoid any microbial contamination during hydrolysis reaction, 40 μL of 2 M sodium azide (Ajax Finechem, Brooklyn, MA, USA) was supplemented to the reaction. The enzymatic reaction was incubated at 45 °C for 72 h with 200 rpm agitation in a shaking incubator (JSSI-100C, JS Research Inc., Gongju, Republic of Korea). Any progress in the hydrolysis reaction was stopped by incubating at 100 °C for 5 min in a water bath (Model: WB-22, Daihan Scientific, Gangwon-do, Korea). The hydrolyzed sample was separated into a solid and liquid fraction by centrifugation at 8000× g for 10 min. The amount of reducing sugars in liquid hydrolysates was quantified using the 3,5-dinitrosalisylic acid (DNS) method [30]. The hydrolyzed solid fraction was collected and oven-dried at 80 °C until moisture was completely removed and constant weight was recorded. All the experiments were performed in triplicates.

2.6. Fermentation and Analysis of Ethanol Yield

The impact of IL pretreatment at the conversion of rice straw to bioethanol was observed by a fermentation experiment using *Saccharomyces cerevisiae* with the procedure used in our previous studies [10,33]. Fermentation was carried out using a liquid hydrolysate fraction obtained from enzymatic hydrolysis without the addition of sodium azide (Section 2.4), eliminating the supplementation of sodium azide. Yeast inoculum (1 mL at the concentration of about 10^8 cells/mL) was inoculated into 19 mL biomass liquid hydrolysate supplemented with glucose (1%, w/v) and yeast extract (1%, w/v). The fermentation mixture containing the yeast culture was incubated in a rotary shaker (Model: JSSI-100C, JS Research Korea) at 30 °C for 72 h at 150 rpm. Then, the supernatant from the fermentation mixture was recovered by centrifugation at 8000× g for 10 min. The ethanol concentration in the supernatant of yeast culture was analyzed using gas chromatography-mass spectrometry (GC-MS) (Shimadzu, Tokyo, Japan) [19].

GC-MS analysis for ethanol quantification was carried out in Shimadzu GC-MS (Shimadzu, Tokyo, Japan) fitted with DB-wax column (Agilent J & W GC column, CA, USA). The analysis used helium as carrier gas with a column flow rate of 1.22 mL/min. The GC inlet was set in a split ratio of 30.0, and the temperature of the injector was maintained at 230 °C. The column oven temperature was controlled in a program with the initial holding temperature at 50 °C for 1 min. The temperature was then increased to 200 °C for a hold time of 5 min, at a ramping rate of 20 °C/min. The MS program was fixed with an ion source temperature of 200 °C and mass range from m/z 30 to 600 [34,35]. The ethanol concentration after fermentation was estimated from the obtained peak area compared to diluted absolute ethanol (99.8% v/v). The analysis was repeated three times.

2.7. Analysis of Chemical Changes in Biomass

Chemical compositions and chemical structural alterations of untreated and pretreated biomass were examined with an FTIR spectrometer (Spectrum 2000, Perkin Elmer, Waltham, MA, USA) to evaluate the impact of pretreatment on biomass. The analysis was conducted at a resolution ranging from 400 cm^{-1} to 1700 cm^{-1}, and the spectral data were analyzed using Spectrum 2.00 software (Perkin Elmer, Waltham, MA, USA).

2.8. Analysis of Morphological Changes

The morphological and structural changes in the biomass before and after pretreatment was visualized through scanning electron microscopy (Model: JSM—5410LV, Jeol, Tokyo, Japan) at an accelerating voltage of 20 kV. Untreated and pretreated samples were prepared by attaching them to a specimen stub and coating them with gold before the inspection. The scanning electron microscope (SEM) images were visualized at a magnification of 100 μm and compared with the untreated sample.

3. Results and Discussion

3.1. EMIMAc Pretreatment and Its Optimization

Pretreatment is considered an essential step in a biorefinery in reducing the recalcitrance of the biomass. In the current study, pretreatment was performed using EMIM-Ac IL. Several studies carried out previously reported pretreatment with EMIM-Ac as efficient in various biomass and reported for more than 90% of sugar yield recovery from biomass [36–39]. Moreover, EMIM-Ac has the capability of rearranging hydrogen bonds efficiently in biomass, thus leading to cellulose dissolution [40]. Additionally, this IL is reported to drastically increase the saccharification rate as well. However, pretreatment, if not carried out under optimized conditions, may not increase the sugar yield significantly, emphasizing the importance of optimization.

Considering the necessity of optimized pretreatment condition, response surface methodology (RSM) was conducted to optimize pretreatment conditions for rice straw using IL. BBD was chosen to optimize the pretreatment conditions as it requires fewer runs than that of factorial design [41]. The design considered three pretreatment factors, including loading ratio of rice straw to EMIM-Ac (X_1), pretreatment temperature (X_2), and pretreatment time (X_3) for optimization depending on the reducing sugar yield after pretreatment. These three pretreatment parameters were selected in this optimization experiment, as they are proved to be an important factor to determine the pretreatment efficiency, and they are easily adjusted during pretreatment operation [2]. The model suggested 17 runs, and the details of each run and its sugar yield are depicted in Table 2. It was noted that the smallest and largest reducing sugars obtained from the RSM experiment were 15.34 (Run No. 11) and 54.64 mg (Run No. 3), respectively, which was equivalent to a 3.56 fold-difference. This observation suggested the significance of optimization and efficiency of the RSM method.

Table 2. BBD design to assess the effects of pretreatment factors (loading ratio (X_1, %), temperature (X_2, °C) and time (X_3, min)) on reducing sugar yield (Y, mg) of EMIM-Ac pretreated rice straw.

Run	Pretreatment Condition			Reducing Sugar (Y) (mg)
	Loading Ratio (X_1) (%)	Temperature (X_2) (°C)	Time (X_3) (min)	
1	15	120	30	22.36
2	5	120	90	44.19
3	5	140	60	54.64
4	10	120	60	48.31
5	15	120	90	45.14
6	10	120	60	43.52
7	10	120	60	45.70
8	15	100	60	16.99
9	10	120	60	46.57
10	5	120	30	32.48
11	10	100	30	15.35
12	15	140	60	29.95
13	5	100	60	23.09
14	10	120	60	44.10
15	10	140	90	31.90
16	10	140	30	40.70
17	10	100	90	27.98

The significance of the mathematical model was also statistically analyzed with ANOVA (Table 3). The model significance was confirmed with a *p*-value less than 0.05. Additionally, Table 3 show temperature as another significant term in this pretreatment. A similar optimization study conducted on rice straw pretreatment with IL also reports temperature as a significant factor in pretreatment [42]. Besides this, the model could also show the effect of interaction between factors on sugar yield (Figure 2). The sugar yield in the 3D plot increases with an increase in temperature and time.

Table 3. Optimum pretreatment conditions and predicted sugar yield obtained from RSM experiment.

EMIM-Ac Pretreatment	Mathematical models	Sugar content (mg) = - 429.80831 − (0.99874 × Conc.) + (7.03793 × Temp.) + (0.97107 × Time) − (0.027403 × Temp^2) − (0.00676158 × Time^2)
	Optimal pretreatment condition	5% loading ratio, 128.4 °C temperature, 71.83 min time
	Predicted sugar yield	51.96 mg

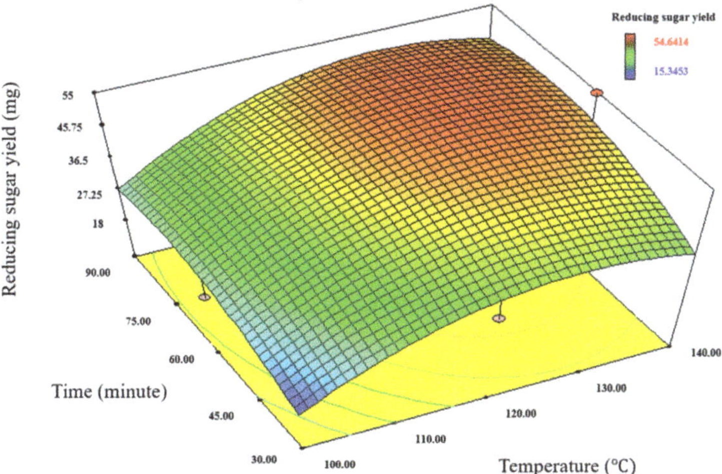

Figure 2. Contour plots representing the relation between the pretreatment factors on sugar yields using EMIM-Ac.

In addition to this, the model was also able to predict the ideal pretreatment conditions for the highest sugar yield based on data from the experimental design (Table 3). Under the optimum pretreatment condition, the model predicted a sugar yield of 51.96 mg. Other experiments in the study followed the optimum pretreatment conditions predicted by the model.

3.2. Analysis of Morphological Changes

Morphological changes in the untreated and pretreated biomass were analyzed using SEM (Figure 3). This analysis facilitates observing any structural changes to the biomass after pretreatment. Hence, to confirm the effect of pretreatment, the biomass was subjected to microscopic analysis to elucidate the physical changes. The untreated biomass had a regular and tough surface. It was more fibrous in appearance and had a smooth surface. The morphology of untreated rice straw was more intact in comparison with the pretreated rice straw. Unlike the untreated biomass, pretreated rice straw appeared to be more disorganized and became more porous. These structural changes enhance the subsequent enzymatic hydrolysis. EMIM-AC was previously reported having the ability to boost the specific surface area of the lignocellulosic biomass by removing lignin and simultaneously reducing its crystallinity, ensuring an enhanced saccharification rate [37]. In the present study, pretreatment with IL generated structural changes in biomass, enhancing the specific surface area for enzymatic accessibility [36]. This porous biomass is more accessible to enzymes and thus can increase the sugar yield [38].

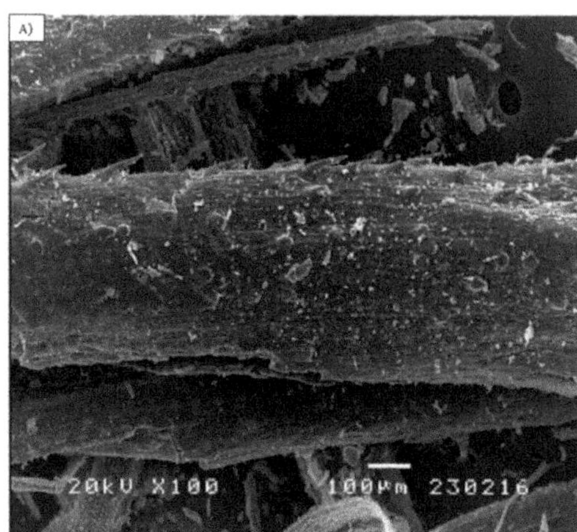
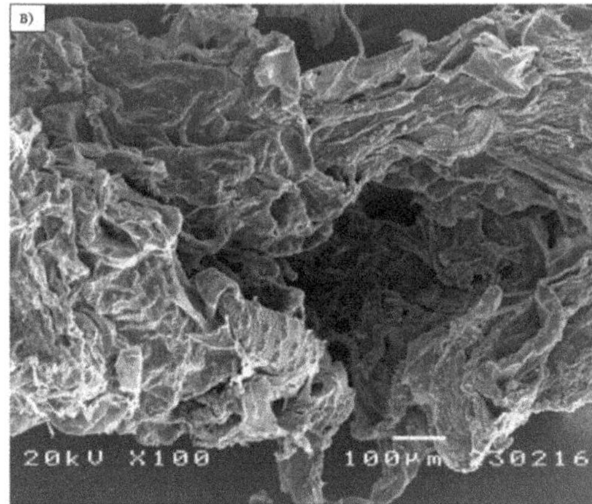

Figure 3. SEM images of biomass portraying the effects of EMIM-Ac pretreatment on biomass morphology. (**A**) Untreated rice straw and (**B**) EMIM-Ac pretreated rice straw using water as an anti-solvent at the optimal condition.

3.3. Effect of Different Anti Solvents and Recycled EMIM-Ac on Reducing Sugar Yields

Anti-solvents are essential in the post pretreatment process. They help to separate IL, in addition to the soluble lignin from pretreated solids. Several anti-solvents such as water, methanol, ethanol, isopropanol, etc., are studied to investigate their performance in delignification and saccharification yield. A study using IL pretreatment on wheat straw investigated the role of methanol, ethanol, and water as anti-solvents to separate pretreated solids [43]. However, this investigation could not find any significant effect of the anti-solvent in increasing sugar yield. A more recent study on investigating different parameters in protic IL pretreatment for ethanol generation studied the effect of anti-solvents in boosting sugar yield [44]. The study used water, ethanol, isopropanol, and isoamyl alcohol as anti-solvents. Even though different anti-solvents did not show significance in sugar yield, they could show a significant difference in delignification. The obtained results infer those anti-solvents with a low number of carbons in the alkyl chain lead to more interaction between hydrophobic lignin and alcohols and thereby allow us to enhance delignification [44].

In the present study, the effect of recycled EMIM-Ac, along with the usage of different anti solvents and their combinations in the washing step, was analyzed. The recyclability of IL is an important consideration as it can help in reducing the production cost [45]. There were previous studies on recycling EMIM-Ac. The effectiveness of recycled EMIM-Ac and different anti-solvents were evaluated based on saccharification yield (Figure 4). Anti-solvents, namely, water, acetone, methanol, and their combinations in 1:1 and 1:5 ratio (A1M5—Acetone:Methanol (1:5), M1A5—Methanol:Acetone (1:5)) were used in this study. Recycling studies were carried out for up to five recycles. The results showed that, in the initial pretreatment (R0), the saccharification yield was high when washing was carried out with solvents other than water. In the subsequent pretreatments carried out with recycled IL, the saccharification yield decreased with an increase in recycling number (R1 to R5) compared with R0. This is in line with the previous study, where recycled EMIM-Ac showed a decrease in sugar conversion after each recycle [26]. Sugar conversion decreased to <5 wt% between the 1st and 5th recycle, and it further decreased to 10–50 wt% between the 5th to 10th recycle. This reduction in sugar conversion was related to the accumulated lignin content in the solvent after each recycle and also to the presence of water in the solvent as the study used water in the washing process [26]. However, in this study, the sugar yield from the untreated rice straw was only 8.30 ± 1.061 mg/100 mg biomass, which clearly indicates the effectiveness of pretreatment even after five recycles.

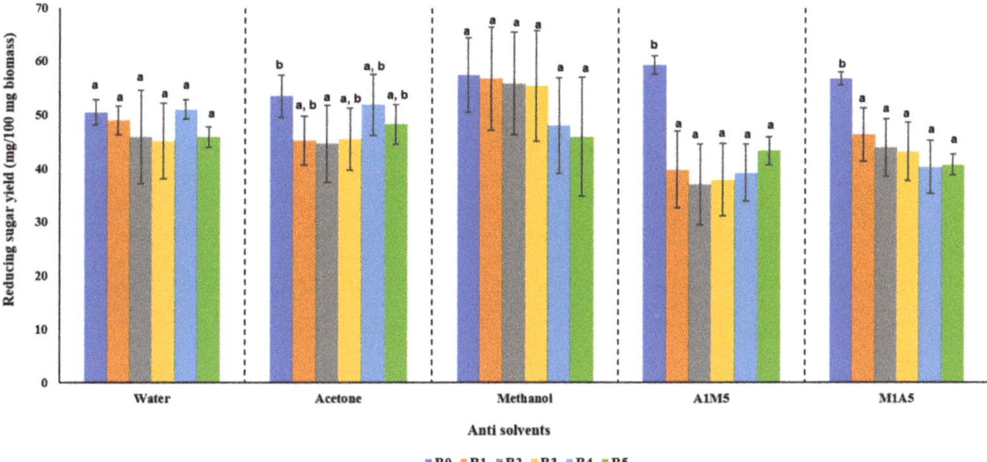

Figure 4. Reducing sugar yield attained from enzymatic saccharification of pretreated biomass with fresh and recycled EMIM-Ac and using various types of anti-solvent (A1M5—Acetone:Methanol (1:5), M1A5—Methanol:Acetone (1:5)). Alphabets indicate the results of ANOVA analysis, and different alphabet means significantly different ($p < 0.05$).

The use of a suitable anti-solvent can also help in increasing the sugar yield. A previous study using water and water:acetone (1:1, v/v) as antisolvent for EMIM-Ac pretreated Miscanthus showed water:acetone (1:1, v/v) as a suitable anti-solvent in increasing sugar yield [46]. However, the exact mechanism correlating the anti-solvent and increased sugar yield was not very clear, and it was assumed that the anti-solvent helped in washing away lignin and inhibitors in regenerated biomass, leading to enhanced sugar yield [47]. Among the different anti-solvents used in this study, the biomass washed by pure methanol yielded the maximum reducing sugar (Figure 4). It could maintain the sugar yield between 57.31 to 45.88 mg even after five recycles with no statistical difference (p-value less than 0.05). The statistical studies also showed that recycling is efficient even after five recycles when methanol was used as an anti-solvent. This could be attributed to the polarity of the solvent. The reported polarity of the solvent boosted its effectiveness to remove IL [44]. Methanol being more polar than other solvents, could facilitate in removing more IL and lignin solubilization after pretreatment. This could enhance the enzyme to access cellulose without any non-specific binding [44]. Moreover, methanol, possessing a lower boiling point, can be easily separated from pretreatment slurry by rotary evaporation. Hence, ethanol fermentation studies were only conducted on pretreated biomass washed with methanol as an anti-solvent.

3.4. Analysis of Chemical Changes in Biomass

The chemical conformational changes that could occur in biomass after IL pretreatment were analyzed by Fourier transform infrared spectroscopy (FT-IR) analysis (Figure 5). The FTIR spectrum of raw rice straw, which corresponds to carbohydrate and aromatic derivatives before and after pretreatment under the optimal condition (5% loading ratio at 128.4 °C for 71.83 min), were comparatively evaluated (Table 4). The FTIR analysis confirms changes in the chemical structures of the components in lignocellulosic biomass when comparing the intensities of peaks. The band appearing near 897 cm^{-1} matches with the typical peak of β-glucosidic bonds in cellulose, increasing its intensity and confirming its exposure and cellulose swelling post pretreatment [46]. The alteration in peak intensity near 1060 cm^{-1} of the pretreated and untreated biomass suggests that the IL treatment process enhanced the ratio of cellulose in the biomass [48]. The reduction in the intensity of peak at 1246 cm^{-1} is indicative of lignin removal in rice straw [10]. The intensity changes in 1321 and 1460 cm^{-1} (syringyl, guaiacyl, and methoxy groups in lignin) imply that the IL pretreatment caused a breakage in the links within and between the lignin–carbohydrate complex. The strong bands at approximately 1373 cm^{-1} were the characteristics of the cellulose biosorption peak that might be due to substituted aromatic components of lignin on the rice straw surface [38]. The decrease in intensity near 1430 cm^{-1} (associated with bending vibration of CH$_2$ group) represents cellulose suggesting the IL pretreatment could reduce crystalline cellulose [46]. The peaks at 1510 and 1637 cm^{-1} represent the damages caused to lignin after EMIM-Ac pretreatment. In fact, these changes depicted in FTIR data discloses the ability of IL in degrading the lignin content in the biomass [46,49]. These changes could also be noted in rice straw samples pretreated with recycled IL, even after five recycles, confirming the potential of EMIM-Ac recyclability (Figure 6).

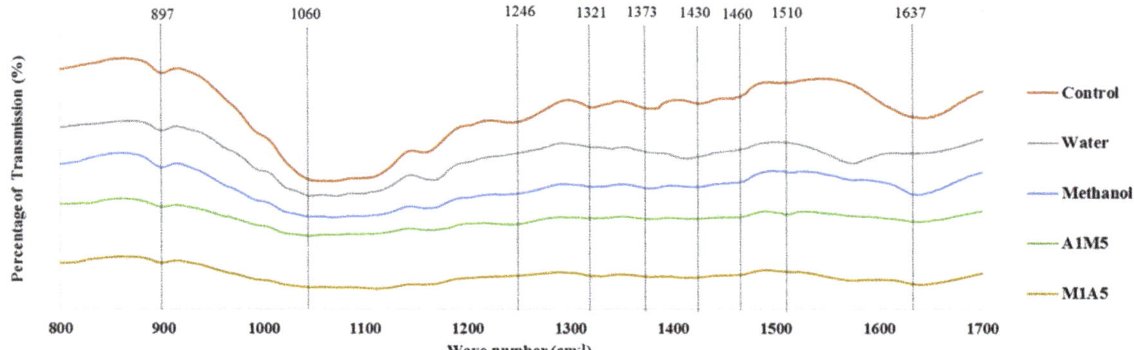

Figure 5. FT-IR spectra of untreated and pretreated rice straw for wavenumbers between 400–1700 cm^{-1} using EMIM-Ac pretreatment with various types of anti-solvents.

Table 4. The FTIR spectrum with different wavenumbers representing different functional groups in chemical derivatives of lignocellulosic biomass.

Peak, cm^{-1}	Functional Group Assignment	References
897	β-glycosidic linkage; vibration of amorphous cellulose	[48,50]
1060	Bond Stretching in C–O of homo and heteropolysaccharide	[38]
1246	C–O stretching of phenolics in lignin	[46]
1321	Stretching vibration of C=O in syringyl, guaiacyl group	[33]
1373	Deformation of C–H in homo and heteropolysaccharide	[38]
1430	C–H$_2$ bending of cellulose	[51]
1460	Deformations in C–H bonds of lignin	[52]
1510	Vibration in aromatic skeleton of lignin	[48]
1637	Phenolics in lignin	[46]

3.5. Effect of Recycled EMIM-AC on Sugar Yield and Fermentation

The efficacy of the IL pretreatment with recycled EMIM-AC was assessed based on the enzymatic hydrolysis and bioethanol fermentation after pretreatment. Fermentation studies were conducted on both pretreated and untreated rice straws that use recycled EMIM-Ac and pure methanol as an anti-solvent. *Saccharomyces cerevisiae* TISTR 5606 was used for the fermentation of biomass. Figure 7 represent the ethanol production and fermentable sugar yield from the pretreated biomass after five recycles. The untreated rice straw could produce only 0.46 ± 0.07% (v/v) ethanol. However, after pretreatment, the ethanol production was increased to 0.75 ± 0.02%, which increased after using recycled IL for pretreatment. The ethanol production was 0.87 ± 0.03% even after the fifth recycle. This implies that the biomass pretreated with recycled IL was able to produce almost 1.9-fold more ethanol than the untreated biomass even after five recycles. This could be due to the presence of some residual anti-solvent in recycled IL, which facilitated more lignin removal [45]. Table 5 show a comparison of the present study with some previously reported ethanol yields from various biomass using EMIM-Ac. In this study, pretreated rice straw at R0 and R5 produced 63% and 89% higher ethanol, respectively, compared to untreated rice straw. The results clearly demonstrate the efficiency of recycled IL in pretreating the biomass and the importance of suitable anti-solvent. It also indicates the potential of IL to be used in industries for reducing the cost of production. However, further studies are required to confirm this effect when upscaling for industrial purposes.

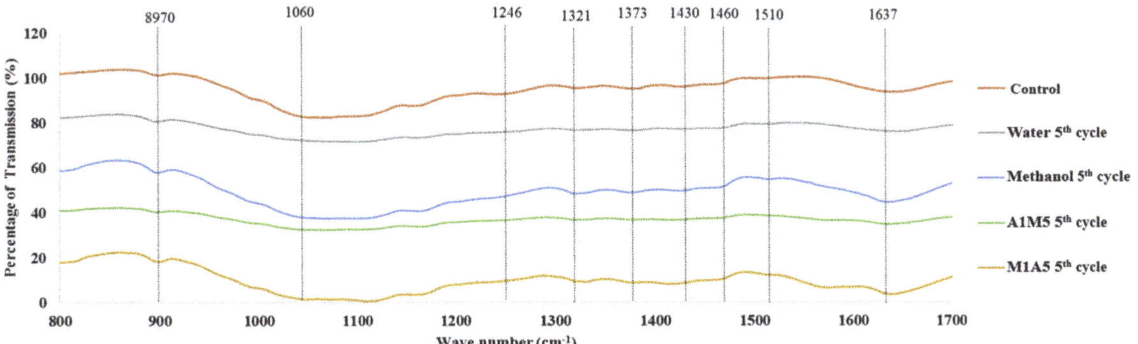

Figure 6. FT-IR spectra of untreated and pretreated rice straw for wavenumbers between 400–1700 cm^{-1} using EMIM-Ac pretreatment after five times recycle with various types of anti-solvents.

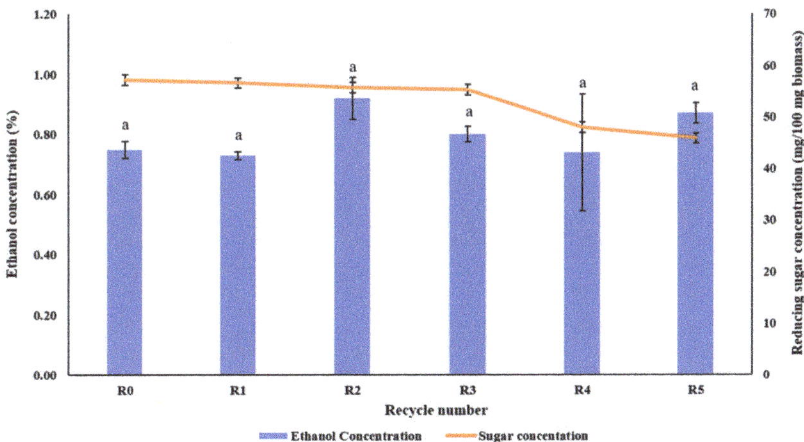

Figure 7. Ethanol concentration and reducing sugar yield obtained from enzymatic hydrolysis of pretreated rice straw by recycled EMIM-Ac and utilizing methanol as an antisolvent. Small alphabet indicate the results of ANOVA analysis, and different alphabet means significantly different ($p < 0.05$).

Table 5. Comparison with previous studies to produce ethanol from IL-pretreated lignocellulosic biomass with various types of ionic liquids and anti-solvents.

Biomass	Pretreatment Conditions	Anti-Solvent	Sugar Concentration	Ethanol Concentration	References
Wood powder	15% solid loading, 120 °C, 2 h	Dimethyl formamide	Glucose: 31 g/100 g biomass Xylose: 314.4 g/100 g biomass	3 g/L	[53]
Yellow pine wood	5% solid loading, 140 °C, 45 min	-	26.89 g/100 g biomass	2.6 g/L	[54]
Barely straw	5.26% solid loading, 105 °C, 7.5 h	Water	36.3 g glucose/100 g biomass 13.2 Xylose/100 g biomass	18.5 g/L	[55]
Water hyacinth	5.89% solid loading, 120 °C, 180 min	Water	4.5 g/100 g biomass	1.3 g/L	[56]
Rice straw	15% solid loading, 120 °C, 5 h	Water	44.3 g glucose/100 g biomass	1.92 g/L	[51]
Rice straw	R0, 5% solid loading, 128.4 °C, 71.83 min	Methanol	57.3 mg/100 mg biomass	5.9 g/L	This study

4. Conclusions

The current study emphasized the requirement of an appropriate anti-solvent and recyclability of ionic liquid in pretreating rice straw for an objective of cost reduction. The optimum pretreatment conditions for rice straw were obtained using RSM. Besides, RSM predicted a pretreatment with a solid loading of 5% wt at a temperature of 128.4 °C for about 71.83 min as the optimum pretreatment condition for rice straw. Despite this, the study also identified pure methanol as an appropriate anti-solvent that can not only enhance lignin removal but also increase sugar yield. In addition to this, methanol, when used as an antisolvent, can be easily recycled and reused for further washing processes. Moreover, it provides an added advantage of reducing wastewater generation in biorefineries. The recycling studies of ionic liquid revealed that a higher ethanol yield (1.9 fold more than untreated biomass) could be achieved even after five recycles of ionic liquid. Altogether, this work demonstrates the potential of recycled ionic liquid and the importance of an appropriate anti-solvent in pretreating biomass, particularly pretreatment cost reduction. Additionally, the recyclability of ionic liquid could consequently cause a reduction in production costs for industrial-scale applications.

Author Contributions: S.C.: conceptualization, investigation, reviewing and editing; E.J.P.: investigation, methodology, writing an original draft; T.R.: investigation, writing—reviewing and editing; K.C.: reviewing and editing; S.K.: research design, data analysis; Y.-S.C.: reviewing and editing; M.S.: conceptualization, data curation, writing—reviewing and editing, funding acquisition, project administration. All authors have read and agreed to the published version of the manuscript.

Funding: The authors would like to thank King Mongkut's University of Technology North Bangkok for their financial support (Contract No. KMUTNB-FF-65-37, KMUTNB-PHD-64-02) in conducting the research.

Institutional Review Board Statement: Not applicable.

Informed Consent Statement: Not applicable.

Data Availability Statement: The authors confirm that the data supporting the findings of this study are available within the article.

Acknowledgments: The authors would like to thank King Mongkut's University of Technology North Bangkok for all the facilities used to conduct the research.

Conflicts of Interest: The authors declare no conflict of interest.

References

1. Haq, I.U.; Qaisar, K.; Nawaz, A.; Akram, F.; Mukhtar, H.; Zohu, X.; Xu, Y.; Mumtaz, M.W.; Rashid, U.; Ghani, W.A.W.A.K.; et al. Advances in Valorization of Lignocellulosic Biomass towards Energy Generation. *Catalysts* **2021**, *11*, 309. [CrossRef]
2. Sriariyanun, M.; Heitz, J.H.; Yasurin, P.; Asavasanti, S.; Tantayotai, P. Intaconic Acid: A Promising and Sustainable Platform Chemical? *Appl. Sci. Eng. Prog.* **2019**, *12*, 75–82. [CrossRef]
3. Saravanan, A.; Kumar, P.S.; Jeevanantham, S.; Karishma, S.; Vo, D.-V.N. Recent advances and sustainable development of biofuels production from lignocellulosic biomass. *Bioresour. Technol.* **2022**, *344*, 126203. [CrossRef] [PubMed]
4. Baruah, J.; Nath, B.K.; Sharma, R.; Kumar, S.; Deka, R.C.; Baruah, D.C.; Kalita, E. Recent Trends in the Pretreatment of Lignocellulosic Biomass for Value-Added Products. *Front. Energy Res.* **2018**, *6*, 141. [CrossRef]
5. Yoo, C.G.; Meng, X.; Pu, Y.; Ragauskas, A.J. The critical role of lignin in lignocellulosic biomass conversion and recent pretreatment strategies: A comprehensive review. *Bioresour. Technol.* **2020**, *301*, 122784. [CrossRef]
6. Kim, J.S.; Lee, Y.Y.; Kim, T.H. A review on alkaline pretreatment technology for bioconversion of lignocellulosic biomass. *Bioresour. Technol.* **2016**, *199*, 42–48. [CrossRef]
7. Tian, S.-Q.; Zhao, R.-Y.; Chen, Z.-C. Review of the pretreatment and bioconversion of lignocellulosic biomass from wheat straw materials. *Renew. Sustain. Energy Rev.* **2018**, *91*, 483–489. [CrossRef]
8. Qin, L.; Li, W.-C.; Liu, L.; Zhu, J.-Q.; Li, X.; Li, B.-Z.; Yuan, Y.-J. Inhibition of lignin-derived phenolic compounds to cellulase. *Biotechnol. Biofuels* **2016**, *9*, 70. [CrossRef]
9. Aftab, M.N.; Iqbal, I.; Riaz, F.; Karadag, A.; Tabatabaei, M. Different Pretreatment Methods of Lignocellulosic Biomass for Use in Biofuel Production. In *Biomass for Bioenergy-Recent Trends and Future Challenges*; IntechOpen: London, UK, 2019.
10. Panakkal, E.J.; Cheenkachorn, K.; Gundupalli, M.P.; Kitiborwornkul, N.; Sriariyanun, M. Impact of sulfuric acid pretreatment of durian peel on the production of fermentable sugar and ethanol. *J. Indian Chem. Soc.* **2021**, *98*, 100264. [CrossRef]
11. Norrrahim, M.N.F.; Ilyas, R.A.; Nurazzi, N.M.; Rani, M.S.A.; Atikah, M.S.N.; Shazleen, S.S. Chemical Pretreatment of Lignocellulosic Biomass for the Production of Bioproducts: An Overview. *Appl. Sci. Eng. Prog.* **2021**, *14*, 588–605. [CrossRef]
12. Areepak, C.; Jiradechakorn, T.; Chuetor, S.; Phalakornkule, C.; Sriariyanun, M.; Raita, M.; Champreda, V.; Laosiripojana, N. Improvement of lignocellulosic pretreatment efficiency by combined chemo-Mechanical pretreatment for energy consumption reduction and biofuel production. *Renew. Energy* **2021**, *182*, 1094–1102. [CrossRef]
13. Mankar, A.R.; Pandey, A.; Modak, A.; Pant, K. Pretreatment of lignocellulosic biomass: A review on recent advances. *Bioresour. Technol.* **2021**, *334*, 125235. [CrossRef] [PubMed]
14. Aguilar, D.L.; Rodríguez-Jasso, R.M.; Zanuso, E.; Lara-Flores, A.A.; Aguilar, C.N.; Sanchez, A.; Ruiz, H.A. Operational Strategies for Enzymatic Hydrolysis in a Biorefinery. In *Biorefining of Biomass to Biofuels: Opportunities and Perception*; Kumar, S., Sani, R.K., Eds.; Springer International Publishing: Cham, Switzerland, 2018; pp. 223–248.
15. Zhao, X.; Liu, D. Multi-products co-production improves the economic feasibility of cellulosic ethanol: A case of Formiline pretreatment-based biorefining. *Appl. Energy* **2019**, *250*, 229–244. [CrossRef]
16. Tayyab, M.; Noman, A.; Islam, W.; Waheed, S.; Arafat, Y.; Ali, F.; Zaynab, M.; Lin, S.; Zhang, H.; Khan, D. Bioethanol production from lignocellulosic biomass by environment-friendly pretreatment methods: A review. *Appl. Ecol. Environ. Res.* **2018**, *16*, 225–249. [CrossRef]
17. Abushammala, H.; Mao, J. A Review on the Partial and Complete Dissolution and Fractionation of Wood and Lignocelluloses Using Imidazolium Ionic Liquids. *Polymers* **2020**, *12*, 195. [CrossRef]
18. Casas, A.; Palomar, J.; Alonso, M.V.; Oliet, M.; Omar, S.; Rodriguez, F. Comparison of lignin and cellulose solubilities in ionic liquids by COSMO-RS analysis and experimental validation. *Ind. Crop. Prod.* **2012**, *37*, 155–163. [CrossRef]
19. Nguyen, T.-A.D.; Kim, K.-R.; Han, S.J.; Cho, H.Y.; Kim, J.W.; Park, S.M.; Park, J.C.; Sim, S.J. Pretreatment of rice straw with ammonia and ionic liquid for lignocellulose conversion to fermentable sugars. *Bioresour. Technol.* **2010**, *101*, 7432–7438. [CrossRef]
20. Reddy, P. A critical review of ionic liquids for the pretreatment of lignocellulosic biomass. *S. Afr. J. Sci.* **2015**, *111*, 1–9. [CrossRef]
21. Xu, J.; Wang, X.; Hu, L.; Xia, J.; Wu, Z.; Xu, N.; Dai, B.; Wu, B. A novel ionic liquid-tolerant Fusarium oxysporum BN secreting ionic liquid-stable cellulase: Consolidated bioprocessing of pretreated lignocellulose containing residual ionic liquid. *Bioresour. Technol.* **2015**, *181*, 18–25. [CrossRef]
22. Trivedi, N.; Gupta, V.; Reddy, C.; Jha, B. Detection of ionic liquid stable cellulase produced by the marine bacterium Pseudoalteromonas sp. isolated from brown alga Sargassum polycystum C. Agardh. *Bioresour. Technol.* **2013**, *132*, 313–319. [CrossRef]
23. Auxenfans, T.; Buchoux, S.; Larcher, D.; Husson, G.; Husson, E.; Sarazin, C. Enzymatic saccharification and structural properties of industrial wood sawdust: Recycled ionic liquids pretreatments. *Energy Convers. Manag.* **2014**, *88*, 1094–1103. [CrossRef]
24. Lopes, A.S.D.C.; Joao, K.G.; Morais, A.R.C.; Lukasik, E.B.; Lukasik, R.B. Ionic liquids as a tool for lignocellulosic biomass fractionation. *Sustain. Chem. Processes* **2013**, *1*, 1–31.
25. Auxenfans, T.; Buchoux, S.; Djellab, K.; Avondo, C.; Husson, E.; Sarazin, C. Mild pretreatment and enzymatic saccharification of cellulose with recycled ionic liquids towards one-batch process. *Carbohydr. Polym.* **2012**, *90*, 805–813. [CrossRef] [PubMed]
26. Weerachanchai, P.; Lee, J.-M. Recyclability of an ionic liquid for biomass pretreatment. *Bioresour. Technol.* **2014**, *169*, 336–343. [CrossRef] [PubMed]
27. Hamidah, U.; Arakawa, T.; H'Ng, Y.Y.; Nakagawa-Izumi, A.; Kishino, M. Recycled ionic liquid 1-ethyl-3-methylimidazolium acetate pretreatment for enhancing enzymatic saccharification of softwood without cellulose regeneration. *J. Wood Sci.* **2017**, *64*, 149–156. [CrossRef]

28. Van Soest, P.J.; Robertson, J.B.; Lewis, B.A. Methods for Dietary Fiber, Neutral Detergent Fiber, and Nonstarch Polysaccha-rides in Relation to Animal Nutrition. *Diary Sci.* **1991**, *74*, 3583–3597. [CrossRef]
29. Box, G.E.P.; Wilson, K.B. On the Experimental Attainment of Optimum Conditions on JSTOR. *J.R. Stat. Soc.* **1951**, *13*, 1–45.
30. Miller, G.L. Use of Dinitrosalicylic Acid Reagent for Determination of Reducing Sugar. *Anal. Chem.* **1959**, *31*, 426–428. [CrossRef]
31. Amnuaycheewa, P.; Hengaroonprasan, R.; Rattanaporn, K.; Kirdponpattara, S.; Cheenkachorn, K.; Sriariyanun, M. Enhancing enzymatic hydrolysis and biogas production from rice straw by pretreatment with organic acids. *Ind. Crop. Prod.* **2016**, *87*, 247–254. [CrossRef]
32. Rattanaporn, K.; Tantayotai, P.; Phusantisampan, T.; Pornwongthong, P.; Sriariyanun, M. Organic acid pretreatment of oil palm trunk: Effect on enzymatic saccharification and ethanol production. *Bioprocess Biosyst. Eng.* **2018**, *41*, 467–477. [CrossRef] [PubMed]
33. Panakkal, E.J.; Sriariyanun, M.; Ratanapoompinyo, J.; Yasurin, P.; Cheenkachorn, K.; Rodiahwati, W.; Tantayotai, P. Influence of Sulfuric Acid Pretreatment and Inhibitor of Sugarcane Bagasse on the Production of Fermentable Sugar and Ethanol. *Appl. Sci. Eng. Prog.* **2021**, *15*. [CrossRef]
34. Gundupalli, M.P.; Cheng, Y.-S.; Chuetor, S.; Bhattacharyya, D.; Sriariyanun, M. Effect of dewaxing on saccharification and ethanol production from different lignocellulosic biomass. *Bioresour. Technol.* **2021**, *339*, 125596. [CrossRef] [PubMed]
35. Gundupalli, M.P.; Chuetor, S.; Cheenkachorn, K.; Rattanaporn, K.; Show, P.-L.; Cheng, Y.-S.; Sriariyanun, M. Interferences of Waxes on Enzymatic Saccharification and Ethanol Production from Lignocellulose Biomass. *Bioengineering* **2021**, *8*, 171. [CrossRef] [PubMed]
36. Da Silva, A.S.; Lee, S.-H.; Endo, T.; Bon, E.P. Major improvement in the rate and yield of enzymatic saccharification of sugarcane bagasse via pretreatment with the ionic liquid 1-ethyl-3-methylimidazolium acetate ([Emim] [Ac]). *Bioresour. Technol.* **2011**, *102*, 10505–10509. [CrossRef]
37. Lee, S.H.; Doherty, T.V.; Linhardt, R.J.; Dordick, J.S. Ionic liquid-mediated selective extraction of lignin from wood leading to enhanced enzymatic cellulose hydrolysis. *Biotechnol. Bioeng.* **2008**, *102*, 1368–1376. [CrossRef]
38. Li, C.; Knierim, B.; Manisseri, C.; Arora, R.; Scheller, H.; Auer, M.; Vogel, K.P.; Simmons, B.; Singh, S. Comparison of dilute acid and ionic liquid pretreatment of switchgrass: Biomass recalcitrance, delignification and enzymatic saccharification. *Bioresour. Technol.* **2010**, *101*, 4900–4906. [CrossRef]
39. Tantayotai, P.; Gundupalli, M.P.; Panakkal, E.J.; Sriariyanun, M.; Rattanaporn, K.; Bhattacharyya, D. Differential Influence of Imidazolium Ionic Liquid on Cellulase Kinetics in Saccharification of Cellulose and Lignocellulosic Biomass Substrate. *Appl. Sci. Eng. Prog.* **2021**, *15*, 5510. [CrossRef]
40. Rao, J.; Kumar, B. 3D Blade root shape optimization. In Proceedings of the 10th International Conference on Vibrations in Rotating Machinery, London, UK, 11–13 September 2012.
41. Poy, H.; Lladosa, E.; Gabaldón, C.; Loras, S. Optimization of rice straw pretreatment with 1-ethyl-3-methylimidazolium acetate by the response surface method. *Biomass-Convers. Biorefinery* **2021**, 1–16. [CrossRef]
42. Raeisi, S.M.; Tabatabaei, M.; Ayati, B.; Ghafari, A.; Mood, S.H. A Novel Combined Pretreatment Method for Rice Straw Using Optimized EMIM[Ac] and Mild NaOH. *Waste Biomass-Valorization* **2015**, *7*, 97–107. [CrossRef]
43. Nakasu, P.Y.S.; Pin, T.C.; Hallett, J.P.; Rabelo, S.C.; Costa, A.C. In-depth process parameter investigation into a protic ionic liquid pretreatment for 2G ethanol production. *Renew. Energy* **2021**, *172*, 816–828. [CrossRef]
44. Wei, H.L.; Wang, Y.T.; Hong, Y.Y.; Zhu, M.J. Pretreatment of rice straw with recycled ionic liquids by phase-separation process for low-cost biorefinery. *Biotechnol. Appl. Biochem.* **2020**, *68*, 871–880. [CrossRef] [PubMed]
45. Gao, D.; Haarmeyer, C.; Balan, V.; Whitehead, T.A.; Dale, B.E.; Chundawat, S.P. Lignin triggers irreversible cellulase loss during pretreated lignocellulosic biomass saccharification. *Biotechnol. Biofuels* **2014**, *7*, 175. [CrossRef] [PubMed]
46. Kahar, P. Synergistic Effects of Pretreatment Process on Enzymatic Digestion of Rice Straw for Efficient Ethanol Fermentation. In *Environmental Biotechnology-New Approaches and Prospective Applications*; BoD–Books on Demand: Norderstedt, Germany, 2013.
47. Dash, M.; Mohanty, K. Effect of different ionic liquids and anti-solvents on dissolution and regeneration of Miscanthus towards bioethanol. *Biomass-Bioenergy* **2019**, *124*, 33–42. [CrossRef]
48. Dai, Y.; Si, M.; Chen, Y.; Zhang, N.; Zhou, M.; Liao, Q.; Shi, D.; Liu, Y.-N. Combination of biological pretreatment with NaOH/Urea pretreatment at cold temperature to enhance enzymatic hydrolysis of rice straw. *Bioresour. Technol.* **2015**, *198*, 725–731. [CrossRef]
49. Zhang, Q.; Huang, H.; Han, H.; Qiu, Z.; Achal, V. Stimulatory effect of in-situ detoxification on bioethanol production by rice straw. *Energy* **2017**, *135*, 32–39. [CrossRef]
50. Cichosz, S.; Masek, A. IR Study on Cellulose with the Varied Moisture Contents: Insight into the Supramolecular Structure. *Materials* **2020**, *13*, 4573. [CrossRef]
51. Poornejad, N.; Karimi, K.; Behzad, T. Ionic Liquid Pretreatment of Rice Straw to Enhance Saccharification and Bioethanol Production. *J. Biomass-Biofuel* **2014**, *1*, 8–15. [CrossRef]
52. Kubovský, I.; Kačíková, D.; Kačík, F. Structural Changes of Oak Wood Main Components Caused by Thermal Modification. *Polymers* **2020**, *12*, 485. [CrossRef]
53. Han, S.Y.; Park, C.W.; Park, J.B.; Ha, S.J.; Kim, N.H.; Lee, S.H. Ethanol Fermentation of the Enzymatic Hydrolysates from the Products Pretreated using [EMIM]Ac and Its Co-Solvents with DMF. *J. For. Environ. Sci.* **2020**, *36*, 62–66. [CrossRef]
54. Elmacı, S.B.; Ozcelik, F. Ionic liquid pretreatment of yellow pine followed by enzymatic hydrolysis and fermentation. *Biotechnol. Prog.* **2018**, *34*, 1242–1250. [CrossRef]

55. Lara-Serrano, M.; Angulo, F.S.; Negro, M.J.; Morales-Delarosa, S.; Campos-Martin, J.M.; Fierro, J.L.G. Second-Generation Bioethanol Production Combining Simultaneous Fermentation and Saccharification of IL-Pretreated Barley Straw. *ACS Sustain. Chem. Eng.* **2018**, *6*, 7086–7095. [CrossRef]
56. Rezania, S.; Alizadeh, H.; Park, J.; Md Din, M.F.; Darajeh, N.; Shafiei Ebrahimi, S.; Saha, B.B.; Kamyab, H. Effect of various pretreatment methods on sugar and ethanol production from cellulosic water hyacinth. *BioResources* **2019**, *14*, 592–606.

Article

Vascular Relaxation and Blood Pressure Lowering Effects of *Prunus mume* in Rats

Cheolmin Jo [1], Bumjung Kim [2], Kyungjin Lee [1] and Ho-Young Choi [1,*]

[1] Department of Herbal Pharmacology, College of Korean Medicine, Kyung Hee University, Seoul 02447, Republic of Korea
[2] Department of Oriental Health Management, Kyung Hee Cyber University, Seoul 02447, Republic of Korea
* Correspondence: hychoi@khu.ac.kr; Tel.: +82-2-961-9372

Abstract: *Prunus mume* Siebold et Zuccarini is mainly consumed as processed fruits in beverages, vinegar, alcohol, or fruit syrup; studies have reported various functional effects. Many pharmacological and functional studies exist on fruit extracts or processed foods using fruits, however, efficacy studies on various parts of *P. mume*, including the bark, branches, flowers, and leaves, have not been sufficiently conducted. A previous study revealed that a 70% ethanol extract of *P. mume* branches induced vascular endothelium-dependent vasorelaxant effects in rat thoracic aortic rings. Therefore, we hypothesized that various parts (the fruits, flowers, leaves, and bark) might have vasorelaxant effects. We evaluated the effects of *P. mume* extracts on the vascular relaxation of isolated rat thoracic aorta and hypotensive effects in spontaneous hypertensive rats (SHR). A 70% ethanol extract of *P. mume* bark (PBaE) was the most effective, thus, we investigated its vasorelaxant mechanisms and hypotensive effects. PBaE lowered the blood pressure in SHR and induced the vascular endothelium-dependent relaxation of isolated rat aortic rings via the NO/sGC/cGMP and the PGI$_2$ pathways in the vascular smooth muscle. Potassium channels, such as K_{Ca}, K_{ATP}, K_V, and K_{ir}, were partially associated with a PBaE-induced vasorelaxation. Therefore, PBaE might help prevent and treat hypertension.

Keywords: *Prunus mume*; vasorelaxant; hypertension; angiotensin receptor blockers

Citation: Jo, C.; Kim, B.; Lee, K.; Choi, H.-Y. Vascular Relaxation and Blood Pressure Lowering Effects of *Prunus mume* in Rats. *Bioengineering* **2023**, 10, 74. https://doi.org/10.3390/bioengineering10010074

Academic Editors: Minaxi Sharma, Kandi Sridhar and Zeba Usmani

Received: 16 December 2022
Revised: 4 January 2023
Accepted: 4 January 2023
Published: 6 January 2023

1. Introduction

Hypertension (high blood pressure) is a major cause of premature death worldwide, affecting one in four men and one in five women (over 1 billion people) [1]. The main causes of a high blood pressure are unhealthy eating habits, a lack of exercise, smoking, drinking, and obesity. Therefore, reducing these modifiable risk factors most effectively prevent and control high blood pressure [1]. However, in uncontrolled high blood pressure, despite these lifestyle changes, antihypertensive drugs, such as angiotensin-converting enzyme inhibitors, angiotensin receptor blockers, calcium channel blockers, renin inhibitors, thiazide diuretics, α-adrenergic blockers, β-adrenergic blockers, sympatholytic agents, and vasodilators are used [2]. Despite the research and discovery of these various drugs, the number of hypertensive patients is not decreasing, and there is an increasing demand for more efficient and reliable approaches to prevent and treat hypertension.

Natural products have been used to treat various diseases, including cancer and cardiovascular disease [3,4]. From 1946 to date, approved nature-derived cancer treatments account for over half of all anti-cancer drugs. In hypertension, natural products accounted for 20% of all approved antihypertensive drugs between 1981 and 2019 [5]. Therefore, natural products can treat and prevent hypertension.

Prunus mume Siebold et Zuccarini is a deciduous tree in the Rosaceae family with over 3000 years of cultivation history [6]. They are mainly used as landscape or fruit trees in Asia, including Korea. The fumigated fruit of *P. mume* is used as the traditional medicine, "Omae," in Korea [7]. Additionally, various parts (the fruits, flowers, leaves, branches, seeds, and roots) of *P. mume* have been used in traditional Chinese medicine [8]. *P. mume* is mainly consumed as fruits processed into beverages [9–11], vinegar [12], alcohol [13,14], or fruit syrup [15]. Studies have identified their various functions. Additionally, flowers are used as tea [16].

Phytochemical research on *Prunus mume* revealed compounds such as phenols, organic acids, steroids, terpenes, lignans, benzyl glycosides, furfural, cyanogenic glycosides, and alkaloids, mainly in flowers and fruits [17]. In basic research, various activities, such as the antioxidant [18,19], anti-inflammatory [20], anti-cancer [21,22], anti-osteoporosis [23], anti-obesity [24], anti-helicobacter [25], blood flow improvement [26], anti-allergic [27], anti-fatigue [12], hepatoprotective [28], and immune-enhancing effects [29] of the fruits of *P. mume* have been identified. There are pharmacological and functional studies on fruit extracts or processed foods using fruits; nonetheless, the efficacy studies on various parts of *P. mume*, such as the bark, branches, flowers, and leaves, have not been sufficiently studied. Some processed fruit products help control blood pressure; however, the scientific evidence is lacking. Studies have revealed that *P. mume* fruit did not significantly impact the control of blood pressure in patients with hypertension [30].

However, we demonstrated in a previous study that a 70% ethanol extract of *P. mume* branches induced vascular endothelium-dependent vasorelaxant effects in rat thoracic aortic rings [7]. Therefore, we hypothesized that various parts of *P. mume* might have vasorelaxant effects. Therefore, to further validate the benefits of *P. mume* products in pharmaceutical and nutraceutical applications, the vasodilatory activity of various *P. mume* parts was investigated. Additionally, one extract with vasorelaxant effects was selected, and its vascular relaxation mechanism and hypotensive effects were evaluated.

2. Materials and Methods

2.1. Chemicals

Angiotensin II (Ang II), calcium chloride ($CaCl_2$), phenylephrine hydrochloride (PE), NG-nitro-L-arginine methyl ester (L-NAME), 1H-[1,2,4]Oxadiazolo[4,3-a]quinoxalin-1-one (ODQ), methylene blue (MB), indomethacin, and ethylene glycol-bis(2-aminoethylether)-N, N, N′, N′-tetraacetic acid (EGTA) were purchased from Sigma Aldrich, Inc. (St. Louis, MO, USA). Tetraethylammonium (TEA), 4-aminopyridine (4-AP), and glibenclamide were purchased from Wako Pure Chemical Industries, Ltd. (Osaka, Japan). Magnesium sulfate ($MgSO_4$), potassium chloride (KCl), and potassium phosphate monobasic (KH_2PO_4) were purchased from Duksan Pure Chemicals Co., Ltd. (Ansan, Korea). Barium chloride ($BaCl_2$), glucose, sodium chloride (NaCl), sodium hydrogen carbonate ($NaHCO_3$), and urethane were purchased from Daejung Chemicals & Metals Co., Ltd. (Siheung, Korea).

2.2. Plant Material and Extraction

The fresh fruits, flowers, leaves, branches, and bark of *Prunus mume* were collected from Dangjin-si, Chungcheongnam-do, the Republic of Korea, and the taxonomic identities of the plant were authenticated by a professor in the Department of Herbology, the University of Kyung Hee, the Republic of Korea. The collected plant parts were washed with water to remove contaminants, cut into small pieces, and dried in a convection oven. The dried samples are mixed with water or 70% ethanol and boiled for 2 h. After vacuum filtration, the filtrate was frozen at $-20\ °C$ and freeze-dried to obtain 10 extract powders (Table 1).

2.3. Animals

Male Sprague Dawley rats (SD, 220–250 g, 8 weeks old) were obtained from Daehanbiolink Co., Ltd. (Eumseong, Korea). Male spontaneously hypertensive rats (SHR, 200–250 g, 8 weeks old) were purchased from Charles River Laboratories (Yokohama, Japan). All animal procedures were conducted according to the animal welfare guidelines and were approved (KHSASP-21-050) by the Kyung Hee University Institutional Animal Care and Use Committee. The animals were maintained under controlled environmental conditions (12/12 h light/dark cycle, $22 \pm 2\ °C$). Food and water were available ad libitum.

Table 1. List of *Prunus mume* extracts used in this study.

Plant Part	Collection Date	Extracts	Yield (%)	Abbreviation
Fruit	June 2020	Water	21.5	PFrW
Fruit	June 2020	70% Ethanol	31.5	PFrE
Flower	March 2020	Water	25.0	PFlW
Flower	March 2020	70% Ethanol	9.3	PFlE
Leaf	June 2020	Water	31.0	PLW
Leaf	June 2020	70% Ethanol	23.0	PLE
Branch	February 2020	Water	8.4	PBrW
Branch	February 2020	70% Ethanol	9.3	PBrE
Bark	February 2020	Water	23.0	PBaW
Bark	February 2020	70% Ethanol	20.0	PBaE

2.4. Measurement of Vasorelaxant Activity

2.4.1. Preparation of Rat Aortic Rings

SD rats were anesthetized using urethane (1.2 g/kg, i.p.). After the abdominal incision was made to expose the aorta, the thoracic aorta was separated. The fat and connective tissues were removed while immersed in Krebs–Henseleit buffer (KH, composition (mM): NaCl, 118.0; KCl, 4.7; $MgSO_4$, 1.2; KH_2PO_4, 1.2; $CaCl_2$, 2.5; $NaHCO_3$, 25.0; and glucose, 11.1; pH 7.4). The tissue bath solution bubbled continuously with 95% O_2 and 5% CO_2 at 37 °C. Aortic rings were made by cutting the thoracic aorta approximately 3 mm long, placing it between two stainless steel hooks in organ bath chambers, and connecting it to isometric force transducers. After incubation without tension for 20 min, the vessel segments were allowed to equilibrate for 40 min at a resting tension of 1.2 g. The KH was replaced every 20 min during the equilibrium period. Changes in the tension were recorded via the isometric transducers connected to a data acquisition system (PowerLab, ADI instrument Co., Ltd., New South Wales, Australia). Ca^{2+}-free KH buffer was prepared by replacing $CaCl_2$ with 1 mM EGTA.

2.4.2. Vasorelaxant Effects of *Prunus mume* Extract on Isolated Aortic Rings

The aortic rings were pre-contracted with PE (1 μM). When the degree of contraction reached the maximum, *Prunus mume* flower water extract (PFlW), *Prunus mume* flower 70% ethanol extract (PFlE), *Prunus mume* fruit water extract (PFrW), *Prunus mume* fruit 70% ethanol extract (PFrE), *Prunus mume* leaf water extract (PLW), *Prunus mume* leaf 70% ethanol extract (PLE), *Prunus mume* branch water extract (PBrW), *Prunus mume* branch 70% ethanol extract (PBrE), *Prunus mume* bark water extract (PBaW), or *Prunus mume* bark 70% ethanol extract (PBaE) was added cumulatively (10–1000 μg/mL). To compare the effect of each extract using the concentration, the minimum concentration was 10 μg/mL, and the maximum concentration was 1000 μg/mL. However, the concentration is flexible and can be changed based on the degree of the blood vessel relaxation. Mechanism studies were conducted by selecting the extract with the greatest vasorelaxant effect.

The equation for calculating the degree of vasorelaxation is:

$$\text{Relaxation (\%)} = [\{(B - A) - (C - A)\}/(B - A)] \times 100$$

where A = the resting tension of aortic rings before pre-contraction with PE; B = the maximum contraction of aortic rings after pre-contraction using PE; and C = the contraction of the aortic rings after the drug treatment.

2.4.3. Effect of PBaE on Endothelium-Intact and Endothelium-Denuded Aortic Rings

To investigate whether the vascular endothelium participates in the vasorelaxant mechanism of PBaE, we measured the vasorelaxant effect of PBaE (10 μg/mL) with or without vascular endothelium on aortic rings pre-contracted with PE (1 μM) KH buffer.

2.4.4. Effect of PBaE on Endothelium-Intact Aortic Rings Pre-Incubated with L-NAME, Indomethacin, or Combination of L-NAME and Indomethacin

To determine the effect of PBaE on nitric oxide (NO), cyclooxygenase (COX), and prostacyclin (PGI_2), the endothelium-intact aortic rings were pre-incubated with an inhibitor, such as L-NAME (NO synthase inhibitor, 100 μM), indomethacin (COX inhibitor, 10 μM), and L-NAME (100 μM) + indomethacin (10 μM), for 20 min before pre-contraction using PE (1 μM). The cumulative concentration–response of PBaE (0.5–10 μg/mL) on the aortic ring was compared to that of the control (not treated with inhibitors).

2.4.5. Effect of PbaE on Endothelium-Intact Aortic Rings Pre-Incubated with ODQ or MB

To determine the effect of PbaE on soluble guanylate cyclase (sGC) or cyclic guanosine monophosphate (cGMP), the endothelium-intact aortic rings were pre-incubated with inhibitors, such as ODQ (sGC inhibitor, 10 μM) or MB (cGMP inhibitor, 10 μM), for 20 min before pre-contraction using PE (1 μM). The cumulative concentration–response of PbaE (0.5–10 μg/mL) on the aortic ring was compared to that of the control (not treated with inhibitors).

2.4.6. Effect of PbaE on Endothelium-Intact Aortic Rings Pre-Incubated with TEA, Glibenclamide, 4-AP, or $BaCl_2$

To examine the effect of PbaE on the non-selective calcium-activated K^+ (K_{Ca}), non-specific adenosine triphosphate-sensitive K^+ (K_{ATP}), voltage-dependent K^+ (K_V), and inwardly rectifying K^+ (K_{ir}) channel, the endothelium-intact aortic rings were pre-incubated with inhibitors, such as TEA (K_{Ca} blocker, 1 mM), glibenclamide (K_{ATP} blocker, 10 μM), 4-AP (K_V blocker, 1 mM), and $BaCl_2$ (K_{ir} blocker, 10 μM), for 20 min before pre-contraction using PE (1 μM). The cumulative concentration–response of PbaE (0.5–10 μg/mL) on the aortic ring was compared to that of the control group (not treated with inhibitors).

2.4.7. Effects of PbaE on Extracellular Ca^{2+}-Induced Contraction

To investigate the mechanism of the vasorelaxant effects through the receptor-operated calcium channel (ROCC), the rat thoracic aortic ring was pretreated using PBaE (10 μg/mL) in Ca^{2+}-free KH buffer, and PE was administered 10 min later to activate ROCC in the aortic rings. $CaCl_2$ (0.3–10 mM) was administered to the aortic ring in which the calcium channel was activated, and the inhibitory effect of PBaE on the vasoconstriction induced by Ca^{2+} was measured.

2.4.8. Inhibitory Effect of PBaE Pre-Treatment on Ang II-Induced Contraction

To investigate the vasorelaxant mechanism related to the angiotensin receptor, the aortic rings were pre-incubated with PBaE (10 μg/mL) for 20 min. Then, Ang II (10^{-9}–10^{-7} M) was cumulatively administered to measure the inhibitory effect of PBaE on the Ang II-induced vasoconstriction.

2.5. Blood Pressure Measurement

The systolic blood pressure (SBP) and diastolic blood pressure (DBP) of the SHRs were measured using the non-invasive tail-cuff method (CODA 8-Channel High Throughput Noninvasive Blood Pressure System, Kent Scientific Co., Ltd., Torrington, CT, USA). Measurements were taken after restraining the animals with an adjustable nose cone holder to restrict excessive movement and a rear gate with access to the base of the animal's tail. SBP and DBP of SHR were measured and recorded using an occlusion cuff and a volume pressure recording (VPR) cuff sensor (Figure 1). The 12 animals were randomly divided into three groups. Each group was orally administered PBaE (100 mg/kg), PBaE (300 mg/kg), and distilled water (control group). The blood pressure of the SHRs was measured before the administration and 1, 2, 4, and 8 h after the drug administration. During the experiment, the surface temperature of the animals was maintained at 32−35 °C using a heating pad.

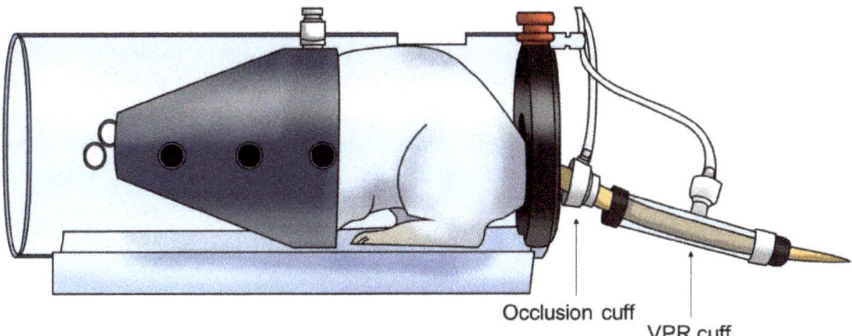

Figure 1. Schematic of the non-invasive tail-cuff method in rat.

2.6. Data Analysis

The values are expressed as the mean ± standard error of the mean (SEM) of n animals (for in vivo studies) or n aortic rings (for ex vivo studies). All data analyses were performed using GraphPad Prism 8 (GraphPad Software, San Diego, CA, USA). The concentration–response relationships were analyzed using an ordinary two-way analysis of variance followed by the Bonferroni's test. Unpaired Student's t-test was used for two group comparisons. A $p < 0.05$ was considered significant.

3. Results

3.1. Vasorelaxant Effects of PFrW and PFrE

The effects of PFrW (10–1000 μg/mL) and PFrE (10–1000 μg/mL) were compared to evaluate the vasorelaxant effect of the *P. mume* fruit extracts. PFrW did not significantly affect the aortic rings pre-contracted with PE (1 μM). Among them, PFrE caused a concentration-dependent relaxation on the endothelium-intact aortic ring. The half maximal effective concentration (EC_{50}) and maximal relaxation (R_{max}) were 363.8 ± 20.8 μg/mL and 37.0 ± 6.5%, respectively (Figure 2).

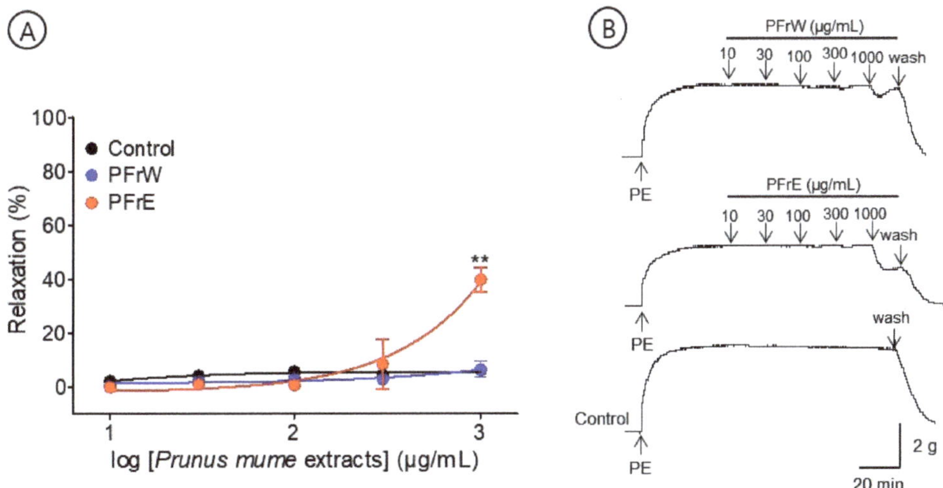

Figure 2. Cumulative concentration–response curves (**A**) and representative original traces (**B**) of *Prunus mume* fruit water extract (PFrW) and *Prunus mume* fruit 70% ethanol extract (PFrE) on endothelium-intact aortic rings. Relaxation was expressed as a percentage of phenylephrine hydrochloride (PE, 1 μM)-induced contraction. Values are expressed as mean ± SEM (n = 4–6). ** $p < 0.01$ vs. control.

3.2. Vasorelaxant Effect of PFlW and PFlE

The effects of PFlW (10–1000 µg/mL) and PFlE (10–1000 µg/mL) were compared to evaluate the vasorelaxant effect on the *P. mume* flower extracts. PFlW did not cause a significant effect on the aortic rings pre-contracted with PE (1 µM). Among them, PFlE caused the concentration-dependent relaxation of the endothelium-intact aortic ring. The EC_{50} and R_{max} were 96.5 ± 1.2 µg/mL and 33.5 ± 10.1%, respectively (Figure 3).

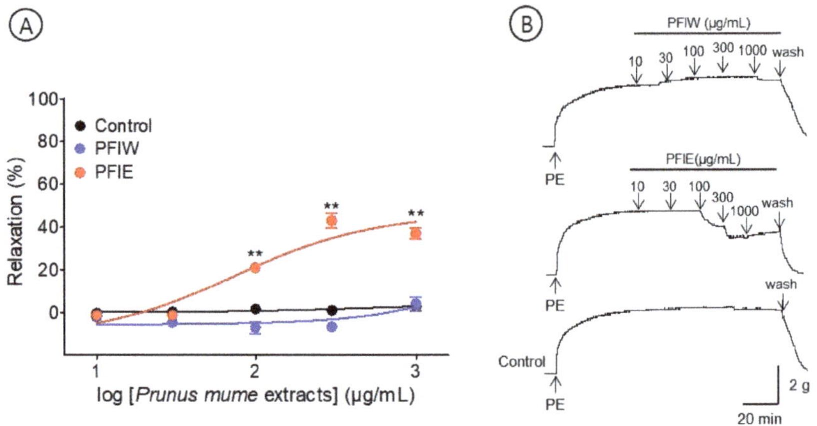

Figure 3. Cumulative concentration–response curves (**A**) and representative original traces (**B**) of *Prunus mume* flower water extract (PFlW) and *Prunus mume* flower 70% ethanol extract (PFlE) on endothelium-intact aortic rings. Relaxation was expressed as a percentage of phenylephrine hydrochloride (PE, 1 µM)-induced contraction. Values are expressed as mean ± SEM (n = 4–6). ** $p < 0.01$ vs. control.

3.3. Vasorelaxant Effects of PLW and PLE

The effects of PLW (10–1000 µg/mL) and PLE (10–1000 µg/mL) were compared to examine the vasorelaxant effect on the *P. mume* leaf extracts. PLW and PLE did not relax the pre-contracted aortic rings but caused a constriction at all concentrations (10–1000 µg/mL) (Figure 4).

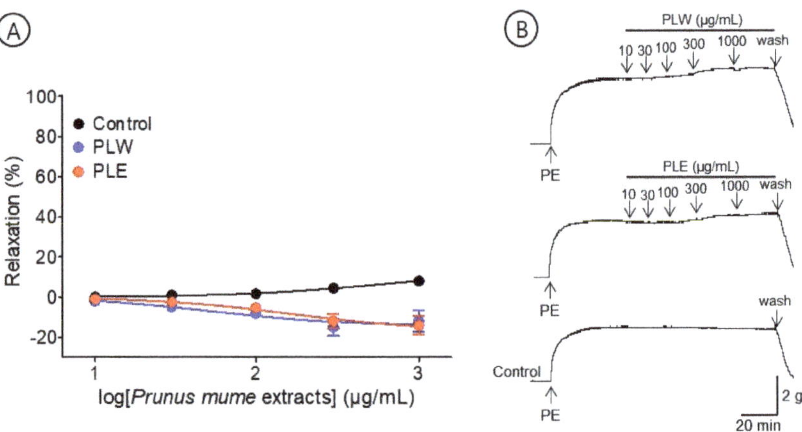

Figure 4. Cumulative concentration–response curves (**A**) and representative original traces (**B**) of *Prunus mume* leaf water extract (PLW) and *Prunus mume* leaf 70% ethanol extract (PLE) on endothelium-intact aortic rings. Relaxation was expressed as a percentage of phenylephrine hydrochloride (PE, 1 µM)-induced contraction. Values are expressed as mean ± SEM (n = 4–6).

3.4. Vasorelaxant Effects of PBrW and PBaW

To evaluate the effects of the branch and bark water extracts, we compared the vasorelaxant effects of PBrW (10–1000 µg/mL) and PBaW (10–1000 µg/mL). PBrW and PBaW caused the vasorelaxation of the endothelium-intact aortic rings pre-contracted with PE (1 µM). The EC_{50} and R_{max} for PBrW and PBaW were 78.4 ± 1.5 µg/mL and 55.0 ± 4.4% and 28.9 ± 1.0 µg/mL and 48.8 ± 4.1%, respectively (Figure 5).

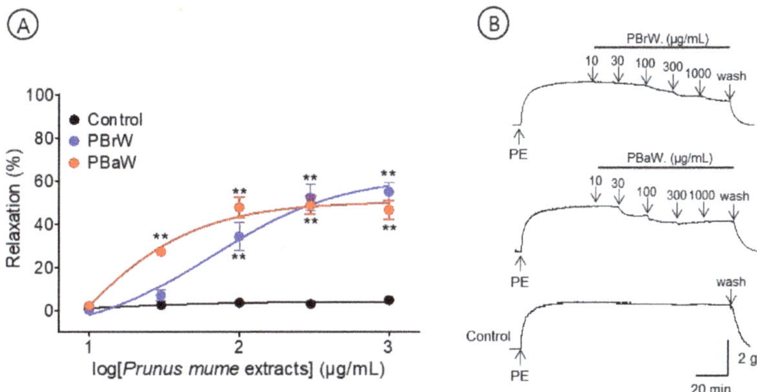

Figure 5. Cumulative concentration–response curves (**A**) and representative original traces (**B**) of *Prunus mume* branch water extract (PBrW) and *Prunus mume* bark water extract (PBaW) on endothelium-intact aortic rings. Relaxation was expressed as a percentage of phenylephrine hydrochloride (PE, 1 µM)-induced contraction. Values are expressed as mean ± SEM (n = 4–6). ** $p < 0.01$ vs. control.

3.5. Vasorelaxant Effects of PBrE and PBaE

To evaluate the effects of the branch and bark water extracts, we compared the vasorelaxant effects of PBrE (0.5–10 µg/mL) and PBaE (0.5–10 µg/mL). PBrE and PBaE caused the concentration-dependent relaxation of the endothelium-intact aortic ring pre-contracted with PE (1 µM). The EC_{50} and R_{max} for PBrE and PBaE were 4.0 ± 1.1 µg/mL and 42.8 ± 3.4% and 3.2 ± 1.0 µg/mL and 81.5 ± 2.7%, respectively (Figure 6).

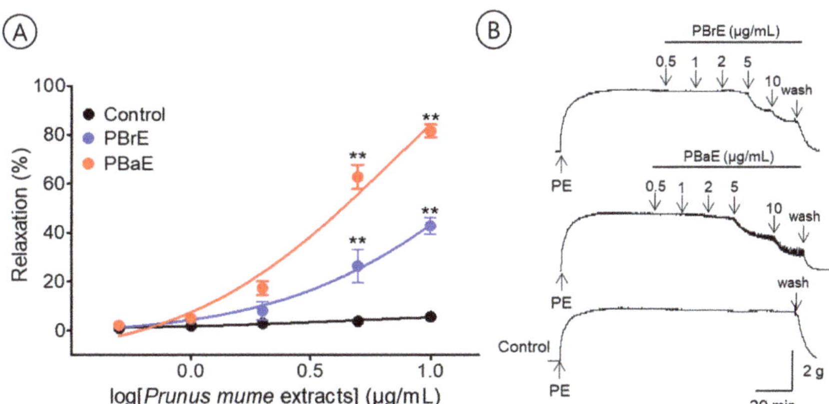

Figure 6. Cumulative concentration–response curves (**A**) and representative original traces (**B**) of *Prunus mume* branch 70% ethanol (PBrE) and *Prunus mume* bark 70% ethanol (PBaE) on endothelium-intact aortic rings. Relaxation was expressed as a percentage of phenylephrine hydrochloride (PE, 1 µM)-induced contraction. Values are expressed as mean ± SEM (n = 5–6). ** $p < 0.01$ vs. control.

3.6. Vasorelaxant Mechanism of PBaE

The PBaE was the most effective, therefore, it was investigated further for the mechanism of its vasorelaxant effect (Table 2). Mechanism studies were designed to evaluate whether the vasorelaxant effects of PBaE are related to the endothelium-dependent pathway, NO/sGC/cGMP pathway, PGI_2 pathway, potassium channel, calcium channel, or angiotensin receptor.

Table 2. EC_{50} and R_{max} of *Prunus mume* extract-induced vasorelaxation.

Samples	R_{max} (%)	EC_{50} (µg/mL)
PFrW	6.5 ± 2.9	-
PFrE	37.0 ± 6.5 **	363.8 ± 20.8
PFlW	4.0 ± 3.1	-
PFlE	33.5 ± 10.1 **	96.5 ± 1.2
PLW	−12.0 ± 5.2	-
PLE	−14.0 ±4.6	-
PBrW	55.0 ± 4.4 **	78.4 ± 1.5
PBrE	42.8 ± 3.4 **	4.0 ± 1.1
PBaW	48.8 ± 4.1 **	28.9 ±1.0
PBaE	81.5 ± 2.7 **	3.2 ± 1.0

Values are expressed as mean ± SEM (n = 4–6). ** $p < 0.01$ vs. control. EC_{50}, half maximal effective concentration; R_{max}, maximal relaxation; PFrW, *Prunus mume* fruit water extract; PFrE, *Prunus mume* fruit 70% ethanol extract; PFlW, *Prunus mume* flower water extract; PFlE, *Prunus mume* flower 70% ethanol extract; PLW, *Prunus mume* leaf water extract; PLE, *Prunus mume* flower 70% ethanol extract; PBrW, *Prunus mume* branch water extract; PBaW, *Prunus mume* bark water extract; PBrE, *Prunus mume* branch 70% ethanol extract; PBaE, *Prunus mume* bark 70% ethanol extract.

3.6.1. Vasorelaxant Effects of PBaE on Endothelium-Intact or Endothelium-Denuded Aortic Rings

The maximum relaxation effect concentration of PBaE, 10 µg/mL, was used in this experiment. PBaE (10 µg/mL) caused the vascular relaxation of the endothelium-intact aortic rings but did not induce the vascular relaxation of the endothelium-denuded aortic rings. The vasorelaxant effect in the PE-induced contraction was 84.5 ± 5.6% and 1.3 ± 0.3% for the endothelium-intact and endothelium-denuded aortic rings using 10 µg/mL, respectively (Figure 7).

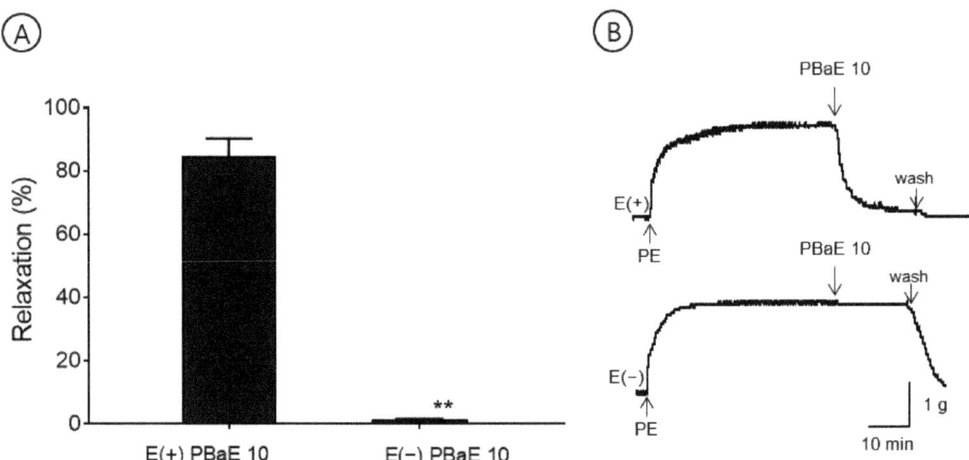

Figure 7. The vasorelaxant effect of *Prunus mume* bark in 70% ethanol (PBaE, 10 µg/mL) on intact [E(+)] or denuded [E(−)] endothelium rat aortic rings pre-contracted with phenylephrine hydrochloride (PE, 1 µM) (**A**). Representative traces under the indicated conditions (**B**). Values are expressed as mean ± SEM (n = 4–6). ** $p < 0.01$ vs. control.

3.6.2. Vasorelaxant Effect of PBaE on Endothelium-Intact Aortic Rings Pre-Incubated with L-NAME, Indomethacin, or L-NAME and Indomethacin Combined

A pre-incubation with L-NAME (100 µM) significantly decreased the PBaE-induced relaxation of endothelium-intact aortic rings pre-contracted using PE (1 µM). In the presence and absence of L-NAME, the maximum relaxation effect was 4.5 ± 2.3% and 81.5 ± 2.7%, respectively (Figure 8). A pre-incubation with indomethacin (10 µM) significantly decreased the PBaE-induced relaxation of the endothelium-intact aortic rings pre-contracted using PE (1 µM). With and without indomethacin, the maximum relaxation effect was 54.0 ± 1.9% and 81.5 ± 2.7%, respectively (Figure 8). A pre-incubation with L-NAME (100 µM) + indomethacin (10 µM) significantly decreased the PBaE-induced relaxation of endothelium-intact aortic rings pre-contracted with PE (1 µM). In the presence and absence of L-NAME (100 µM) + indomethacin (10 µM), the maximum relaxation effect was 3.2 ± 0.9% and 81.5 ± 2.7%, respectively (Figure 8).

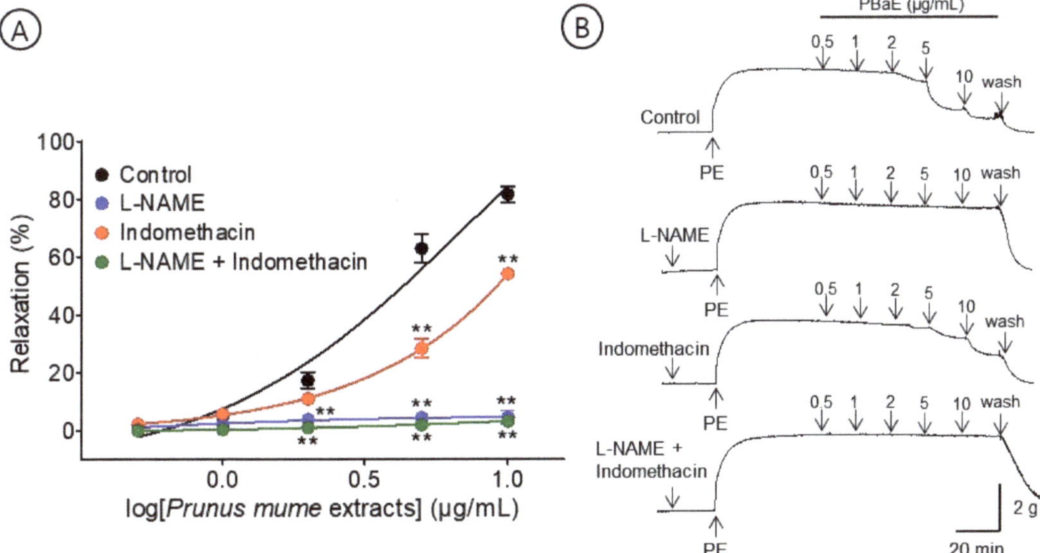

Figure 8. Cumulative concentration–response curves (**A**) and representative original traces (**B**) of vasorelaxant effect for *Prunus mume* bark in 70% ethanol (PBaE, 0.5–10 µg/mL) in the absence (control) or presence of NG-nitro-L-arginine methyl ester (L-NAME (100 µM), indomethacin (10 µM), or L-NAME (100 µM) + indomethacin (10 µM). Values are expressed as mean ± SEM (n = 4–6). ** $p < 0.01$ vs. control.

3.6.3. Vasorelaxant Effect of PBaE on Endothelium-Intact Aortic Rings Pre-Incubated with ODQ or MB

A pre-incubation with MB (10 µM) significantly decreased the PBaE-induced relaxation of endothelium-intact aortic rings pre-contracted using PE (1 µM). In the presence and absence of ODQ, the maximum relaxation effect was 4.7 ± 0.4% and 81.5 ± 2.7%, respectively. A pre-incubation with MB (10 µM) significantly decreased the PBaE-induced relaxation of endothelium-intact aortic rings pre-contracted with PE (1 µM). In the presence and absence of MB, the maximum relaxation effect was 17.0 ± 3.4% and 81.5 ± 2.7%, respectively (Figure 9).

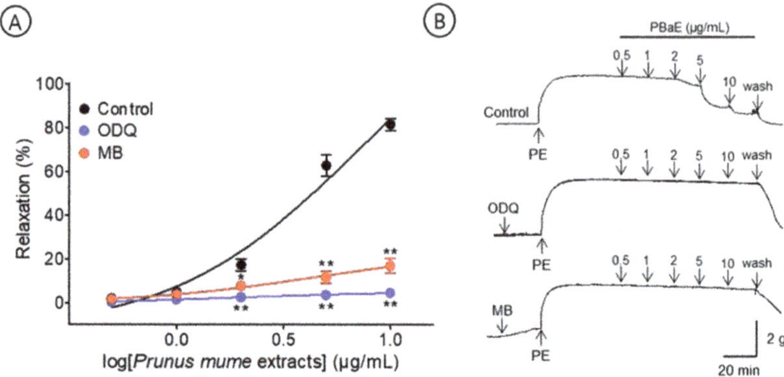

Figure 9. Cumulative concentration–response curves (**A**) and representative original traces (**B**) of vasorelaxant effect for *Prunus mume* bark in 70% ethanol (PBaE, 0.5–10 μg/mL) with (control) or without 1H-[1,2,4]Oxadiazolo [4,3-a]quinoxalin-1-one (ODQ, 10 μM) or methylene blue (MB, 10 μM). Values are expressed as mean ± SEM (n = 4–6). * $p < 0.05$, ** $p < 0.01$ vs. control.

3.6.4. Vasorelaxant Effect of PBaE on Endothelium-Intact Aortic Rings Pre-Incubated with TEA, Glibenclamide, 4-AP, or BaCl₂

A pre-incubation with potassium channel blockers, such as TEA, glibenclamide, 4-AP, and BaCl₂, significantly decreased the PBaE-induced relaxation on endothelium-intact aortic rings pre-contracted with PE (1 μM). Using TEA (1 mM), glibenclamide (10 μM), 4-AP (1 mM), or BaCl₂ (10 μM), the maximum relaxation effects were $42.5 \pm 2.3\%$, $71.7 \pm 2.5\%$, 38.0 ± 6.4, or 43.2 ± 6.1 at the 10 μg/mL, respectively (Figure 10).

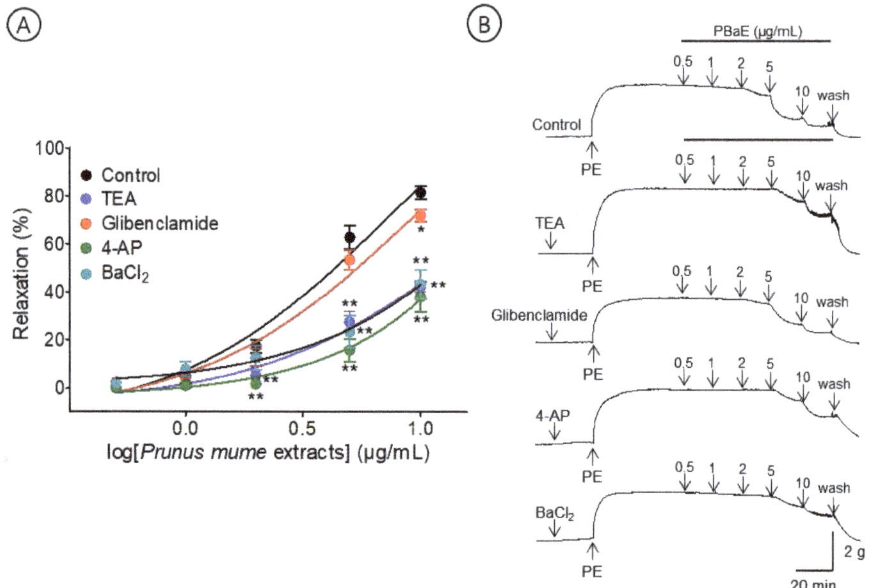

Figure 10. Cumulative concentration–response curves (**A**) and representative original traces (**B**) of vasorelaxant effect for *Prunus mume* bark in 70% ethanol (PBaE, 0.5–10 μg/mL) without (control) or with tetraethylammonium (TEA; 1 mM), glibenclamide (10 μM), 4-aminopyridine (4-AP; 1 mM), or barium chloride (BaCl₂, 10 μM). Values are expressed as mean ± SEM (n = 4–6). * $p < 0.05$, ** $p < 0.01$ vs. control.

3.6.5. Vasorelaxant Effect of PBaE on Extracellular Ca^{2+}-Induced Contraction

The cumulative addition of CaCl$_2$ (0.3–10 mM) gradually contracted the tension of the aortic rings pretreated with PE (1 μM) in the Ca^{2+}-free KH buffer. However, the PBaE (10 μg/mL) pre-treatment did not significantly differ from the control group (Figure 11).

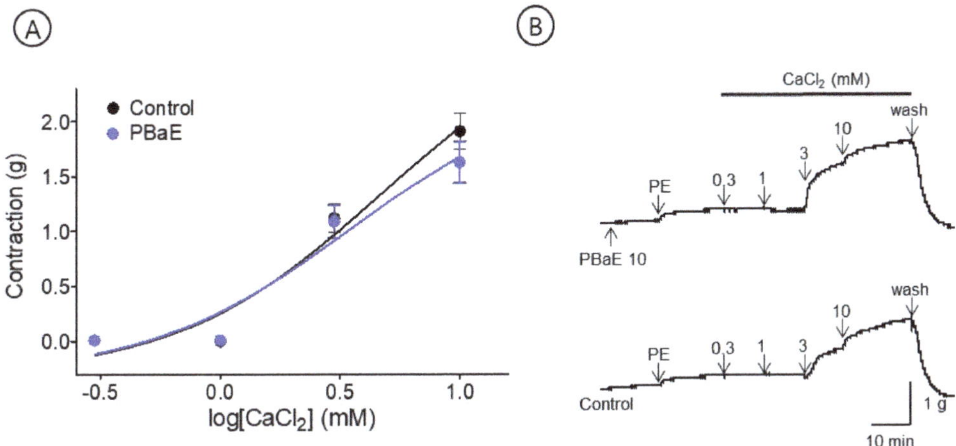

Figure 11. Inhibitory effect of *Prunus mume* bark in 70% ethanol (PBaE, 10 μg/mL) on the contraction induced by extracellular calcium chloride (CaCl$_2$, 0.3–10 mM) on aortic rings that were pre-contracted with phenylephrine hydrochloride (PE, 1 μM) (**A**). Representative traces under the indicated conditions (**B**). Values are expressed as mean ± SEM (n = 4).

3.6.6. Inhibitory Effect of PBaE Pre-Treatment on Ang II-Induced Contraction

An experiment was performed to evaluate the inhibitory effect of the PBaE (10 μg/mL) on Ang II (10^{-9}–10^{-7} M)-induced vasoconstriction in the endothelium-intact aortic rings. The PBaE pre-treatment significantly reduced the Ang II-induced contractions. The degree of contraction decreased to 0.60 ± 0.11 g compared to the control group and 1.28 ± 0.11 g at Ang II 10^{-7} M concentration, respectively (Figure 12).

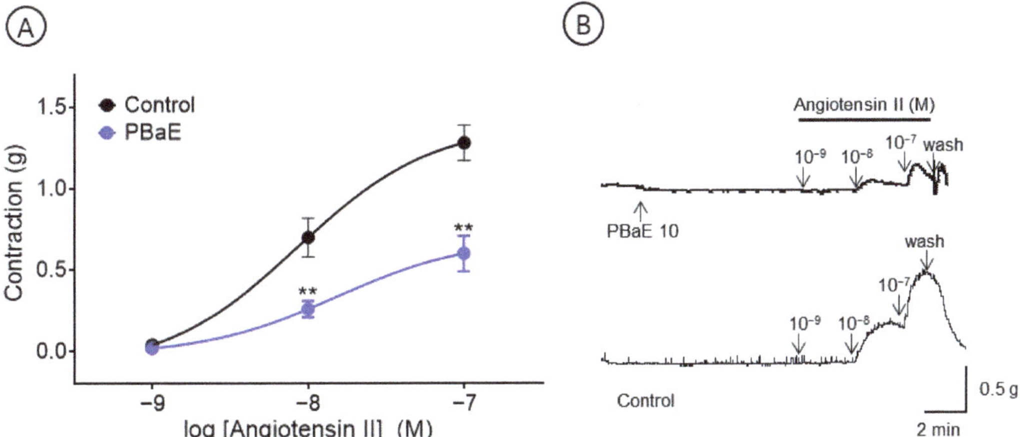

Figure 12. Inhibitory effect (**A**) and original representative traces (**B**) of *Prunus mume* bark in 70% ethanol (PBaE, 10 μg/mL) in the contraction induced by angiotensin II (Ang II, 10^{-9}–10^{-7} M) on endothelium-intact aortic rings. Values are expressed as mean ± SEM (n = 6). ** $p < 0.01$ vs. control.

3.7. Hypotensive Effect of PBaE on Blood Pressure in SHR

To investigate the hypotensive effect of PBaE, SBP and DBP were measured 1, 2, 4, and 8 h after administering 100 or 300 mg/Kg of PBaE orally to SHR. At 4 h after administering PBaE 300 mg/kg, the SBP was significantly lowered from 210.0 ± 2.4 mmHg to 187.6 ± 8.7 mmHg, and the DBP decreased from 164.1 ± 3.2 mmHg to 133.0 ± 5.8 mmHg (Figure 13). Due to the characteristics of SHR, there was a difference in the blood pressure for each rat. A significant trend was confirmed by comparing the individual blood pressure values (Table 3).

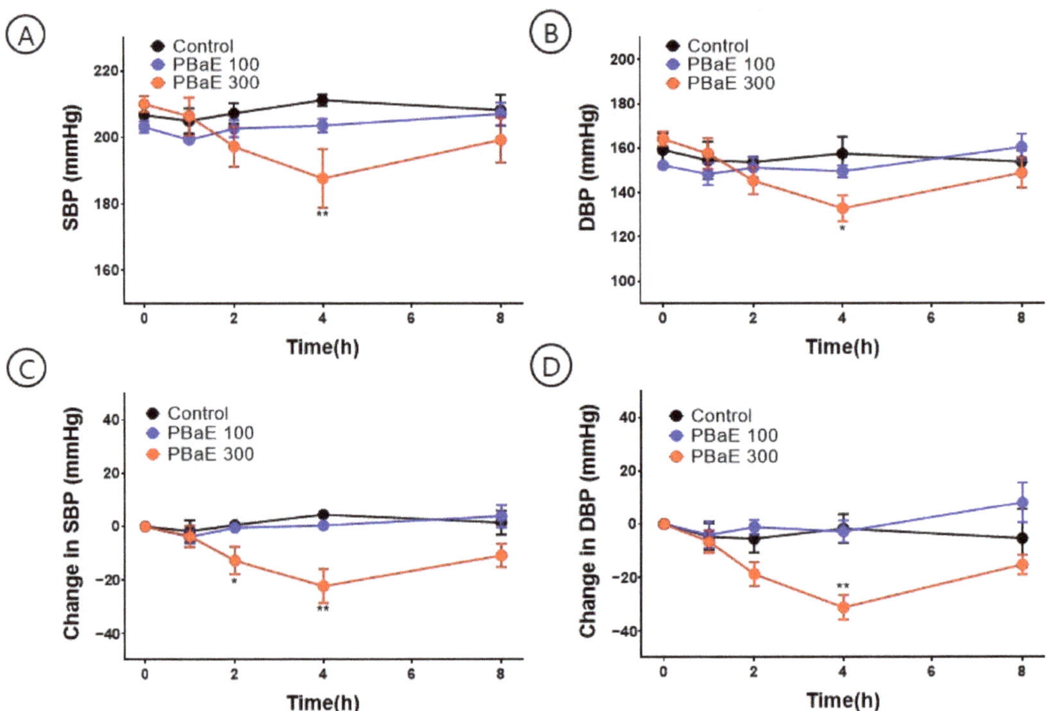

Figure 13. Hypotensive effect of *Prunus mume* in 70% ethanol extract (PBaE) in a spontaneously hypertensive rat (SHR). Systolic blood pressure (SBP) and diastolic blood pressure (DBP) were measured using the non-invasive tail-cuff system (**A,B**). Changes in SBP (**C**) and DBP (**D**) caused SHR by administering water (control), PBaE (100 mg/kg), or PBaE (300 mg/kg). The values are expressed as the mean $\pm$ SEM (n = 4). * $p < 0.05$, ** $p < 0.01$ vs. control.

Table 3. Effect of PBaE on blood pressure in SHR.

	Time (h)				
	0	**1**	**2**	**4**	**8**
	Systolic blood pressure (mmHg)				
Control	206.6 ± 2.8	204.9 ± 3.7	207.1 ± 3.0	211.1 ± 1.7	208.1 ± 4.6
PBaE 100 mg/kg	203.1 ± 1.7	199.2 ± 1.3	202.6 ± 2.5	203.5 ± 2.0	207.0 ± 3.4
PBaE 300 mg/kg	210.0 ± 2.4	206.3 ± 5.7	197.2 ± 6.0	187.6 ± 8.7 **	199.1 ± 6.7
	Diastolic blood pressure (mmHg)				
Control	159.2 ± 7.3	154.5 ± 8.4	153.7 ± 2.5	157.5 ± 7.5	153.8 ± 4.9
PBaE 100 mg/kg	152.3 ± 1.9	148.3 ± 4.9	151.2 ± 4.6	149.6 ± 2.6	160.4 ± 5.9
PBaE 300 mg/kg	164.1 ± 3.2	157.6 ± 6.9	145.4 ± 6.1	133.0 ± 5.8 *	148.8 ± 6.8

Values are expressed as mean $\pm$ SEM (n = 4). * $p < 0.05$, ** $p < 0.01$ vs. control. PBaE, *Prunus mume* bark 70% ethanol extract; SHR, spontaneously hypertensive rat.

4. Discussion

In this study, the fruits, flowers, leaves, branches, and bark of the *P. mume* were collected, and the extracts were prepared using two solvents: water and 70% ethanol. According to the results of the investigations on its vasorelaxant effects, 70% ethanol extracts of fruit, flower, branches, and bark had a vasorelaxant activity. Among them, branch and bark water extracts also had a vasorelaxant effect. However, the leaf water extract and 70% ethanol extract caused a vasoconstriction. The results reveal that the solvents of different polarities can extract various biologically active compounds, demonstrating a difference in the biological activity of each part, even within the same plant. Except for PBrW and PBaW, PFrE, PFlE, PBrE, and PBaE exhibited a vasorelaxant activity, suggesting that using an organic solvent, such as ethanol, to extract specific active ingredients that induce vasodilation may be advantageous. Among them, the vasorelaxant effects of PBrE and PBaE were 42.8 ± 3.4% and 81.5 ± 2.7%, respectively, at a relatively low concentration of 10 μg/mL, exhibiting strong vasorelaxant effects. Considering that the branch and bark of *P. mume* were not a single compound but a natural product consisting of a mixture of various compounds, it was a very effective vasodilator for regulating the tone of the blood vessels. In a previous study, amlodipine, a representative calcium channel blocker for treating hypertension, had a vasorelaxant effect of up to 48.6 ± 3.5% at 10 μg/mL [31]. Based on the vasodilation screening, PBaE was the most potent vasorelaxant, and a vasorelaxant mechanism and hypotensive effect study were performed on PBaE. The mechanism studies evaluated whether the vasorelaxant effect of PBaE was related to the endothelium-dependent pathway, NO/sGC/cGMP pathway, PGI$_2$ pathway, potassium channel, calcium channel, or angiotensin receptor.

The vascular endothelium lies at the border between circulating blood cells and vascular smooth muscle cells and is essential in regulating the blood flow and vascular tone [32]. Vascular endothelium can synthesize and release different vasodilators, such as NO, PGI$_2$, and the endothelium-derived hyperpolarizing factor [33,34]. The release of these substances causes an endothelium-dependent vasorelaxation in the rat thoracic aorta [35,36]. NO is produced from L-arginine in vascular endothelial cells under the catalysis of NO synthase and activates sGC to induce a cGMP-mediated vasodilation [37,38]. Additionally, PGI$_2$ is generated from arachidonic acid by the catalytic action of COX and increases the cyclic adenosine monophosphate levels through an adenylate cyclase activation to induce vasodilation [39]. In the present study, PBaE induced vasorelaxation in endothelium-intact aortic rings pre-contracted with PE; however, this relaxation was significantly abrogated by removing the vascular endothelium. These results indicate that PBaE acts on vascular endothelial cells to stimulate vasodilators to mediate its endothelium-dependent vasorelaxation. Additionally, the endothelium-dependent vasorelaxation of PBaE was investigated using inhibitors, such as L-NAME (NO synthase inhibitor), ODQ (sGC inhibitor), MB (cGMP inhibitor), or indomethacin (COX inhibitor). The vasorelaxant effects of PBaE were significantly reduced by indomethacin and significantly inhibited by L-NAME, ODQ, MB, or L-NAME and indomethacin combined. Therefore, the results revealed that the PBaE vasodilation was mainly exerted through the NO/sGC/cGMP and PGI$_2$ pathways.

The potassium channels also play a vital role in regulating the muscle contraction and vascular tone [40]. Four types of potassium channels exist in the arterial smooth muscle: K_{Ca}, K_{ATP}, K_V, and K_{ir} channels. The activation of the potassium channels in vascular smooth muscle cells causes vasodilation by hyperpolarizing the cell membrane due to the efflux of K^+ [41]. The results of this study revealed that the PBaE-induced relaxation in endothelium-intact aortic rings was reduced by the treatment with potassium channel blockers, including TEA, glibenclamide, 4-AP, and BaCl$_2$. These data indicate that potassium channel activation in the vascular smooth muscle and endothelium, including K_{Ca}, K_{ATP}, K_V, and K_{ir} channels, which may involve a PBaE-induced vasorelaxation.

Ang II is a final product of the renin-angiotensin system, which causes a vasoconstriction and increases the blood pressure by binding to the angiotensin receptor type 1

(AT-1) [42]. Therefore, the blood pressure and vascular tone can be controlled by using Ang II receptor blockers (ARB) that block Ang II from binding to AT-1 [43] or using angiotensin-converting enzyme inhibitors that inhibit the production of Ang II by directly acting on the converting enzyme [44]. Our results revealed that PBaE significantly reduced the degree of contractility induced by Ang II (10^{-9}–10^{-7} M) by almost 50% in the endothelium-intact aortic rings. This suggests that PBaE replaces the function of ARB, inhibiting Ang II from binding to the angiotensin receptor. However, future studies are needed for more precise mechanisms by which PBaE contributes to regulating the vasoconstriction.

To evaluate the hypotensive effect of PBaE, the SBP and DBP of SHR were measured using the non-invasive tail-cuff method. They significantly decreased 4 h after the oral administration of PBaE (300 mg/kg).

In the present study, PBaE lowered the blood pressure in SHR and induced the vascular endothelium-dependent relaxation of isolated rat aortic rings via the NO/sGC/cGMP and PGI$_2$ pathway mechanisms in the vascular smooth muscle. In addition, the potassium channels, such as K_{Ca}, K_{ATP}, K_V, and K_{ir} channels, were partially associated with a PBaE-induced vasorelaxation. Therefore, PBaE can be developed as food or medicine to help prevent or treat high blood pressure. However, in this study, the changes in the blood pressures of the SHR were only measured for 8 h using a tail-cuff experiment to assess the antihypertensive effect of PBaE.

5. Conclusions

In conclusion, the vasorelaxant effect of PBaE was endothelium-dependent and was related to the NO/sGC/cGMP vascular prostacyclin pathway. In addition, potassium channels, such as K_{Ca}, K_{ATP}, K_V, and K_{ir}, were partially related to the PBaE-induced vasorelaxation. PBaE was effective in relaxing the contraction induced by Ang II, and the vasorelaxant effects of PBaE were unassociated with the influx of extracellular Ca^{2+} via ROCC. Furthermore, the SBP and DBP of SHR significantly decreased 4 h after the oral administration of PBaE (300 mg/kg). Our findings provide a basis for the use of the bark of *Prunus mume* as a medicinal and food resource. In future studies, the comparative evaluation of non-polar solvent extracts and safety and stability analyses, including the identification and standardization of the active ingredients, the determination of the appropriate dose, and a toxicity evaluation, should be conducted.

6. Patents

On 10 June 2021, a patent was registered for composition for preventing and/or treating a hypertensive disease comprising an extract of *Prunus mume* Siebold et Zuccarini or a fraction thereof as an active ingredient (Registration number: 10-2265786).

Author Contributions: Conceptualization, C.J. and H.-Y.C.; methodology, C.J.; software, C.J.; validation, B.K., K.L., and C.J.; investigation, B.K. and C.J.; resources, H.-Y.C.; data curation, H.-Y.C.; writing—original draft preparation, C.J.; writing—review and editing, C.J. and H.-Y.C.; visualization, K.L.; supervision, H.-Y.C. All authors have read and agreed to the published version of the manuscript.

Funding: This study was carried out with the support of 'R&D Program for Forest Science Technology (Project No. 2020250A00-2021-0001)' provided by Korea Forest Service (Korea Forestry Promotion Institute).

Institutional Review Board Statement: The study was conducted according to the guidelines of the Declaration of Helsinki and approved by the Institutional Animal Care and Use Committee of Kyung Hee University (KHSASP-21-050).

Informed Consent Statement: Not applicable.

Data Availability Statement: Not applicable.

Acknowledgments: We thank Changseuk Cho and Yangsoon Lim for their assistance in collecting plant materials.

Conflicts of Interest: The authors declare no conflict of interest.

References

1. World Health Organization. Available online: https://www.who.int/news-room/fact-sheets/detail/hypertension (accessed on 16 December 2022).
2. World Health Organization. *Guideline for the Pharmacological Treatment of Hypertension in Adults: Web Annex A: Summary of Evidence*; World Health Organization: Geneva, Switzerland, 2021.
3. Atanasov, A.G.; Zotchev, S.B.; Dirsch, V.M.; International Natural Product Sciences Taskforce; Supuran, C.T. Natural products in drug discovery: Advances and opportunities. *Nat. Rev. Drug Discov.* **2021**, *20*, 200–216. [CrossRef] [PubMed]
4. Ajebli, M.; Eddouks, M. Phytotherapy of hypertension: An updated overview. *Endocr. Metab. Immune Disord. Drug Targets* **2020**, *20*, 812–839. [CrossRef] [PubMed]
5. Newman, D.J.; Cragg, G.M. Natural products as sources of new drugs over the nearly four decades from 01/1981 to 09/2019. *J. Nat. Prod.* **2020**, *83*, 770–803. [CrossRef] [PubMed]
6. Zhang, Q.; Chen, W.; Sun, L.; Zhao, F.; Huang, B.; Yang, W.; Tao, Y.; Wang, J.; Yuan, Z.; Fan, G. The genome of *Prunus mume*. *Nat. Commun.* **2012**, *3*, 1318. [CrossRef]
7. Jo, C.; Kim, B.; Lee, S.; Ham, I.; Lee, K.; Choi, H.Y. Vasorelaxant Effect of *Prunus mume* (Siebold) Siebold & Zucc. Branch through the endothelium-dependent pathway. *Molecules* **2019**, *24*, 3340.
8. Shi, J.; Gong, J.; Liu, J.; Wu, X.; Zhang, Y. Antioxidant capacity of extract from edible flowers of *Prunus mume* in China and its active components. *LWT Food Sci. Technol.* **2009**, *42*, 477–482. [CrossRef]
9. Kim, J.H.; Won, Y.S.; Cho, H.D.; Hong, S.M.; Moon, K.D.; Seo, K.I. Protective Effect of *Prunus mume* Fermented with Mixed lactic acid Bacteria in dextran sodium sulfate-Induced Colitis. *Foods* **2020**, *10*, 58. [CrossRef]
10. Yingsakmongkon, S.; Miyamoto, D.; Sriwilaijaroen, N.; Fujita, K.; Matsumoto, K.; Jampangern, W.; Hiramatsu, H.; Guo, C.T.; Sawada, T.; Takahashi, T.; et al. In vitro inhibition of human influenza A virus infection by fruit-juice concentrate of Japanese plum (*Prunus mume* SIEB. et ZUCC). *Biol. Pharm. Bull.* **2008**, *31*, 511–515. [CrossRef]
11. Bae, J.-H.; Kim, K.-J.; Kim, S.-M.; Lee, W.-J.; Lee, S.-J. Development of the functional beverage containing the *Prunus mume* extracts. *Korean J. Food Sci. Technol.* **2000**, *32*, 713–719.
12. Kim, J.-H.; Cho, H.-D.; Won, Y.-S.; Hong, S.-M.; Moon, K.-D.; Seo, K.-I. Anti-fatigue effect of *Prunus mume* vinegar in high-intensity exercised rats. *Nutrients* **2020**, *12*, 1205. [CrossRef]
13. Park, L.-Y.; Chae, M.-H.; Lee, S.-H. Effect of ratio of maesil (*Prunus mume*) and alcohol on quality changes of maesil liqueur during leaching and ripening. *Korean J. Food Preserv.* **2007**, *14*, 645–649.
14. Kim, N.-Y.; Eom, M.-N.; Do, Y.-S.; Kim, J.-B.; Kang, S.-H.; Yoon, M.-H.; Lee, J.-B. Determination of ethyl carbamate in maesil wine by alcohol content and ratio of maesil (*Prunus mume*) during ripening period. *Korean J. Food Preserv.* **2013**, *20*, 429–434. [CrossRef]
15. Mun, K.-H.; Lee, H.-C.; Jo, A.-H.; Lee, S.-H.; Kim, N.-Y.-S.; Park, E.-J.; Kang, J.-Y.; Kim, J.-B. Effect of sugared sweeteners on quality characteristics of *Prunus mume* fruit syrup. *Korean J. Food Nutr.* **2019**, *32*, 161–166.
16. Kim, Y.-D.; Jeong, M.-H.; Koo, I.-R.; Cho, I.-K.; Kwak, S.-H.; Na, R.; Kim, K.-J. Analysis of volatile compounds of *Prunus mume* flower and optimum extraction conditions of *Prunus mume* flower tea. *Korean J. Food Preserv.* **2006**, *13*, 180–185.
17. Gong, X.-P.; Tang, Y.; Song, Y.-Y.; Du, G.; Li, J. Comprehensive Review of Phytochemical Constituents, Pharmacological Properties, and Clinical Applications of *Prunus mume*. *Front. Pharmacol.* **2021**, *12*, 679378. [CrossRef] [PubMed]
18. Imahori, Y.; Takemura, M.; Bai, J. Chilling-induced oxidative stress and antioxidant responses in mume (*Prunus mume*) fruit during low temperature storage. *Postharvest Biol. Technol.* **2008**, *49*, 54–60. [CrossRef]
19. Hwang, J.-Y.; Ham, J.-W.; Nam, S.-H. The antioxidant activity of maesil (*Prunus mume*). *Korean J. Food Sci. Technol.* **2004**, *36*, 461–464.
20. Jin, H.-L.; Lee, B.-R.; Lim, K.-J.; Debnath, T.; Shin, H.-M.; Lim, B.-O. Anti-inflammatory effects of *Prunus mume* mixture in colitis induced by dextran sodium sulfate. *Korean J. Med. Crop Sci.* **2011**, *19*, 16–23. [CrossRef]
21. Jeong, J.T.; Moon, J.H.; Park, K.H.; Shin, C.S. Isolation and characterization of a new compound from *Prunus mume* fruit that inhibits cancer cells. *J. Agric. Food Chem.* **2006**, *54*, 2123–2128. [CrossRef]
22. Bailly, C. Anticancer properties of *Prunus mume* extracts (Chinese plum, Japanese apricot). *J. Ethnopharmacol.* **2020**, *246*, 112215. [CrossRef]
23. Yan, X.T.; Lee, S.H.; Li, W.; Sun, Y.N.; Yang, S.Y.; Jang, H.D.; Kim, Y.H. Evaluation of the antioxidant and anti-osteoporosis activities of chemical constituents of the fruits of *Prunus mume*. *Food Chem.* **2014**, *156*, 408–415. [CrossRef]
24. Xia, D.; Wu, X.; Yang, Q.; Gong, J.; Zhang, Y. Anti-obesity and hypolipidemic effects of a functional formula containing Prumus mume in mice fed high-fat diet. *Afr. J. Biotechnol.* **2010**, *9*, 2463–2467.
25. Babarikina, A.; Nikolajeva, V.; Babarykin, D. Anti-Helicobacter activity of certain food plant extracts and juices and their composition in vitro. *Food Nutr. Sci.* **2011**, *2*, 7889.
26. Chuda, Y.; Ono, H.; Ohnishi-Kameyama, M.; Matsumoto, K.; Nagata, T.; Kikuchi, Y. Citric acid derivative improving blood fluidity from fruit-juice concentrate of Japanese apricot. *J. Agric. Food Chem.* **1999**, *47*, 828–831. [CrossRef] [PubMed]
27. Kono, R.; Nakamura, M.; Nomura, S.; Kitano, N.; Kagiya, T.; Okuno, Y.; Inada, K.I.; Tokuda, A.; Utsunomiya, H.; Ueno, M. Biological and epidemiological evidence of anti-allergic effects of traditional Japanese food ume (*Prunus mume*). *Sci. Rep.* **2018**, *8*, 11638. [CrossRef] [PubMed]

28. Khan, A.; Pan, J.H.; Cho, S.; Lee, S.; Kim, Y.J.; Park, Y.H. Investigation of the hepatoprotective effect of *Prunus mume* Sieb. et Zucc extract in a mouse model of alcoholic liver injury through high-resolution metabolomics. *J. Med. Food* **2017**, *20*, 734–743. [CrossRef]
29. Jung, B.G.; Ko, J.H.; Cho, S.J.; Koh, H.B.; Yoon, S.R.; Han, D.U.; Lee, B.J. Immune-enhancing effect of fermented maesil (*Prunus mume* Siebold & Zucc.) with probiotics against Bordetella bronchiseptica in mice. *J. Vet. Med. Sci.* **2010**, *72*, 1195–1202. [CrossRef]
30. Takemura, S.; Yoshimasu, K.; Fukumoto, J.; Mure, K.; Nishio, N.; Kishida, K.; Yano, F.; Mitani, T.; Takeshita, T.; Miyashita, K. Safety and adherence of Umezu polyphenols in the Japanese plum (*Prunus mume*) in a 12-week double-blind randomized placebo-controlled pilot trial to evaluate antihypertensive effects. *Environ. Health Prev. Med.* **2014**, *19*, 444–451. [CrossRef] [PubMed]
31. Kim, B.; Jo, C.; Choi, H.Y.; Lee, K. Vasorelaxant and hypotensive effects of Cheonwangbosimdan in SD and SHR rats. *Evid. Based Complement. Altern. Med.* **2018**, *2018*, 6128604. [CrossRef]
32. Furchgott, R.F.; Vanhoutte, P.M. Endothelium-derived relaxing and contracting factors. *FASEB J.* **1989**, *3*, 2007–2018. [CrossRef]
33. Vanhoutte, P.M. Endothelium and control of vascular function. State of the Art lecture. *Hypertension* **1989**, *13*, 658–667. [CrossRef] [PubMed]
34. Moncada, S.; Gryglewski, R.; Bunting, S.; Vane, J.R. An enzyme isolated from arteries transforms prostaglandin endoperoxides to an unstable substance that inhibits platelet aggregation. *Nature* **1976**, *263*, 663–665. [CrossRef]
35. Hongo, K.; Nakagomi, T.; Kassell, N.F.; Sasaki, T.; Lehman, M.; Vollmer, D.G.; Tsukahara, T.; Ogawa, H.; Torner, J. Effects of aging and hypertension on endothelium-dependent vascular relaxation in rat carotid artery. *Stroke* **1988**, *19*, 892–897. [CrossRef]
36. Di Wang, H.D.; Pagano, P.J.; Du, Y.; Cayatte, A.J.; Quinn, M.T.; Brecher, P.; Cohen, R.A. Superoxide anion from the adventitia of the rat thoracic aorta inactivates nitric oxide. *Circ. Res.* **1998**, *82*, 810–818. [CrossRef]
37. Archer, S.L.; Huang, J.M.; Hampl, V.; Nelson, D.P.; Shultz, P.J.; Weir, E.K. Nitric oxide and cGMP cause vasorelaxation by activation of a charybdotoxin-sensitive K channel by cGMP-dependent protein kinase. *Proc. Natl Acad. Sci. USA* **1994**, *91*, 7583–7587. [CrossRef]
38. Rapoport, R.M.; Murad, F. Agonist-induced endothelium-dependent relaxation in rat thoracic aorta may be mediated through cGMP. *Circ. Res.* **1983**, *52*, 352–357. [CrossRef] [PubMed]
39. Yang, S.; Xu, Z.; Lin, C.; Li, H.; Sun, J.; Chen, J.; Wang, C. Schisantherin A causes endothelium-dependent and -independent vasorelaxation in isolated rat thoracic aorta. *Life Sci.* **2020**, *245*, 117357. [CrossRef]
40. Jackson, W.F. Ion channels and vascular tone. *Hypertension* **2000**, *35*, 173–178. [CrossRef]
41. Nelson, M.T.; Quayle, J.M. Physiological roles and properties of potassium channels in arterial smooth muscle. *Am. J. Physiol.* **1995**, *268*, C799–C822. [CrossRef] [PubMed]
42. Griendling, K.K.; Murphy, T.J.; Alexander, R.W. Molecular biology of the renin-angiotensin system. *Circulation* **1993**, *87*, 1816–1828. [CrossRef]
43. Burnier, M.; Brunner, H. Angiotensin II receptor antagonists. *Lancet* **2000**, *355*, 637–645. [CrossRef] [PubMed]
44. Clozel, M.; Kuhn, H.; Hefti, F. Effects of angiotensin converting enzyme inhibitors and of hydralazine on endothelial function in hypertensive rats. *Hypertension* **1990**, *16*, 532–540. [CrossRef] [PubMed]

 *bioengineering*

Article

Japanese Knotweed Rhizome Bark Extract Inhibits Live SARS-CoV-2 In Vitro

Urška Jug [1,*,†], Katerina Naumoska [1,*,†] and Tadej Malovrh [2,*,†]

1   Laboratory for Food Chemistry, Department of Analytical Chemistry, National Institute of Chemistry, Hajdrihova 19, 1000 Ljubljana, Slovenia
2   Veterinary Faculty, University of Ljubljana, Gerbičeva ulica 60, 1000 Ljubljana, Slovenia
*   Correspondence: urska.jug@ki.si (U.J.); katerina.naumoska@ki.si (K.N.); tadej.malovrh@vf.uni-lj.si (T.M.); Tel.: +386-1-4760-521 (U.J. & K.N.); +386-1-4779-824 (T.M.)
†   These authors contributed equally to this work.

Abstract: Coronavirus disease 2019 (COVID-19), a viral infectious respiratory disease, is caused by highly contagious severe acute respiratory syndrome coronavirus 2 (SARS-CoV-2) and is responsible for the ongoing COVID-19 pandemic. Since very few drugs are known to be effective against SARS-CoV-2, there is a general need for new therapeutics, including plant-based drugs, for the prophylaxis and treatment of infections. In the current study, the activity of a 70% ethanolic$_{(aq)}$ extract of the rhizome bark of Japanese knotweed, an invasive alien plant species, was tested for the first time against the wild-type SARS-CoV-2 virus using a specific and robust virus neutralization test (VNT) on Vero-E6 cells, which best mimics the mechanism of real virus–host interaction. A statistically significant antiviral effect against SARS-CoV-2 (p-value < 0.05) was observed for the 50.8 µg mL^{-1} extract solution in cell medium. A suitable extract preparation was described to avoid loss of polyphenols throughout filtration of the extract, which was dissolved in cell medium containing fetal bovine serum (FBS). The significance of the differences between the sums of the test and control groups in the incidence of cytopathic effects (CPE) was determined using the one-way ANOVA test. A dose–response relationship was observed, with the cytotoxic effect occurring at higher concentrations of the extract ($\geq$101.6 µg mL^{-1}). The obtained results suggest possible use of this plant material for the production of various products (e.g., packaging, hygiene products, biodisinfectants, etc.) that would be useful against the spread of and for self-protection against COVID-19.

Keywords: Japanese knotweed rhizome bark; SARS-CoV-2; COVID-19; virus neutralization test

Citation: Jug, U.; Naumoska, K.; Malovrh, T. Japanese Knotweed Rhizome Bark Extract Inhibits Live SARS-CoV-2 In Vitro. *Bioengineering* **2022**, *9*, 429. https://doi.org/10.3390/bioengineering9090429

Academic Editors: Kandi Sridhar, Zeba Usmani, Minaxi Sharma and Qi Zhang

Received: 20 July 2022
Accepted: 23 August 2022
Published: 1 September 2022

Publisher's Note: MDPI stays neutral with regard to jurisdictional claims in published maps and institutional affiliations.

1. Introduction

Severe acute respiratory syndrome coronavirus 2 (SARS-CoV-2) [1] is a virus that is contagious to humans and some animal species and causes coronavirus disease 2019 (COVID-19), a viral respiratory illness responsible for the ongoing COVID-19 pandemic [2] declared by the World Health Organization on 11 March 2020 [3]. SARS-CoV-2 belongs to the *Coronaviridae* family, represented by enveloped viruses containing a positive-sense, single-stranded ribonucleic acid (+ssRNA) [4–6]. SARS-CoV-2 and other coronaviruses have four structural proteins known as spike, envelope, membrane, and nucleocapsid proteins. During viral infection, the spike protein promotes binding and fusion between the virus and the cell membrane [7]. The virus invades human cells by binding to the cell surface receptor angiotensin converting enzyme 2 (ACE2), a membrane glycoprotein [8,9]. The virus is transmitted from person to person mainly by close contact via aerosols and droplets from the respiratory air [10]. The nasal cavity respiratory epithelium is probably the predominant site of initial infection and virus replication [11]. Common symptoms and signs of a person infected with coronavirus include fever, cough, dyspnea, lymphopenia, abnormal results of chest computed tomography (CT), myalgia or fatigue, sputum production, headache, hemoptysis, diarrhea, etc. In more severe cases, the infection leads to acute respiratory distress syndrome, RNAemia, cardiac damage, secondary infections, and also death [12].

Depending on the target, there are two potential approaches to anti-coronavirus therapy, one targeting the coronavirus itself and the other targeting the support of the human immune system or human cells [7]. To date, few drugs are known to effectively inhibit SARS-CoV-2. Among specific drugs against SARS-CoV-2, Veklury (remdesivir) and Olumiant (baricitinib) have been approved by the FDA [13]. The EMA has approved several treatments for COVID-19: Evusheld (tixagevimab/cilgavimab), Kineret (anakinra), Paxlovid (PF-07321332/ritonavir), Regkirona (regdanvimab), RoActemra (tocilizumab), Ronapreve (casirivimab/imdevimab), Veklury (remdesivir), and Xevudy (sotrovimab) [14].

Plant-based drugs, including Japanese knotweed, have been used to treat various viral infections. The identification of phytochemicals with health-beneficial effects would be crucial for the development of new drugs against SARS-CoV-2 infection.

Japanese knotweed (*Fallopia japonica* Houtt., *Reynoutria japonica* Houtt., *Polygonum cuspidatum* Siebold & Zucc.) is an alien plant species native to East Asia that has become invasive in Europe and North America [15]. The rhizome extracts of Japanese knotweed have been tested in various biological studies [16], and the extracts themselves or their compounds showed antioxidant [17–24], antiproliferative [17], estrogenic [25], antiatherosclerotic [26], anti-inflammatory [27], antibacterial [28], and antiviral activities. Japanese knotweed ethanolic extract inhibits the lytic cycle and reduces the production of Epstein –Barr (EBV) viral particles [29]. Ethanolic and water extracts (the latter at a higher dose) inhibit the production of hepatitis B virus (HBV) [30]. Methanolic extract inhibits infection with the mosquito-borne pathogen dengue virus (DENV) in the early viral entry phases and also reduces the infectivity of hepatitis C virus (HCV) and Zika virus (ZIKV) [31]. The antiviral effect of Japanese knotweed water extract was investigated in a model of acquired immunodeficiency syndrome in mice, and immunodeficiency was partially inhibited [32]. In addition, a 70% ethanolic$_{(aq)}$ extract was tested against human immunodeficiency virus type 1 (HIV-1) and inhibited HIV-1-induced syncytium formation [33]. Bioactivity-guided fractionation enabled the isolation of 20 phenolic compounds with anti-HIV potential, and potent antiviral activity against HIV-1 was demonstrated for resveratrol, (+)-catechin, emodin-8-*O*-glucoside, and 5,7-dimethoxyphthalide [33]. Japanese knotweed water extract and its bioactive components resveratrol and emodin inhibit the replication of influenza A (H1N1) virus via interference with the mechanisms of the Toll-like receptor-9 pathway [34]. Furthermore, bioactivity-guided fractionation of the ethyl acetate extract was performed, and of the seven compounds isolated, resveratrol, (*E*)-3,5,12-trihydroxystilbene-3-*O*-beta-D-glucopyranoside-2′-(3″,4″,5″-trihydroxybenzoate) and catechin-3-*O*-gallate showed an inhibitory effect on neuraminidase activity, while the last two compounds also showed inhibitory activity against H1N1 [35].

Two important substances of Japanese knotweed rhizomes with well-studied and confirmed antiviral activity are emodin and resveratrol [36–45], phenolic compounds belonging to the anthraquinone and stilbene groups, respectively.

For emodin extracted from the rhizome of Japanese knotweed, antiviral activity against human simplex virus type 1 (HSV-1) was observed in guinea pigs [36]. Emodin inhibits the DNA replication of HBV [44]. Emodin and an ethyl acetate subfraction of Japanese knotweed rhizome, which contained 68.2% emodin, inhibited the expression of EBV immediate-early proteins and DNA replication [43]. Emodin isolated from Japanese knotweed inhibited the entry and replication of Coxsackie B4 virus (CVB$_4$) and improved the survival of infected mice when administered orally [45].

Resveratrol inhibited the induced expression of early EBV antigen in Raji cells [41] and was proven useful in preventing the proliferation of EBV [37]. In addition, resveratrol inhibited the replication of human cytomegalovirus (HCMV) [38], varicella-zoster virus (VZV) [40], HSV-1 [39], and even HIV-1 [42].

Due to the beneficial effects of some traditional Chinese medicine (TCM) plants on SARS-CoV virus infection and other coronaviruses, these plants and their secondary metabolites have attracted attention and have been studied also within the current pandemic [46].

Emodin inhibited the interaction of SARS-CoV spike protein and ACE2 [47] and blocked coronavirus SARS-CoV and HCoV-OC43 ion channel 3a, which impaired virus release [48].

Resveratrol inhibited Middle East respiratory syndrome coronavirus (MERS-CoV) infection and prolonged cellular survival after infection. The expression of nucleocapsid protein essential for MERS-CoV replication decreased, and apoptosis induced by MERS-CoV in vitro was downregulated, after resveratrol treatment [49].

The aim of the current study was to test the antiviral activity of the 70% ethanol$_{(aq)}$ extract of Japanese knotweed rhizome bark extract against live SARS-CoV-2 in vitro for the first time. A search in the literature shows a lack of in vitro studies as opposed to in silico studies to test the activity of this plant extract; the exceptions are Nawrot-Hadzik et al., 2021 [50], reporting the in vitro SARS-CoV-2 Mpro enzyme inhibitory activity of 70% acetone extract of the plant whole rhizome and Lin et al., 2022 [51], reporting the blocked entry of the SARS-CoV-2 pseudotyped virus into fibroblasts by water and ethanol extracts of the rhizome and root. The studies discussing Japanese knotweed or its compounds in connection with SARS-CoV are elaborated in detail in Results and Discussion.

2. Materials and Methods

2.1. Preparation of the Japanese Knotweed Rhizome Bark Extract

Rhizomes of Japanese knotweed were harvested in Ljubljana, Slovenia (Vrhovci, Mali Graben river bank; N 46°02′33.9″, E 14°27′00.9″). A voucher specimen was deposited in the Herbarium LJU (LJU10143477). The rhizomes were cleaned with tap water, the bark was peeled and lyophilized at $-50\ ^{\circ}$C for 24 h (Micro Modulyo, IMAEdwards, Bologna, Italy), and the obtained dry material was frozen with liquid N_2 and pulverized by a Mikro-Dismembrator S (Sartorius, Goettingen, Germany; 1 min, 1700 min^{-1}). The rhizome bark powder (1 g) was extracted with 20 mL of 70% ethanol$_{(aq)}$ (ethanol absolute anhydrous was purchased from Carlo Erba Reagents (Val de Reuil, France), and a Milli-Q water purification system (18 MΩ cm^{-1}; Millipore, Bedford, MA, USA) was used to obtain ultrapure water). Vortexing (5 min), sonication (15 min) and centrifugation (5 min, 6700× g) were executed and the supernatant was transferred to a pre-weighted vial, while the extraction of the solid residue was repeated with 10 mL 70% ethanol$_{(aq)}$. The solvent of the pooled supernatants was evaporated under N_2 flow. The dry extract (414.17 mg) was dissolved in 127.37 mL of cell medium (extract stock solution) in a sterile hood and was used for antiviral activity assays without filtration.

2.2. Native SARS-CoV-2 Virus Neutralization Test (VNT)

Different concentrations of the extract from the stock solution (concentration 3.25 mg mL^{-1}) in the culture media (ATCC; E-MEM with addition of 10% of FBS and 1% of standard antibiotic and antimycotic (Gibco, Grand Island, NY, USA; Anti-Anti 100x) were prepared in volume ratios of 1:64, 1:128, and 1:256. The number of infectious viral particles was quantified using the Median Tissue Culture Infectious Dose (TCID50) test. The assay is based on adding a serial dilution of the virus sample to the susceptible cells in a 96-well plate. The dilution at which 50% of the wells show CPE is used to mathematically calculate the TCID50 of the virus sample as generally described.

Extract dilutions (25 μL) and medium (25 μL) containing 6.2 TCID50 SARS-CoV-2 as virus working concentrations (SARS-CoV-2, 4265/20; EVAg) were mixed and incubated at 37 °C and 5% CO$_2$ for 7 h. Each dilution of the extract and virus was incubated in 48 replicates in 96-well tissue culture plates (tissue culture test-plate 96F, TPP Techno Plastic Products AG, Trasadingen, Swiss). After the incubation period, 3.5×10^5 Vero-E6 (African green monkey kidney cell line; ATCC CRL-1587) cell suspension was added in a volume of 100 μL to reach a final volume of 150 μL/well. To standardize the test procedures, a positive control with a working dilution of virus that was not incubated with the extract under the same conditions and volumes was included. After 110 h, the microplates were observed by inverted light microscope (ECLIPSE Ts2R, NIKON Instruments, Melville, NY, USA) at 400-fold magnification under a microscope equipped with an LED display. Characteristic morphological changes in the cells (rounding of adherent infected cells) in the culture that were the target of the SARS-CoV-2 replication were labelled as CPE. The other cellular transformation (cell dissolution and loss of cellular ultrastructure) was declared as a cytotoxic, necrotic effect. Each well in which the CPE was detected was labelled as positive for virus replication.

Before the main experiment, combinations of screening tests with different virus concentrations (virus working solution 100 TCID50 and subsequent dilutions in volume ratios of 1:2, 1:4, 1:8, 1:16, 1:32, 1:64, and 1:128) were performed in combination with different concentrations of extract (dilutions of stock solution in volume ratios of 1:2, 1:4, 1:8, 1:16, 1:32, 1:64, 1:128, 1:256, 1:512, 1:1024, and 1:2048) to determine the optimal concentration range for determining the cytotoxic properties of the extract and its potential antiviral activity under the same conditions as those under which the main experiment was later performed.

2.3. Statistical Analysis

The significance of the differences between the sums of the test and control groups in CPE incidence was determined using the one-way ANOVA test. A p-value of <0.05 was considered statistically significant. All the data were analyzed using Microsoft Excel 2016 (v16.0, Microsoft, Redmond, WA, USA).

3. Results and Discussion

A green solvent, 70% ethanol$_{(aq)}$, suitable for tincture preparation was selected as an extraction medium for Japanese knotweed rhizome bark due to the obtained higher extraction yield (44.3%) in comparison to other solvents [52]. It is also considered less harmful when present as a residual solvent in pharmaceutical formulations compared to other organic solvents [53]. An additional advantage of ethanol is its commercial availability as a food-grade solvent. Japanese knotweed rhizome bark 70% ethanolic$_{(aq)}$ extract was tested for its antioxidant activity in our previous study and was found to possess potent and stable time-dependent antioxidant activity [52].

In the search for an herbal antiviral drug and supported by already proven antiviral properties against other viruses, Japanese knotweed is one of several candidates to be tested against SARS-CoV-2. Within the current study we investigated the effect of the 70% ethanol$_{(aq)}$ Japanese knotweed rhizome bark extract on the antiviral properties of a wild-type SARS-CoV-2 using the virus neutralization test (VNT), which is routinely used in laboratory diagnostics of viral diseases or in the detection of specific antibodies [54]. The test is also a suitable research tool for studying all potential antiviral substances, such as antiviral peptides, drugs, or disinfectants. The test detects prevention or inhibition of cytopathic effect (CPE) and viral replication if neutralizing antibodies or substances with antiviral properties are detected [54].

In the current study, the extract was incubated together with SARS-CoV-2 for 7 h before cell exposure.

It is illusory to expect a neutralizing effect in the range of the specific neutralizing antibodies (NAbs), making the partial reduction in the cytopathic effect in such an experimental model a realistic expectation.

One of the biological effects of NAbs is to reduce the infectivity of the virus by their specific binding to viral ligands, which block the attachment of the virus to the cell receptor, which is the principle of virus neutralization [54]. Other antiviral mechanisms include a wide spectrum of possibly undefined interactions at the molecular level between the virus and the host cell, leading to impaired virus infectivity or even to the destruction of the virus. The gold standard for evaluating the in vitro neutralization ability or antiviral properties of substances is the virus neutralization test with native, infectious virus, which most closely mimics the mechanism of virus–host interaction in natural infection. Accuracy is its key feature, requiring manipulation of wild-type live virus, and is therefore performed in level III biosafety standards laboratories. The closest alternative to this type of testing is the pseudovirus neutralization test, which uses a noninfectious pseudovirus and is therefore a safer test performed in level II laboratories, but the results must be interpreted with a certain degree of caution [54].

Modifications of VNT have been described in the scientific literature to investigate the antiviral properties of such herbal extracts [51]. However, only laboratory testing models with a wild-type virus can most suitably mimic the natural viral infection of the host by allowing long-term virus–cell contact with prolonged antiviral contact of the virus extract in the experiment. The obtained results indicate that the tested extract has a certain level of inhibitory properties on the viral infectivity cycle; however, these are not as potent as the specific neutralizing antibody. CPE or antiviral effect of SARS-CoV-2 was studied in 48 parallels at each virus' and extract's concentration, taking into consideration only dilutions of the extract stock solution that did not cause cytotoxicity (1:64; 1:128; 1:256; *v/v*) in combination with the optimal working concentration of the virus (6.2 TCID50 SARS-CoV-2) (Figures 1 and 2) as previously determined by screening tests.

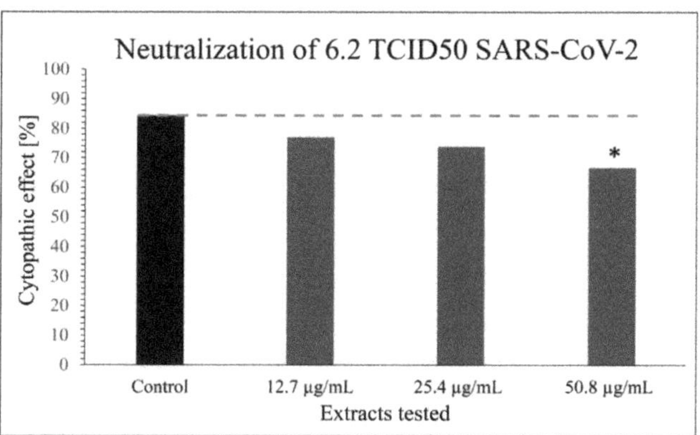

Figure 1. Graph showing the reduction in CPE by 70% ethanol$_{(aq)}$ Japanese knotweed rhizome bark extract on SARS-CoV-2 performed in 48 parallels at extract concentrations that did not cause cytotoxicity in comparison with the optimal working concentration of the virus (6.2 TCID50 SARS-CoV-2). The results are presented as a percentage of wells with cytopathic effect on Vero-E6 cells (*n* = 48); * $p < 0.05$. The result for the positive control (cells exposed to the virus only) is presented for comparison.

A statistically significant antiviral effect (reduction in CPE) against SARS-CoV-2 was observed for the stock solution diluted 1:64 (*v/v*) (i.e., 50.8 µg mL^{-1}) compared to the incidence of CPE in the positive control, where the virus-working concentration was inoculated on Vero-E6 cells only; $p < 0.05$ (Figure 1). Besides the statistically significant result for the CPE reduction effect of the extract at 50.8 µg mL^{-1} compared to the positive control, a dose–response relationship was generally noticed for all the extract concentrations (Figure 1).

The interactions between Japanese knotweed flavan-3-ols and proanthocyanidins [52,55] and FBS albumin in the cell medium should be considered when evaluating Japanese knotweed extract using in vitro cell culture systems. Namely, among various biological activities, Japanese knotweed polyphenols possess astringent activity [56]. The VNT executed with the filtered extract dissolved in cell medium did not show promising results. Therefore, final experiments were performed with the non-filtered extract to avoid possible removal of these complexes. According to the literature, interactions between flavan-3-ols and FBS albumin may induce cytotoxicity at high concentrations of flavan-3-ols [57].

In the current study, the cytotoxic effect was observed at extract concentrations above 50.8 µg mL^{-1}. Although SARS-CoV-2 inhibitory activity was expected to be even higher at higher concentrations of the extract, this could not be evaluated at cytotoxic concentrations.

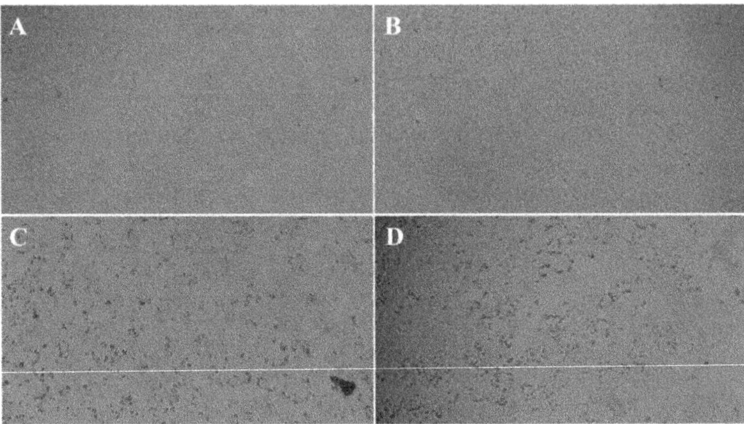

Figure 2. Images of: (**A**) non-treated Vero-E6 cells; (**B**) Vero-E6 cells exposed to SARS-CoV-2 virus (6.2 TCID50 SARS-CoV-2) and Japanese knotweed rhizome bark extract in concentration which prevents a cytopathic effect without provoking cytotoxicity (50.8 µg mL^{-1}); (**C**) cytotoxic effect caused by higher concentrations of Japanese knotweed rhizome bark extract; (**D**) cytopathic effect resulting from viral infection.

The mechanisms of antiviral activity on SARS-CoV-2 inhibitory activity and the main antiviral compounds of the examined extract as well as possible cell protection effect should be investigated on a molecular level in a further study. To support the obtained result, a few studies elaborating the effect of Japanese knotweed extract and its compounds on SARS-CoV virus are described below; however, none of these studies involve live, wild-type SARS-CoV-2 virus.

An herbal drug called Shufeng Jiedu, consisting of eight medicinal plants including Japanese knotweed, is known for its antiviral, anti-inflammatory, and immunomodulatory effects in acute lung diseases and was discussed as a promising candidate for the treatment of COVID-19 [58]. However, the antiviral and anti-inflammatory effects of Shufeng Jiedu were tested in a HCoV-229E (group 1 coronavirus) mouse model [58]. The formulation reduced viral load and decreased inflammatory factors in the lungs while increasing the concentration of CD4$^+$ and CD8$^+$ cells in the blood [58]. Direct binding of polydatin (an anthraquinone from the Japanese knotweed rhizome [52,55]), quercetin, and wogonin to the major protease of SARS-CoV-2 was demonstrated in silico [58]. Clinical data for Shufeng Jiedu added to standard antiviral therapy showed significant reduction in the COVID-19 clinical recovery time. Still, largescale, randomized, placebo-controlled, double-blinded clinical trials are lacking.

In addition, Baidu Jieduan granules containing Japanese knotweed and eleven other herbs have been clinically evaluated for efficacy and safety in the treatment of moderate COVID-19 [59]. Nonetheless, placebo-controlled and double-blinded clinical trial design is lacking.

The results of network pharmacology and bioinformatics analysis suggest that Japanese knotweed material is a promising therapeutic agent against COVID-19 [60].

The antiviral activity of purchased standards of polydatin and resveratrol, typical compounds of Japanese knotweed, has been demonstrated in vitro against HCoV-OC43 strain, an alternative model for SARS-CoV-2 [61]. Furthermore, computer-aided virtual screening was used to predict the binding site, and surface plasmon resonance (SPR) analysis was employed to confirm the interaction. SPR results showed a specific affinity of polydatin and resveratrol toward SARS-CoV 3CLpro and PLpro proteins as well as toward SARS-CoV-2. Moreover, Japanese knotweed was suggested as a potential therapeutic agent for pulmonary fibrosis caused by COVID-19 based on network pharmacology and data mining [62].

In other studies, a decreased infectivity of the SARS-CoV-2 spike pseudovirus in HEK293T-ACE2 cells was observed by polydatin [63], while resveratrol-inhibited SARS-CoV-2 replication (0–99.3%) in a dose-dependent manner (0–25 μM) with an EC90 and EC50 (50% and 90% maximal effective concentration, respectively) 11.42 and 10.66 μM, respectively. Higher concentrations of resveratrol (>50 μM) posed cytotoxicity on Vero-E6 cells [64].

In the study by Nawrot-Hadzik et al., 2021 [50], 25 compounds known to be present in the rhizomes of *Reynoutria japonica* (Japanese knotweed) and *Reynoutria sachalinensis* were docked into the main protease binding site of SARS-CoV-2. Further, 11 of them together with the extracts of both plants were tested in vitro for inhibition of SARS-CoV-2 Mpro enzyme [50]. Vanicosides A and B, isolated from the rhizomes of *Reynoutria sachalinensis*, showed moderate inhibitory activity against SARS-CoV-2 Mpro, while acetone extract and especially the butanol fractions of plants (of tested: dichloromethane, diethyl ether, ethyl acetate, *n*-butanol, and water fractions) containing vanicosides and polymerized procyanidins showed strong inhibition against SARS-CoV-2 Mpro [50]. Water and 90% ethanol$_{(aq)}$ extracts of Japanese knotweed rhizome and root blocked the entry of SARS-CoV-2 pseudotyped virus into HEK293T-ACE2 cells and zebrafish larvae and were shown to inhibit the spike protein–ACE2 receptor interaction and 3CL protease activity. Anti-SARS-CoV-2 activity was confirmed for the extract component gallic acid [51].

Some doubts and counter-opinions about the antiviral activity of Japanese knotweed against SARS-CoV-2 are also present in the literature. For instance, although the frequency of COVID-19 infection cases should be lower in Asian countries where Japanese knotweed occurs as a component of local foods, this has not been the case [65].

The results of the current study could be prospectively used to formulate or at least promote food supplements containing a Japanese knotweed rhizome bark extract for the prevention or complementary treatment of COVID-19 patients. A biofoil enriched with this extract which could be used as active packaging for food, drugs, and cosmetics to protect food or other packed contents from bacteria and oxidation has already been formulated by our group [66]. These foils may potentially prevent the spread of coronavirus (or other virus species), since another possible way for coronavirus to spread, besides through respiratory air (as mentioned in the introduction) is via indirect contact with contaminated surfaces. The virus remains viable and infectious in aerosols for hours and on surfaces for up to a day [67]. It is particularly stable on conventional plastic and stainless steel [67]. Other hygiene products such as soaps (soap destabilizes the lipid bilayer of viruses [68]) and disinfectant solutions (ethanol or isopropanol inactivate human coronavirus [69]) could also be enriched with Japanese knotweed rhizome bark extract to enhance their effect (act as bio-disinfectants). In this way, we would fight two battles at the same time: potential new COVID-19 waves and the invasiveness of Japanese knotweed, as the best way to eradicate or at least restrict this plant from Europe and North America (where it represents a major economic and environmental problem) is via mechanical excavation, which, on the other hand, is a necessary step to obtain extracts from its rhizome bark.

Author Contributions: Conceptualization, U.J., K.N. and T.M.; methodology, K.N., U.J. and T.M.; formal analysis, T.M.; investigation, U.J., K.N. and T.M.; resources, K.N. and T.M.; data curation, K.N. and U.J.; writing—original draft preparation, U.J., K.N. and T.M.; writing—review and editing, K.N. and U.J.; visualization, T.M. and U.J.; funding acquisition, K.N. and T.M. All authors have read and agreed to the published version of the manuscript.

Funding: The study was carried out with financial support from the Slovenian Research Agency (P1-0005 and P4-0092).

Institutional Review Board Statement: Not applicable.

Informed Consent Statement: Not applicable.

Data Availability Statement: All the data generated for this study are included in the article.

Conflicts of Interest: The authors declare no conflict of interest.

References

1. Gorbalenya, A.E.; Baker, S.C.; Baric, R.S.; De Groot, R.J.; Drosten, C.; Gulyaeva, A.A.; Haagmans, B.L.; Lauber, C.; Leontovich, A.M.; Neuman, B.W.; et al. The species Severe acute respiratory syndrome-related coronavirus: Classifying 2019-nCoV and naming it SARS-CoV-2. *Nat. Microbiol.* **2020**, *5*, 536–544. [CrossRef]
2. COVID-19: Epidemiology, Virology, and Prevention. Available online: https://www.uptodate.com/contents/covid-19 -epidemiology-virology-and-prevention (accessed on 14 November 2021).
3. WHO Director-General's opening remarks at the media briefing on COVID-19—11 March 2020. World Health Organization (WHO) (Press Release). Available online: https://www.who.int/director-general/speeches/detail/who-director-general-s-opening-remarks-at-the-media-briefing-on-covid-19---11-march-2020 (accessed on 14 November 2021).
4. Cui, J.; Li, F.; Shi, Z.L. Origin and evolution of pathogenic coronaviruses. *Nat. Rev. Microbiol.* **2019**, *17*, 181–192. [CrossRef] [PubMed]
5. Machhi, J.; Herskovitz, J.; Senan, A.M.; Dutta, D.; Nath, B.; Oleynikov, M.D.; Blomberg, W.R.; Meigs, D.D.; Hasan, M.; Patel, M.; et al. The natural history, pathobiology, and clinical manifestations of SARS-CoV-2 infections. *J. Neuroimmune Pharmacol.* **2020**, *15*, 359–386. [CrossRef]
6. Pal, M.; Berhanu, G.; Desalegn, C.; Kandi, V. Severe acute respiratory syndrome coronavirus-2 (SARS-CoV-2): An update. *Cureus* **2020**, *12*, e7423. [CrossRef] [PubMed]
7. Wu, C.; Liu, Y.; Yang, Y.; Zhang, P.; Zhong, W.; Wang, Y.; Wang, Q.; Xu, Y.; Li, M.; Li, X.; et al. Analysis of therapeutic targets for SARS-CoV-2 and discovery of potential drugs by computational methods. *Acta Pharm. Sin. B* **2020**, *10*, 766–788. [CrossRef]
8. Hoffmann, M.; Kleine-Weber, H.; Schroeder, S.; Krüger, N.; Herrler, T.; Erichsen, S.; Schiergens, T.S.; Herrler, G.; Wu, N.-H.; Nitsche, A.; et al. SARS-CoV-2 Cell entry depends on ACE2 and TMPRSS2 and is blocked by a clinically proven protease inhibitor. *Cell* **2020**, *181*, 271–280. [CrossRef]
9. Zhou, P.; Yang, X.-L.; Wang, X.-G.; Hu, B.; Zhang, L.; Zhang, W.; Si, H.-R.; Zhu, Y.; Li, B.; Huang, C.-L.; et al. A pneumonia outbreak associated with a new coronavirus of probable bat origin. *Nature* **2020**, *579*, 270–273. [CrossRef]
10. Coronavirus Disease (COVID-19): How Is It Transmitted? Available online: https://web.archive.org/web/20201015230546/https://www.who.int/news-room/q-a-detail/coronavirus-disease-covid-19-how-is-it-transmitted (accessed on 14 November 2021).
11. Hou, Y.J.; Okuda, K.; Edwards, C.E.; Martinez, D.R.; Asakura, T.; Dinnon, K.H., 3rd; Kato, T.; Lee, R.E.; Yount, B.L.; Mascenik, T.M.; et al. SARS-CoV-2 Reverse genetics reveals a variable infection gradient in the respiratory tract. *Cell* **2020**, *182*, 429–446. [CrossRef]
12. Huang, C.; Wang, Y.; Li, X.; Ren, L.; Zhao, J.; Hu, Y.; Zhang, L.; Fan, G.; Xu, J.; Gu, X.; et al. Clinical features of patients infected with 2019 novel coronavirus in Wuhan, China. *Lancet* **2020**, *395*, 497–506. [CrossRef]
13. Coronavirus (COVID-19), Drugs. Available online: https://www.fda.gov/drugs/emergency-preparedness-drugs/coronavirus-covid-19-drugs (accessed on 12 August 2022).
14. COVID-19 Treatments, European Medicines Agency. Available online: https://www.ema.europa.eu/en/human-regulatory/overview/public-health-threats/coronavirus-disease-covid-19/treatments-vaccines/covid-19-treatments (accessed on 12 August 2022).
15. Balogh, L. Japanese, Giant and Bohemian knotweed. In *The Most Important Invasive Plants in Hungary*; Botta-Dukat, Z., Balogh, L., Eds.; HAS Institute of Ecology and Botany: Budapest, Hungary, 2008; pp. 13–33.
16. Zhang, H.; Li, C.; Kwok, S.-T.; Zhang, Q.-W.; Chan, S.-W. A review of the pharmacological effects of the dried root of *Polygonum cuspidatum* (Hu Zhang) and its constituents. *Evid.-Based Complementary Altern. Med.* **2013**, *2013*, 208349. [CrossRef]
17. Lin, Y.-W.; Yang, F.-J.; Chen, C.-L.; Lee, W.-T.; Chen, R.-S. Free radical scavenging activity and antiproliferative potential of *Polygonum cuspidatum* root extracts. *J. Nat. Med.* **2010**, *64*, 146–152. [CrossRef] [PubMed]
18. Hsu, C.-Y.; Chan, Y.-P.; Chang, J. Antioxidant activity of extract from *Polygonum cuspidatum. Biol. Res.* **2007**, *40*, 13–21. [CrossRef] [PubMed]
19. Pogačnik, L.; Rogelj, A.; Ulrih, N.P. Chemiluminescence method for evaluation of antioxidant capacities of different invasive knotweed species. *Anal. Lett.* **2015**, *49*, 350–363. [CrossRef]
20. Lachowicz, S.; Oszmianski, J. Profile of bioactive compounds in the morphological parts of wild *Fallopia japonica* (Houtt) and *Fallopia sachalinensis* (F. Schmidt) and their antioxidative activity. *Molecules* **2019**, *24*, 1436. [CrossRef]
21. Ardelean, F.; Moacă, E.A.; Păcurariu, C.; Antal, D.S.; Dehelean, C.; Toma, C.-C.; Drăgan, S. Invasive *Polygonum cuspidatum*: Physico-chemical analysis of a plant extract with pharmaceutical potential. *Studia Univ. Vasile Goldis Arad Ser. Stiintele Vietii* **2016**, *26*, 415–421.
22. Kurita, S.; Kashiwagi, T.; Ebisu, T.; Shimamura, T.; Ukeda, H. Content of resveratrol and glycoside and its contribution to the antioxidative capacity of *Polygonum cuspidatum* (Itadori) harvested in Kochi. *Biosci. Biotechnol. Biochem.* **2014**, *78*, 499–502. [CrossRef]
23. Chan, C.-L.; Gan, R.-Y.; Corke, H. The phenolic composition and antioxidant capacity of soluble and bound extracts in selected dietary spices and medicinal herbs. *Int. J. Food Sci. Technol.* **2016**, *51*, 565–573. [CrossRef]

24. Nawrot-Hadzik, I.; Ślusarczyk, S.; Granica, S.; Hadzik, J.; Matkowski, A. Phytochemical diversity in rhizomes of three *Reynoutria* species and their antioxidant activity correlations elucidated by LC-ESI-MS/MS analysis. *Molecules* **2019**, *24*, 1136. [CrossRef]

25. Zhang, C.; Zhang, X.; Zhang, Y.; Xu, Q.; Xiao, H.; Liang, X. Analysis of estrogenic compounds in *Polygonum cuspidatum* by bioassay and high performance liquid chromatography. *J. Ethnopharmacol.* **2006**, *105*, 223–228. [CrossRef]

26. Xue, Y.; Liang, J. Screening of bioactive compounds in rhizoma *Polygoni cuspidati* with hepatocyte membranes by HPLC and LC-MS. *J. Sep. Sci.* **2014**, *37*, 250–256. [CrossRef]

27. Fan, P.; Zhang, T.; Hostettmann, K. Anti-inflammatory activity of the invasive neophyte *Polygonum cuspidatum* Sieb. and Zucc. (*Polygonaceae*) and the chemical comparison of the invasive and native varieties with regard to resveratrol. *J. Tradit. Complementary Med.* **2013**, *3*, 182–187. [CrossRef] [PubMed]

28. Shan, B.; Cai, Y.-Z.; Brooks, J.D.; Corke, H. Antibacterial properties of *Polygonum cuspidatum* roots and their major bioactive constituents. *Food Chem.* **2008**, *109*, 530–537. [CrossRef]

29. Yiu, C.-Y.; Chen, S.-Y.; Huang, C.-W.; Yeh, D.-B.; Lin, T.-P. Inhibitory effects of *Polygonum cuspidatum* on the Epstein-Barr virus lytic cycle. *J. Food Drug Anal.* **2011**, *19*, 107–113. [CrossRef]

30. Chang, J.-S.; Liu, H.-W.; Wang, K.-C.; Chen, M.-C.; Chiang, L.-C.; Hua, Y.-C.; Lin, C.-C. Ethanol extract of *Polygonum cuspidatum* inhibits hepatitis B virus in a stable HBV-producing cell line. *Antivir. Res.* **2005**, *66*, 29–34. [CrossRef]

31. Kuo, Y.-T.; Liu, C.-H.; Li, J.-W.; Lin, C.-J.; Jassey, A.; Wu, H.-N.; Perng, G.C.; Yen, M.-H.; Lin, L.T. Identification of the phytobioactive *Polygonum cuspidatum* as an antiviral source for restricting dengue virus entry. *Sci. Rep.* **2020**, *10*, 16378. [CrossRef] [PubMed]

32. Yan, J.I.; Hong-Xia, W.A.; Zuo-Yi, B.A.; Guan-Fu, Z.H. Evaluation of antiviral effect of *Polygonum cuspidatum* water extract with a model of murine acquired immunodeficiency syndrome. *Virol. Sin.* **1998**, *13*, 311.

33. Lin, H.-W.; Sun, M.-X.; Wang, Y.-H.; Yang, L.-M.; Yang, Y.-R.; Huang, N.; Xuan, L.-J.; Xu, Y.-M.; Bai, D.-L.; Zheng, Y.-T.; et al. Anti-HIV activities of the compounds isolated from *Polygonum cuspidatum* and *Polygonum multiflorum*. *Planta Med.* **2010**, *76*, 889–892. [CrossRef]

34. Lin, C.J.; Lin, H.J.; Chen, T.H.; Hsu, Y.A.; Liu, C.S.; Hwang, G.Y.; Wan, L. *Polygonum cuspidatum* and its active components inhibit replication of the influenza virus through toll-like receptor 9-induced interferon beta expression. *PLoS ONE* **2015**, *10*, e0117602. [CrossRef]

35. Chen, K.-T.; Zhou, W.-L.; Liu, J.-W.; Zu, M.; He, Z.N.; Du, G.H.; Chen, W.W.; Liu, A.L. Active neuraminidase constituents of *Polygonum cuspidatum* against influenza A(H1N1) influenza virus. *Zhongguo Zhong Yao Za Zhi* **2012**, *37*, 3068–3073. [CrossRef]

36. Wang, Z.; Huang, T.; Guo, S.; Wang, R. Effects of emodin extracted from Rhizoma Polygoni Cuspidati in treating HSV-1 cutaneous infection in guinea pigs. *J. Anhui Tradit. Chin. Med. Coll.* **2003**, *22*, 36–38.

37. Yiu, C.-Y.; Chen, S.-Y.; Chang, L.-K.; Chiu, Y.-F.; Lin, T.-P. Inhibitory effects of resveratrol on the Epstein-Barr virus lytic cycle. *Molecules* **2010**, *15*, 7115–7124. [CrossRef] [PubMed]

38. Evers, D.L.; Wang, X.; Huong, S.-M.; Huang, D.Y.; Huang, E.-S. 3,4',5-Trihydroxy-*trans*-stilbene (resveratrol) inhibits human cytomegalovirus replication and virus-induced cellular signaling. *Antivir. Res.* **2004**, *63*, 85–95. [CrossRef]

39. Docherty, J.J.; Fu, M.M.; Stiffler, B.S.; Limperos, R.J.; Pokabla, C.M.; DeLucia, A.L. Resveratrol inhibition of herpes simplex virus replication. *Antivir. Res.* **1999**, *43*, 145–155. [CrossRef]

40. Docherty, J.J.; Sweet, T.J.; Bailey, E.; Faith, S.A.; Booth, T. Resveratrol inhibition of varicella-zoster virus replication in vitro. *Antivir. Res.* **2006**, *72*, 171–177. [CrossRef] [PubMed]

41. Kapadia, G.J.; Azuine, M.A.; Tokuda, H.; Takasaki, M.; Mukainaka, T.; Konoshima, T.; Nishino, H. Chemopreventive effect of resveratrol, sesamol, sesame oil and sunflower oil in the Epstein-Barr virus early antigen activation assay and the mouse skin two-stage carcinogenesis. *Pharmacol. Res.* **2002**, *45*, 499–505. [CrossRef] [PubMed]

42. Zhang, H.-S.; Zhou, Y.; Wu, M.-R.; Zhou, H.-S.; Xu, F. Resveratrol inhibited Tat-induced HIV-1 LTR transactivation via NAD+-dependent SIRT1 activity. *Life Sci.* **2009**, *85*, 484–489. [CrossRef] [PubMed]

43. Yiu, C.-Y.; Chen, S.-Y.; Yang, T.-H.; Chang, C.-J.; Yeh, D.-B.; Chen, Y.-J.; Lin, T.-P. Inhibition of Epstein-Barr virus lytic cycle by an ethyl acetate subfraction separated from *Polygonum cuspidatum* root and its major component, emodin. *Molecules* **2014**, *19*, 1258–1272. [CrossRef]

44. Shuang-Suo, D.; Zhengguo, Z.; Yunru, C.; Xin, Z.; Baofeng, W.; Lichao, Y.; Yan'an, C. Inhibition of the replication of hepatitis B virus in vitro by emodin. *Med. Sci. Monit.* **2006**, *12*, 302–306.

45. Liu, Z.; Wei, F.; Chen, L.-J.; Xiong, H.-R.; Liu, Y.-Y.; Luo, F.; Hou, W.; Xiao, H.; Yang, Z.-Q. In vitro and in vivo studies of the inhibitory effects of emodin isolated from *Polygonum cuspidatum* on Coxsakievirus B_4. *Molecules* **2013**, *18*, 11842–11858. [CrossRef]

46. Yang, Y.; Islam, M.S.; Wang, J.; Li, Y.; Chen, X. Traditional Chinese medicine in the treatment of patients infected with 2019-new coronavirus (SARS-CoV-2): A review and perspective. *Int. J. Biol. Sci.* **2020**, *16*, 1708–1717. [CrossRef]

47. Ho, T.-Y.; Wu, S.-L.; Chen, J.-C.; Li, C.-C.; Hsiang, C.-Y. Emodin blocks the SARS coronavirus spike protein and angiotensin-converting enzyme 2 interaction. *Antivir. Res.* **2007**, *74*, 92–101. [CrossRef] [PubMed]

48. Schwarz, S.; Wang, K.; Yu, W.J.; Sun, B.; Schwarz, W. Emodin inhibits current through SARS-associated coronavirus 3a protein. *Antivir. Res.* **2011**, *90*, 64–69. [CrossRef] [PubMed]

49. Lin, S.-C.; Ho, C.-T.; Chuo, W.-H.; Li, S.; Wang, T.T.; Lin, C.-C. Effective inhibition of MERS-CoV infection by resveratrol. *BMC Infect. Dis.* **2017**, *17*, 144. [CrossRef] [PubMed]

50. Nawrot-Hadzik, I.; Zmudzinski, M.; Matkowski, A.; Preissner, R.; Kęsik-Brodacka, M.; Hadzik, J.; Drag, M.; Abel, R. *Reynoutria* rhizomes as a natural source of SARS-CoV-2 Mpro inhibitors-molecular docking and in vitro study. *Pharmaceuticals* **2021**, *14*, 742. [CrossRef]

51. Lin, S.; Wang, X.; Tang, R.W.; Lee, H.C.; Chan, H.H.; Choi, S.S.A.; Dong, T.T.; Leung, K.W.; Webb, S.E.; Miller, A.L.; et al. The extracts of *Polygonum cuspidatum* root and rhizome block the entry of SARS-CoV-2 wild-type and omicron pseudotyped viruses via inhibition of the S-protein and 3CL protease. *Molecules* **2022**, *27*, 3806. [CrossRef]

52. Jug, U.; Naumoska, K.; Vovk, I. (−)-Epicatechin—An important contributor to the antioxidant activity of Japanese knotweed rhizome bark extract as determined by antioxidant activity-guided fractionation. *Antioxidants* **2021**, *10*, 133. [CrossRef]

53. Grodowska, K.; Parczewski, A. Organic solvents in the pharmaceutical industry. *Acta Pol. Pharm.* **2010**, *67*, 3–12. Available online: https://www.ptfarm.pl/wydawnictwa/czasopisma/acta-poloniae-pharmaceutica/110/-/12992 (accessed on 10 September 2020).

54. Lu, Y.; Wang, J.; Li, Q.; Hu, H.; Lu, J.; Chen, Z. Advances in neutralization assays for SARS-CoV-2. *Scand. J. Immunol.* **2021**, *94*, e13088. [CrossRef]

55. Jug, U.; Glavnik, V.; Vovk, I.; Makuc, D.; Naumoska, K. *Off-line* multidimensional high performance thin-layer chromatography for fractionation of Japanese knotweed rhizome bark extract and isolation of flavan-3-ols, proanthocyanidins and anthraquinones. *J. Chromatogr. A* **2021**, *1637*, 461802. [CrossRef]

56. Cos, P.; De Bruyne, T.; Hermans, N.; Apers, S.; Vanden Berge, D.; Vlietinck, A.J. Proanthocyanidins in health care: Current and new trends. *Curr. Med. Chem.* **2004**, *11*, 1345–1359. [CrossRef]

57. Fujii, Y.; Suhara, Y.; Sukikara, Y.; Teshima, T.; Hirota, Y.; Yoshimura, K.; Osakabe, N. Elucidation of the interaction between flavan-3-ols and bovine serum albumin and its effect on their in-vitro cytotoxicity. *Molecules* **2019**, *24*, 3667. [CrossRef] [PubMed]

58. Xia, L.; Shi, Y.; Su, J.; Friedemann, T.; Tao, Z.; Lu, Y.; Ling, Y.; Lv, Y.; Zhao, R.; Geng, Z.; et al. Shufeng Jiedu, a promising herbal therapy for moderate COVID-19: Antiviral and anti-inflammatory properties, pathways of bioactive compounds, and a clinical real-world pragmatic study. *Phytomedicine* **2021**, *85*, 153390. [CrossRef] [PubMed]

59. Zhang, W.; Xie, Q.; Xu, X.; Sun, S.; Fan, T.; Wu, X.; Qu, Y.; Che, J.; Huang, T.; Li, H.; et al. Baidu Jieduan granules, traditional Chinese medicine, in the treatment of moderate coronavirus disease-2019 (COVID-19): Study protocol for an open-label, randomized controlled clinical trial. *Trials* **2021**, *22*, 476. [CrossRef] [PubMed]

60. Zhao, J.; Pan, B.; Xia, Y.; Liu, L. Network pharmacology-based analysis reveals the putative action mechanism of *Polygonum cuspidatum* against COVID-19. *Int. J. Clin. Exp. Med.* **2021**, *14*, 1852–1863.

61. Xu, H.; Li, J.; Song, S.; Xiao, Z.; Chen, X.; Huang, B.; Sun, M.; Su, G.; Zhou, D.; Wang, G.; et al. Effective inhibition of coronavirus replication by *Polygonum cuspidatum*. *Front. Biosci.* **2021**, *26*, 789–798. [CrossRef]

62. Yu, M.X.; Song, X.; Ma, X.Q.; Hao, C.X.; Huang, J.J.; Yang, W.H. Investigation into molecular mechanisms and high-frequency core TCM for pulmonary fibrosis secondary to COVID-19 based on network pharmacology and data mining. *Ann. Palliat. Med.* **2021**, *10*, 3960–3975. [CrossRef]

63. Wang, M.; Qin, K.; Zhai, X. Combined network pharmacology, molecular docking, and experimental verification approach to investigate the potential mechanisms of polydatin against COVID-19. *Nat. Prod. Commun.* **2022**, *17*, 1934578X221095352. [CrossRef]

64. Pasquereau, S.; Nehme, Z.; Haidar Ahmad, S.; Daouad, F.; Van Assche, J.; Wallet, C.; Schwartz, C.; Rohr, O.; Morot-Bizot, S.; Herbein, G. Resveratrol inhibits HCoV-229E and SARS-CoV-2 Coronavirus replication in vitro. *Viruses* **2021**, *13*, 354. [CrossRef]

65. Wiwanitkit, V. Polydatin and COVID-19. *Clin. Nutr.*. [CrossRef]

66. Naumoska, K.; Jug, U.; Kõrge, K.; Oberlintner, A.; Golob, M.; Novak, U.; Vovk, I.; Likozar, B. Antioxidant and antimicrobial biofoil based on chitosan and Japanese knotweed (*Fallopia japonica*, Houtt.) rhizome bark extract. *Antioxidants* **2022**, *11*, 1200. [CrossRef]

67. Van Doremalen, N.; Bushmaker, T.; Morris, D.H.; Holbrook, M.G.; Gamble, A.; Williamson, B.N.; Tamin, A.; Harcourt, J.L.; Thornburg, N.J.; Gerber, S.I.; et al. Aerosol and surface stability of SARS-CoV-2 as compared with SARS-CoV-1. *N. Engl. J. Med.* **2020**, *382*, 1564–1567. [CrossRef] [PubMed]

68. Why the Coronavirus Has Been So Successful. Available online: https://www.theatlantic.com/science/archive/2020/03/biography-new-coronavirus/608338/ (accessed on 14 November 2021).

69. Meyers, C.; Kass, R.; Goldenberg, D.; Milici, J.; Alam, S.; Robison, R. Ethanol and isopropanol inactivation of human coronavirus on hard surfaces. *J. Hosp. Infect.* **2021**, *107*, 45–49. [CrossRef] [PubMed]

 *bioengineering*

Article

A Biotechnological Approach for the Production of Pharmaceutically Active Human Interferon-α from *Raphanus sativus* L. Plants

Rashad Kebeish [1,2,*], Emad Hamdy [1], Omar Al-Zoubi [2], Talaat Habeeb [2], Raha Osailan [2] and Yassin El-Ayouty [1]

[1] Botany and Microbiology Department, Faculty of Science, Zagazig University, Zagazig 44519, Egypt
[2] Biology Department, Faculty of Science Yanbu, Taibah University, Yanbu El-Bahr 46423, Saudi Arabia
* Correspondence: rkebeish@taibahu.edu.sa or rkebeish@gmail.com

Abstract: Human interferon (IFN) is a type of cytokine that regulates the immune system's response to viral and bacterial infections. Recombinant IFN-α has been approved for use in the treatment of a variety of viral infections as well as an anticancer medication for various forms of leukemia. The objective of the current study is to produce a functionally active recombinant human IFN-α2a from transgenic *Raphanus sativus* L. plants. Therefore, a binary plant expression construct containing the IFN-α2a gene coding sequence, under the regulation of the cauliflower mosaic virus 35SS promoter, was established. *Agrobacterium*-mediated floral dip transformation was used to introduce the IFN-α2a expression cassette into the nuclear genome of red and white rooted *Raphanus sativus* L. plants. From each genotype, three independent transgenic lines were established. The anticancer and antiviral activities of the partially purified recombinant IFN-α2a proteins were examined. The isolated IFN-α2a has been demonstrated to inhibit the spread of the Vesicular Stomatitis Virus (VSV). In addition, cytotoxicity and cell apoptosis assays against Hep-G2 cells (Human Hepatocellular Carcinoma) show the efficacy of the generated IFN-α2a as an anticancer agent. In comparison to bacterial, yeast, and animal cell culture systems, the overall observed results demonstrated the efficacy of using *Raphanus sativus* L. plants as a safe, cost-effective, and easy-to-use expression system for generating active human IFN-α2a.

Keywords: *Raphanus sativus* L.; recombinant IFN-α2a; apoptosis; antiviral and antitumor activity

Citation: Kebeish, R.; Hamdy, E.; Al-Zoubi, O.; Habeeb, T.; Osailan, R.; El-Ayouty, Y. A Biotechnological Approach for the Production of Pharmaceutically Active Human Interferon-α from *Raphanus sativus* L. Plants. *Bioengineering* 2022, 9, 381. https://doi.org/10.3390/bioengineering9080381

Academic Editors: Minaxi Sharma, Kandi Sridhar and Zeba Usmani

Received: 21 July 2022
Accepted: 8 August 2022
Published: 10 August 2022

Publisher's Note: MDPI stays neutral with regard to jurisdictional claims in published maps and institutional affiliations.

1. Introduction

Some human disorders are caused by protein deficiency and poor performance [1]. Protein medicines are expected to be an armory against illnesses and have enormous commercial value [2]. Many advantages of molecular farming have been discovered in recent years, particularly in terms of cost, practicality, and safety. Transgenic plants have been used in the development of vaccines and therapeutic proteins in the biopharmaceutical field [2,3]. Plants such as rice [4], aloe [5], tobacco [6], and *Raphanus sativus* can be used to make edible vaccines for oral delivery and immunization. Currently, new research is focusing on figuring out how to make them inexpensive, easily produced, and fully functional. Previously, routinely produced proteins were taken from natural sources for use in research, medicine, and industry [7]. The protein generated using this approach rarely met the requirements, and it also posed several risks and challenges in isolation [2]. Sorensen and Mortensen, (2005) [8] believe that biotechnological technologies are the only way to address the demand for pure, soluble, and functional proteins. To meet the capital required as well as the production scale-up and efficacy, numerous expression systems were developed, including bacterial, yeast, animal cell lines, and plants [9]. As a result, deciding on an expression system necessitates a cost breakdown in terms of design, procedure, and other economic factors. Because its genome has been fully sequenced and the organism is easy to handle, grows quickly, and requires an inexpensive, easy-to-prepare medium for

Figure 2. Representative photographs of *Raphanus sativus* L. plants. (**A**) Floral dip transformation of *Raphanus sativus* L. plants. (**B**) Selection of transgenic *Raphanus sativus* L. plants after spraying the foliar leaves with BASTA solution (phosphinothricin; 25 µg/mL) 6 times at 3–day intervals. (**C**) T$_1$ plants of IFN-α2a transgenic *Raphanus Sativus* L. plants grown in soil.

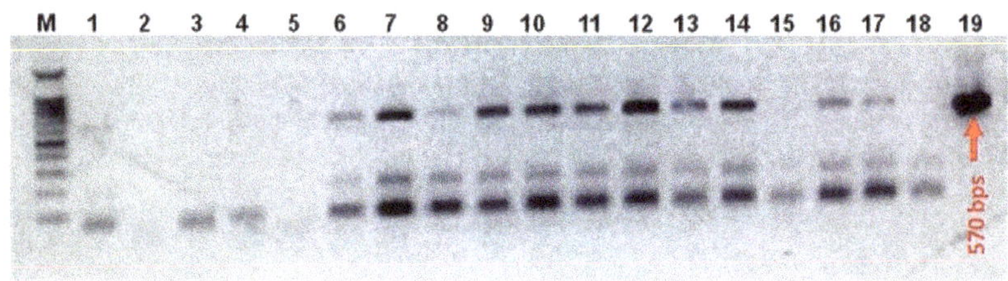

Figure 3. DNA gel electrophoresis of PCR screening for transgenic white and red *Raphanus sativus* L. plants. Shown is the ethidium bromide-mediated fluorescence of DNA fragments after UV excitation. Products of PCR screening of genomic DNA isolated from transgenic white and red *Raphanus sativus* L. plants were separated on a 1 % (*w/v*) agarose gel in 1x TAE for 50 min at 100 V. M: 100 bps DNA ladder, lane 1–12: PCR products of the screened white *Raphanus sativus L* plants, lane 13–17: PCR products of the screened red *Raphanus sativus* L. plants, lane 18: negative PCR water control, and lane 19: pTRA-PT-IFN-α2a plasmid positive control.

2.4. Real-Time RT-PCR

Quantitative real-time PCR was performed as described previously [26]. The BCP (1-bromo-3-chloropropane) protocol [27] was applied to extract RNA from transgenic *Raphanus sativus* L. leaves. The isolated RNA was subjected to DNase-1 digestion before proceeding to cDNA synthesis. First-strand cDNA synthesis was implemented as previously described [28]. RT-PCR was performed using the StepOne qRT-PCR system (Applied Biosystems, Waltham, MA, USA) following the manufacturer's instructions in the presence of SYBR Green (SYBR1 GreenERTM qPCR SuperMixes; Karlsruhe, Germany). Primers were purchased from Intron Biotechnology Inc. (Kyungki-Do, Korea). 5′-CTG AAA CCA TCC CTG TCC TC 3′ and 5′-TCT AGG AGG GTC TCA TCC CA-3′ primers were used to detect IFN-α2a transcripts. For the detection of the housekeeping gene, *ACTIN2* transcripts was performed using 5′-GGTAACATTGTGCTCAGTGGTGG-3′ and 5′-GGTGCAACGACCTTAATCTTCAT-3′ primers. A final concentration of 200 nM primer was applied to the reaction mixture. The amplification program used for both *ACTIN2* and IFN-α2a was 10 min of initial denaturation at 95 °C, followed by 40 cycles each of 15 s denaturation at 95 °C and 1 min of annealing and extension at 60 °C.

2.5. IFN-α2a Protein Extraction and Partial Purification

100 mg frozen green leaves collected from 5-week-old WT and transgenic *Raphanus sativus* L. plants were homogenized in liquid nitrogen. 500 µLof protein extraction buffer (50 mM sodium phosphate buffer (pH 7.4) supplemented with 0.1% Triton X-100 and 10 mM EDTA) was added followed by centrifugation (30,000× *g*/15 min/4 °C). The

supernatant was collected and the centrifugation step was repeated. Partial purification of IFN-α2a protein samples was performed using DEA-Sepharose followed by Sephadex G-50 (Sigma Aldrich, Taufkirchen, Germany). Protein concentration was evaluated based on the Bradford method [29]. Total protein extracts and enriched IFN-α2a protein fractions were kept in freezers at $-20\ ^{\circ}C$ and used for further analyses.

2.6. Enzyme-Linked Immunosorbent Assay (ELISA) for Detection of IFN-α2a Protein

Enzyme-linked Immunosorbent Assay (ELISA) was applied for the detection of recombinant IFN-α2a protein using an IFN alpha Human ELISA Kit (ThermoFisher Scientific, Altrincham, UK). Detection and quantification of IFN-α2a in protein extracts were carried out according to the supplier's protocol. The titer of interferon in test samples was calculated by plotting the optical densities (450 nm) and fitting the standard curve with a 6-parameter fit. Peg-IFN (Pegasys®; Hoffmann-LaRoche, Basel, Switzerland), the commercially available IFN-α, was used as a positive control in this assay. To eliminate the background, the blank sample values were subtracted from the standards and the test samples.

2.7. Antiviral Activity of the Recombinant IFN-α2a

Vesicular Stomatitis Virus (VSV) Indiana strain-ATCC® VR-158 and its host cells (Vero cells, derived from the kidney of a normal, adult African green monkey) were generously provided by the Head of R&D sector VACSERA-Egypt, Prof. Dr. Aly Fahmy. Vero cells were allowed to grow as a monolayer in RPMI medium containing L-glutamine, sodium pyruvate, sodium bicarbonate, non-essential amino acids, 10% fetal bovine serum (FBS), and antibiotics (10 μg/mL streptomycin and 100 μg/mL penicillin) at 37 °C in a 5% CO_2 humidified environment until the formation of confluent cell monolayers. VSV was prepared on Vero cells, aliquoted, and stored at -80 °C. Infectivity titer of VSV stock was determined as previously described [30]. The virus stock titer was 10^7 TCID50/mL. 0.02 mL/well virus dilution was dispensed to pre-cultured Vero cells. In MEM (Sigma Aldrich, Taufkirchen, Germany) supplemented with 5% FCS (GEPCO, Miami, FL, USA), non-cytotoxic concentrations of recombinant IFN-α2a protein fractions were 10-fold serially diluted. To each concentration, equal volumes (~20 μL) of virus at an infective titer of 10^2 TCID50/mLwere added and incubated at 37 °C for 3h. To ensure an even distribution of virus on the cell surface, virus inocula were shaken at 15 min intervals. Culture plates were examined daily using an inverted microscope (Willovert, Helmut Hund GmbH, Wetzlar, Germany) until more than 90% cytopathic effect (CPE) was detected. As described by Lansky and Newman (2007) [31], 50% endpoint-induced CPE was determined. The antiviral activity of *Raphanus sativus* L. recombinant IFN-α2a protein was assessed both directly and indirectly based on its inhibitory effect against the viral cytopathic effect. Extracts from wild-type white and red *Raphanus sativus* L. and peg-IFN were used as negative and positive controls, respectively.

2.8. In Vitro CPE Assay of the Recombinant IFN-α2a on Hep-G2-Cells

Hep-G2 ATCC® HB-8065 (human hepatocellular carcinoma cell line) was used to evaluate the antitumor properties of the recombinant IFN-α2a isolated from transgenic *Raphanus sativus* L. lines. This assay was performed at VACSERA, Cairo, Egypt. The assay is based on the mitochondrial-dependent reduction of the yellow MTT (3-(4,5-dimethylthiazol-2-yl)-2,5-diphenyl tetrazolium bromide) to purple formazan as previously described [32]. In brief, Hep-G2 cells were seeded with 100 μL of culture medium per well in 96-well plates. Cells were cultured alone (control) or with dilution series (final concentration of 10 μg/mL, 25 μg/mL, 50 μg/mL, 75 μg/mL, 100 μg/mL, 200 μg/mL, 300 μg/mL, 500 μg/mL, 750 μg/mL, and 1000 μg/mL) of the recombinant IFN-α2a protein extracts in DMSO. Wild-type *Raphanus sativus* L. protein extracts were used as negative controls. The experiment was performed in quadruplicate. The cells were grown in a 5% CO_2 humidified incubator at 37 °C overnight. After discarding the growth medium, 100 μL of MTT working solution (0.4 mg/mL of PBS) was added and incubated for 3 h. The supernatant was then removed and 100 μL DMSO was added to each sample and absorbance at 550 nm was measured for each sample. The absorbance was calculated as a percentage of the control value. The change in viability percentage was calculated based on the formula (reading of extract/reading of negative control) × 100 CPE was also measured for normal Vero cells and WISH cells in order to determine the IC_{50} value of the recombinant IFN-α2a protein isolated from *Raphanus sativus* L. transgenic lines. IC_{50} values were calculated based on sigmoid concentration–response curve fitting models using Graph Pad Prism computer software (International scientific community, San Diego, CA, USA).

2.9. Hep-G2-Cell Apoptosis Assay

In 25 cm^2 surface area cell culture flasks, Hep-G2 cells were pre-cultured and treated for 24h with the IC_{50} concentrations of the tested samples of recombinant IFN-α2a isolated from *Raphanus sativus* L. transgenic lines in RPMI-1640 and DMEM-media (Sigma Aldrich, Taufkirchen, Germany). For cell cycle analysis, Hep-G2 cells were collected and fixed gently with 70% (*v*/*v*) ethanol in FBS (Sigma Aldrich, Taufkirchen, Germany) overnight at 4 °C and then resuspended in FBS containing 40 μg/mLpenicillin, 0.1% (*v*/*v*) Triton X-100, and 0.1 mg/mL RNase in a dark room. Cells were incubated at 37 °C for 30 min then analyzed using a flowcytometer (Becton Dickinson, San Jose, CA, USA) equipped with an argon ion laser at a wavelength of 488 nm. Multicycle Software (Phoenix Flow Systems, San Diego, CA, USA) was used to evaluate the data, as previously reported [33].

2.10. Statistical Analysis

At least three independent tests were carried out for each experiment. Data were expressed as average values ± standard error (SE) and statistically analyzed using one-way analysis of variance (ANOVA). All statistical analyses were performed using the SPSS-11 program and Excel software (Microsoft Corporation, Redmond, WA, USA).

3. Results and Discussion

3.1. Generation and Selection of Transgenic Raphanus sativus L. Plants

Plant molecular farming is a growing technique used to produce recombinant bio-pharmaceuticals, secondary metabolites, and other industrial proteins all over the world. The majority of this technology is based on introducing a gene or gene clusters into plants and/or plant cell cultures via molecular transformation. In 1986 and 1989, transgenic plants produced human growth hormone, the first recombinant pharmaceutical protein, and the first recombinant antibody, respectively [34,35]. Green plants and/or plant cell culture-based systems can be employed as large-scale biofactories for the manufacture of recombinant proteins [17,36,37]. *Raphanus sativus* L., an edible plant from the Cruciferae family, is an annual vegetable. It is grown mostly for its roots, which are consumed in salads by the majority of people. It can be eaten raw, cooked, or preserved by storage, canning, or drying. Despite this, radish is a valuable source of medicinal compounds [38]. *Raphanus sativus* L. is a promising option for building an efficient molecular farming system for the generation of human IFN-α2a because of its favorable growth properties, ease of genetic transformation [25], economical culture, and increased biosafety issues. As an eukaryotic expression system, *Raphanus sativus* L. plants can perform glycosylation as well as post-translational modification needed for fully active proteins, including human interferon-α [23].

Raphanus sativus L., a white and red rooted radish, was chosen as a test plant in this work with the goal of finding an alternate expression system for the generation of active human IFN-α2a. PCR was used to amplify the human IFN-α2a gene coding sequence (GenBank: JN848522.1). Because the genomic IFN-α2a on chromosome 9 lacks intron sequences, human genomic DNA isolated from blood was employed directly as a template [24]. The PCR product of IFN-α2a was amplified and cloned into the binary expression vector pTRA-PT (gi13508478). In green plants and/or microalgae, CaMV 35S promoter has been demonstrated to enhance high gene expression [20,26,39]. CaMV 35S promoter was therefore employed in the current study to regulate IFN-α2a gene expression in radish plants. The 5′ UTR of tobacco leader peptide (TL) and the 3′ UTR of CaMV 35S (pA35S) surround the IFN-2a expression cassette in the binary expression vector pTRA-PT (Figure 1).

Agrobacterium tumefaciens (GV3101) was transformed with pTRA-PT-IFN-α2a construct which was then utilized to transform wild-type white and red rooted radish plants using the floral dip transformation methodology [25] (Figure 2A). For the selection of transgenic lines, three-week-old T_1 plantlets were sprayed with BASTA herbicide (phosphinothricin 25 μg/mL) six times at three days intervals, see Figure 2B,C. PCR analysis of genomic DNA obtained from radish plants revealed the presence of the IFN-α2a gene in the selected T_1 transgenic radish plants. The result of this PCR screening is shown in Figure 3. Twelve white and six red *Raphanus sativus* L. were tested. Six plant samples from white radish (Lane 6–12 in Figure 3) and four samples of red radish (Lane 13, 14, 16, and 17 in Figure 3) confirm the existence of the IFN-α2a gene. Five transgenic white radishes and four transgenic red radishes were chosen for additional IFN-α2a gene expression analysis.

3.2. Analysis of IFN-α2a Expression in Transgenic White and Red Raphanus sativus L. Plants

3.2.1. RT-PCR Analysis

Quantitative RT-PCR was used to compare the accumulation of IFN-α2a mRNA transcripts in five transgenic white radish lines (named IFN-wRs-1-5) and four transgenic red radish lines (named IFN-rRs-1-4) to the housekeeping gene, Actin2 transcripts (Figure 4A,B respectively and Supplementary Figure S1). Variable IFN-α2a expression levels in both white and red transgenic radish were observed. It is noted that transgenic white radish lines, IFN-wRs-1 & IFN-wRs-2, showed a relatively higher expression level of IFN-α2a compared to transgenic red radish lines. Similar expression levels of the IFN-α2a gene were observed for IFN-wRs-3, 4, 5, and all the transgenic red radish lines under the assay conditions. The higher IFN-α2a expressing transgenic lines from both radish types were chosen for further analyses. A similar expression pattern of human interferon was reported for IFN transgene in green plants [17], and/or microalgae [39]. However, for both bio-systems to elicit large amounts of human interferon production, codon use adaption of the introduced transgene is required.

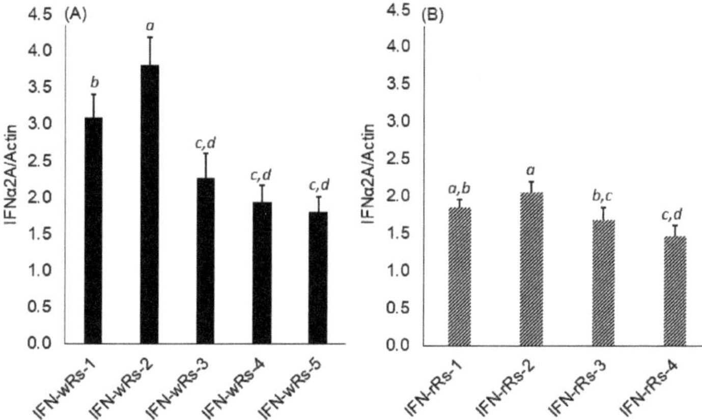

Figure 4. RT-PCR analysis of IFNα2a gene expression in transgenic white (**A**) and red (**B**) *Raphanus sativus* L. plants. IFN-α2a mRNA transcripts was measured by real-time qRT-PCR and calculated in arbitrary units compared to a standard dilution series. IFN-wRs 1-5: samples from five independent IFNα2a transgenic white *Raphanus sativus* L. plants, IFN-rRs 1-4: samples from four independent IFNα2a transgenic red *Raphanus sativus* L. plants. Data are represented as the average of at least three independent measurements of RNA preparations ± SE. Different small letters represent significant differences; $p = 0.05$.

3.2.2. Enzyme-Linked Immunosorbent Assay (ELISA) Detection of Human IFN-α2a and SDS-PAGE Analysis

The presence of the IFN-α2a transgene in transformed *Raphanus sativus* L. plants was validated at the protein level by utilizing an Enzyme-linked Immunosorbent Assay (ELISA) kit to detect IFN-α2a protein (ThermoFisher Scientific, Altrincham, UK). Peg-IFN-α (Pegasys®; Hoffmann-LaRoche, Basel, Switzerland) was used in this experiment as a positive control. The kit includes a double-antibody sandwich that is used to detect the presence of human interferon (IFN-α) in test samples. Total protein extracts isolated from transgenic white and red radish lines, as well as wild-type control samples, were tested to detect IFN-α2a in the transgenic plant protein extracts using a commercially available IFN-α protein standard sample (Figure 5).

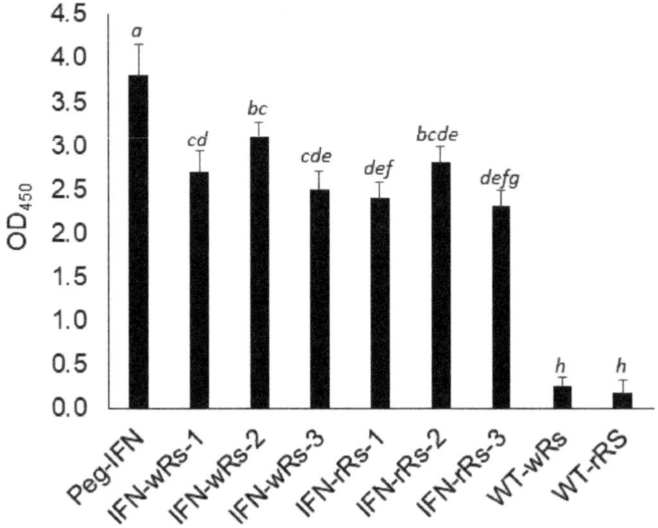

Figure 5. Enzyme-linked Immunosorbent Assay (ELISA) detection for IFN-α2a in wild-type and transgenic *Raphanus sativus* L. plants. Shown are OD$_{450}$ values recorded for recombinant IFN-α2a samples and Peg-IFN (commercially available IFN-α (Pegasys®; Hoffmann-LaRoche, Basel, Switzerland)). IFN-wRs-1–3: total protein extracts isolated from three independent transgenic lines of IFN-α2a- transgenic white *Raphanus sativus* L. plants. IFN-rRs-1–3: total protein extracts isolated from three independent transgenic lines of IFN-α2a-transgenic red *Raphanus sativus* L., WT-wRs: wild-type white *Raphanus sativus* L. protein extracts. WT-rRs: wild-type red *Raphanus sativus* L. protein extracts. Data represent the average OD$_{450}$ values recorded from at least 4 independent measurements ± SE. Different small letters represent significant differences; $p = 0.01$.

OD$_{450}$ values recorded for IFN-α2a proteins in test samples by ELISA assay werefound to be highly correlated with the accumulation of IFN-α2a RNA transcripts (Figures 4A,B and 5). Human IFN-α2a expressed in transgenic *Chlamydomonas reinhardtii* [39] and *E. coli* [40,41] was previously detected using an ELISA assay for human IFN-α2a.

Isolated total protein extracts from IFN-wRs-2 and IFN-rRs-2 transgenic lines were partially purified using DEA-Sepharose and then Sephadex G-50 (Sigma Aldrich, Taufkirchen, Germany). In these purification procedures, total protein extracts isolated from wild-type white and red radish plants were used as controls. The partially purified protein samples were subjected to SDS-PAGE analysis. A protein band in the size range of 19.5 kDa corresponding to IFN-α2a protein was observed for protein extracts isolated from transgenic IFN-wRs-2 and IFN-rRs-2 lines since this protein band was totally absent in wild-type control samples, see Supplementary Figure S2. Moreover, the partially purified and total protein extracts of IFN-α2a were also subjected to UV-spectral scanning and compared to

Peg-IFN control to confirm the existence of IFN-α2a protein in transgenic plant protein extracts (see Supplementary Figure S3). Overall, the results of the qRT-PCR, IFN-α2a ELISA detection assays, SDS-PAGE assay of partially purified IFN-α2a protein isolated from transgenic plants, and UV-spectral scanning assays show that the IFN-α2a gene is efficiently expressed in transgenic *Raphanus sativus* L. lines.

The most commonly used bio-systems for the manufacturing of biopharmaceutical medications are animal cell cultures and microorganisms. Some of the recombinant human proteins created in microbes, including human IFN-α [23], do not have proper activity because microorganisms lack glycosylation and post-translational modifications of proteins. Transgenic animal, mammalian, insect cell cultures, and microbial fermentation systems are probably unsafe and not recommended for the production of biopharmaceuticals [2,42]. As a result, low-cost, simple, and effective technologies are required to meet the demand for large-scale and safe recombinant protein production for human applications. Plant-based systems are particularly cost-effective for producing recombinant proteins, costing around 1% of the cost of mammalian cells and 2–10% of the cost of microbial fermentation. Plants are therefore good candidates for supplying medicinal compounds [42,43]. Radish is known as one of the richest sources of medicinal compounds [38]. Other advantages, including ease of transformation, stability of gene expression, and its safety aspects concerning human consumption, all support the choice of radish as a bio-system for the production of IFN-α2a protein.

Protein extracts obtained from three different lines from each transgenic white and red radish plant were utilized to evaluate the efficiency of the expressed IFN-α2a protein as an antiviral and anticancer agent.

3.3. Antiviral Activity of the Recombinant IFN-α2a

The antiviral activity of recombinant IFN-α2a protein isolated from IFN-α2a transgenic white and red radish lines, total, and partially purified protein extracts was tested against VSV (Vesicular Stomatitis Virus) in two ways: post (direct) and pre-infection (indirect) of Vero cells with VSV. In this test, Peg-IFN and protein extracts from wild-type white and red radish were used as positive and negative controls, respectively. The results of this assay are shown in Table 1.

Table 1. The recombinant IFN-α2a protein antiviral activity.

Sample Code	Reduction in VSV Titer (%)	
	Direct	Indirect
Peg-IFN	7.8 ± 0.25	75.4 ± 6.1
P-IFN-wRs	6.7 ± 0.15	63.2 ± 3.2
IFN-wRs-1	4.8 ± 0.20	50.8 ± 1.2
IFN-wRs-2	5.5 ± 0.14	52.3 ± 1.6
IFN-wRs-3	4.1 ± 0.21	48.2 ± 1.3
P-IFN-rRs	6.2 ± 0.18	60.2 ± 2.2
IFN-rRs-1	4.4 ± 0.16	53.4 ± 1.2
IFN-rRs-2	4.8 ± 0.11	55.6 ± 2.1
IFN-rRs-3	3.8 ± 0.17	49.8 ± 1.8
WT-wRS	0.8 ± 0.06	02.3 ± 0.1
WT-rRS	0.7 ± 0.05	02.0 ± 0.2

Peg-IFN: Commercially available IFN-α2a (Pegasys®; Hoffmann-LaRoche, Basel, Switzerland), P-IFN-wRs: partially purified IFN protein from IFN-α2a transgenic white radish, IFN-wRs-1–3: total protein extracts isolated from three independent IFN-α2a transgenic lines of white radish, P-IFN-rRs: partially purified IFN protein from IFN-α2a-transgenic red radish, IFN-rRs-1–3: total protein extracts isolated from three independent IFN-α2a transgenic lines of red radish, WT-wRs: wild-type white radish protein extracts, WT-rRs: wild-type red radish protein extracts. Data represents the average of the percentage of reduction in VSV titer post-treatment of at least 6 independent measurements $\pm$ SE. *t*-test; $p < 0.001$.

Protein extracts from transgenic white and red *Raphanus sativus* L. are clearly efficacious in inhibiting and decreasing VSV development and reproduction (Table 1). In the case of transgenic white radish protein extracts, it was observed that enriched protein fractions of IFN-wRs-2 transgenic plants (P-IFN-wRs in Table 1) have approximately 8 to and 27-fold more antiviral activity in direct and indirect measurements, respectively than wild-type white radish protein extracts.

Total protein extracts isolated from IFN-wRs-1, 2, and 3 showed approximately 6-fold and 22-fold more suppression of VSV activity compared to the corresponding wild-type protein extracts for direct and indirect measurements, respectively. Similar antiviral activity was observed for recombinant IFN-α2a protein isolated from transgenic red radish plants compared to transgenic white radish plants. Partially purified protein extracts isolated from IFN-rRs-2 transgenic plants (P-IFN-rRs in Table 1) showed 8.8-fold and 27.3-fold more antiviral activity compared to wild-type red radish protein extracts for the direct and indirect measurements, respectively. Total protein extracts isolated from IFN-rRs-1, 2, and 3 recorded an average of 6-fold and 24-fold more suppression of VSV activity compared to the corresponding wild-type protein extracts for direct and indirect assays, respectively. It is also obvious that under the assay conditions, recombinant IFN-α2a protein extracts from both transgenic radish genotypes have equivalent VSV antiviral activity to commercially available peg-IFN.

Previous research has demonstrated that recombinant IFN-α has antiviral and immune-regulatory properties, including the stimulation of class I histocompatibility complex antigens, which causes lymphocytes and macrophages to unleash cytotoxic activities [44]. Moreover, it has been shown that IFN-α acts either by the direct inhibition of cell growth or through the excitation and activation of the immune system [45]. Research implemented for the production of recombinant IFN-α in *Chlamydomonas reinhardtii* [39] and rice [46] or IFN-γ in soybean [17] showed similar inhibitory effects in suppressing the growth and reproduction of VSV.

When recombinant IFN-α2a derived from both types of transgenic radish plants is administered to Vero cells before viral infection, the overall results demonstrate that it functions better as an antiviral agent (indirect assay). As a result, it can be concluded that recombinant IFN-α2a produced in transgenic radish plants is functionally active as an antiviral drug, and its activity is mediated via the cellular immune system's excitation and activation.

3.4. Antitumor Activity of the Recombinant IFN-α2a Isolated from Transgenic White and Red Raphanus sativus L. Plants

3.4.1. Antitumor Effect of the Recombinant IFN-α2a on Hep-G2 Tumor Cell Line (In Vitro Assay)

The research was expanded to assess the impact of recombinant IFN-α2a protein produced in transgenic white and red *Raphanus sativus* L. plants on the growth of the human Hep-G2 (human hepatocellular carcinoma) tumor cell line in vitro. The MTT test was used, as stated in the materials and methods section (Section 2.8). In this test, total and enriched fractions of recombinant IFN-α2a protein extracts were employed. In addition to the commercially available peg-IFN, protein extracts from wild-type white and red radish plants were employed as positive and negative controls, respectively. Partially purified recombinant IFN-α2a protein and protein extracts derived from three independent lines from each IFN-α2a transgenic white and red *Raphanus sativus* L. plant demonstrate a strong antiproliferative effect on the Hep-G2 tumor cell line (Table 2). In total protein extracts isolated from transgenic white and red radish plants, the average IC_{50} values for recombinant IFN-α2a are 8.8 μg/mLand 10.7 μg/mL, respectively. However, the IC_{50} values measured for wild-type white and red radish plants are ~531 μg/mL and ~468 μg/mL, respectively. Thus, the observed IC_{50} values for transgenic white and red radish protein extracts showed ~60-fold and ~43-fold more antitumor activity compared to wild-type protein extracts, respectively. Moreover, it is obvious that IC_{50} values recorded for recombinant IFN-α2a protein are comparable to IC_{50} values recorded for the commercially available Peg-IFN (see Table 2). This assay result demonstrates the functionality of the recombinant IFN-α2a produced in transgenic radish plants against Hep-G2 tumor cells.

Table 2. Cytotoxicity effect of recombinant INFα2a against Hep-G2 cell line.

Sample Name	IC$_{50}$ (µg Protein)	SE
Peg-IFN	6.2	0.62
P-IFN-wRs	7.4	0.77
IFN-wRs-1	8.9	0.47
IFN-wRs-2	8.5	0.55
IFN-wRs-3	9.2	0.81
P-IFN-rRs	8.2	0.93
IFN-rRs-1	10.3	0.65
IFN-rRs-2	10.1	0.53
IFN-rRs-3	11.7	0.68
WT-wRS	531.1	3.62
WT-rRS	468.5	4.23

P-IFN: commercially available IFN-α2a (Pegasys®; Hoffmann-LaRoche, Basel, Switzerland), P-IFN-wRs: partially purified IFN protein from IFN-α2a-transgenic white radish, IFN-wRs-1–3: total protein extracts from three independent transgenic lines of IFN-α2a-transgenic white radish, P-IFN-rRs: partially purified IFN protein from IFN-α2a-transgenic red radish, IFN-rRs-1–3: total protein extracts from three independent transgenic lines of IFN-α2a-transgenic red radish, WT-wRS: wild-type white radish protein extracts, WT-rRS: wild-type red radish protein extracts. Data represents the average of the IC$_{50}$ values recorded from at least 3 independent measurements ± SE. *t*-test; *p*-value < 0.001.

Antitumor properties of plant-, microalgae-, and microorganism-derived protein extracts have been widely tested using Hep-G2 tumor cell lines. Many proteins were tested utilizing Hep-G2 tumor cell lines, including *Helicobacter pylori* CCUG 17874 [47], *E. coli* L-asparaginase (IC$_{50}$: 46 µg/mL) [48], and *Penicillium brevicompactum* NRC 829 (IC$_{50}$: 43.3 µg/mL) [49]. Furthermore, many kinds of human interferon, such as hIFN-α, hIFN-β, and hIFN-γ, have been widely researched for their potential to suppress the development and proliferation of Hep-G2 and Hep3B cells [39,50,51]. All of these reports supported the findings of the current investigation. Thus, IFN-α2a expressed in transgenic white and red radish plants has effective antitumor properties against Hep-G2 tumor cell lines.

3.4.2. Effect of Recombinant IFN-α2a on Hep-G2 Cell Apoptosis

In addition to the CPE assay (Section 3.4.1), the antitumor properties of the recombinant IFN-α2a produced in transgenic white and red radish plants on Hep-G2 cell apoptosis were studied. This assay is based on the application of the IC$_{50}$ concentration of the prepared recombinant IFN-α2a to growing Hep-G2 cells. In addition to wild-type (negative control) and Peg-IFN (positive control), partially purified IFN-α2a proteins and IFN-α2a total protein extracts isolated from transgenic white and red radish plants were employed for this purpose. In the same assay, untreated Hep-G2 cells were employed as an untreated control. Table 3 shows the results of this test. When P-IFN-wRs, IFN-wRs-1, IFN-wRs-2, and IFN-wRs-3 protein extracts are applied to Hep-G2 cells, the percentage of normal cells is greatly reduced (5.40%, 8.47%, 8.58%, and 7.82 %, respectively). This was accompanied by a significant increase in necrotic cell percentage (82.6%, 63.8%, 79.6%, and 57.1%, respectively) as shown in Table 3. A relatively and significantly lower effect was observed for transgenic red radish plants, 5.10%, 7.17%, 7.51%, and 6.82% (normal cells) and 81.2, 60.7, 76.2, and 54.1% (necrotic cells) were observed for P-IFN-rRs, IFN-rRs-1, IFN-rRs-2, and IFN-rRs-3, respectively. Negligible inhibitory effects were observed for white and red radish wild-type controls, the recorded values for WT protein extracts (~0.67% necrosis), and untreated Hep-G2 cells (0.59% necrosis). For all of the measured parameters, normal, early apoptosis, late apoptosis, and necrotic cells, it is evident that recombinant IFN-α2a isolated from white radish has relatively higher effects than the recombinant IFN-α2a isolated from transgenic red radish under the assay conditions.

Table 3. Effect of recombinant IFN-α2a on Hep-G2 cell apoptosis at various stages.

Sample Name	Normal Cells (%)	Early Apoptosis (%)	Late Apoptosis (%)	Necrotic Cells (%)
Control	99.4	0.03	0.07	0.59
Peg-IFN	4.70	1.02	54.55	87.8
P-IFN-wRs	5.40	0.14	37.42	82.6
IFN-wRs-1	8.47	0.07	26.4	63.8
IFN-wRs-2	8.58	0.08	28.7	79.6
IFN-wRs-3	7.82	0.06	25.2	57.1
P-IFN-rRs	5.10	0.12	39.52	81.2
IFN-rRs-1	7.17	0.12	24.3	60.7
IFN-rRs-2	7.51	0.18	26.7	76.2
IFN-rRs-3	6.82	0.16	23.4	54.1
WT-wRS	96.7	0.72	2.21	0.62
WT-rRS	95.9	0.83	2.11	0.72

Control: untreated Hep-G2 cells; *t*-test; *p*-value < 0.001.

IFN has been utilized in the treatment of hepatocellular carcinoma (HCC), albeit with some mixed results [33,52–54]. The current antitumor assay results, on the other hand, demonstrate that recombinant IFN-α2a generated in transgenic radish plants is capable of stopping cell growth and inducing cell apoptosis. Despite all of the research done to figure out how IFNs cause apoptosis, it is still unknown how a tumor cell chooses between cell death and growth arrest. The antitumor characteristics of recombinant IFN-α2a produced in transgenic white and red *Raphanus sativus* L. plants against Hep-G2 cell line assays, revealed in the current investigation, suggest recombinant IFN-α2a's usefulness as an antitumor agent and support the use of radish plants as a promising bio-system for the production of functionally active human IFN-α2a.

4. Conclusions

Producing active biopharmaceuticals in *Raphanus sativus* L. plants has several advantages, including the ability to create proteins with glycosylation and/or post-translational modifications in addition to the absence of toxins and human pathogens. This last feature has the potential to reduce the number of purification steps required in downstream manufacturing. In addition, *Raphanus sativus* L. plants grow quickly and require little time from transformation to protein production. *Raphanus sativus* is a suitable system for the manufacture of pharmaceutically active human interferon because of these properties, as well as the findings of the current investigation. The generated IFN-α2a expressed in radish plants was proven to be functionally active as an anticancer and antiviral drug according to molecular and biochemical investigations. However, more research is needed, such as optimizing transgenic codon usage, finding strong constitutive promoters to boost gene expression, and optimizing both the production and administration of the produced proteins.

Supplementary Materials: The following supporting information can be downloaded at: https://www.mdpi.com/article/10.3390/bioengineering9080381/s1, Figure S1: qRT-PCR analysis of Actin and IFNα2a RNA transcripts; Figure S2: SDS-PAGE of the partially purified IFNα2a protein isolated from IFNα2a transgenic white and red *Raphanus sativus* L. plants; Figure S3: Scanning spectra of peg-IFN protein (A), partially purified protein fractions (B) and un-purified total protein (C) isolated from IFN.wRs-2 transgenic *Raphanus sativus* L. plants.

Author Contributions: Conceptualization, R.K. and Y.E.-A.; methodology, E.H.; software, R.K and O.A.-Z.; validation, R.K., T.H. and R.O.; formal analysis, R.K., O.A.-Z., T.H., R.O. and E.H.; investigation, R.K., Y.E.-A., E.H. and O.A.-Z.; resources, T.H., R.O., O.A.-Z. and R.K.; data curation, R.K., E.H. and O.A.-Z.; writing—original draft preparation, R.K. and O.A.-Z.; writing—review and editing, R.K., T.H., R.O. and O.A.-Z.; visualization, R.K., R.O. and T.H.; supervision, R.K. and Y.E.-A.; project administration, R.K.; funding acquisition, T.H., R.O. and E.H. All authors have read and agreed to the published version of the manuscript.

Funding: This research received no external funding.

Institutional Review Board Statement: Not applicable.

Informed Consent Statement: Not applicable.

Data Availability Statement: Not applicable.

Acknowledgments: Ali Fahmy (Head of R&D Sector at VACSERA, Egypt) was extremely helpful in developing the anticancer and antiviral assays.

Conflicts of Interest: The authors declare no conflict of interest.

References

1. Sato, H.; Tsukamoto-Yasui, M.; Takado, Y.; Kawasaki, N.; Matsunaga, K.; Ueno, S.; Kanda, M.; Nishimura, M.; Karakawa, S.; Isokawa, M.; et al. Protein Deficiency-Induced Behavioral Abnormalities and Neurotransmitter Loss in Aged Mice Are Ameliorated by Essential Amino Acids. *Front. Nutr.* **2020**, *7*, 1–8. [CrossRef] [PubMed]
2. Ma, J.K.; Drake, P.M.; Christou, P. The Production of Recombinant Pharmaceutical Proteins in Plants. *Nat. Rev. Genet.* **2003**, *4*, 794–805. [CrossRef] [PubMed]
3. Ma, S.; Huang, Y.; Davis, A.; Yin, Z.; Mi, Q.; Menassa, R.; Brandle, J.E.; Jevnikar, A.M. Production of Biologically Active Human Interleukin-4 in Transgenic Tobacco and Potato. *Plant Biotechnol. J.* **2005**, *3*, 309–318. [CrossRef] [PubMed]
4. Shirono, H.; Morita, S.; Miki, Y.; Kurita, A.; Morita, S.; Koga, J.; Tanaka, K.; Masumura, T. Highly Efficient Production of Human Interferon-A by Transgenic Cultured Rice Cells. *Plant Biotechnol.* **2006**, *23*, 283–289. [CrossRef]
5. Lowther, W.; Lorick, K.; Lawrence, S.D.; Yeow, W.S. Expression of Biologically Active Human Interferon Alpha 2 in *Aloe Vera*. *Transgenic Res.* **2012**, *21*, 1349–1357. [CrossRef]
6. Nurjis, F.; Khan, M.S. Expression of Recombinant Interferon A-2a in Tobacco Chloroplasts Using Micro Projectile Bombardment. *Afr. J. Biotechnol.* **2011**, *10*, 17016–17022.
7. Rotheim, P. The Biochemotechnology Revolution. *Chemtech* **1997**, *27*, 20–25.
8. Sørensen, H.P.; Mortensen, K.K. Soluble Expression of Recombinant Proteins in the Cytoplasm of *Escherichia coli*. *Microb. Cell Factories* **2005**, *4*, 1–8. [CrossRef]
9. Gasdaska, J.R.; Spencer, D.; Dickey, L. Advantages of Therapeutic Protein Production in the Aquatic Plant Lemna. *Bioprocess J.* **2003**, *2*, 49–56. [CrossRef]
10. Maeda, S.; McCandliss, R.; Gross, M.; Sloma, A.; Familletti, P.C.; Tabor, J.M.; Evinger, M.; Levy, W.P.; Pestka, S. Construction and Identification of Bacterial Plasmids Containing Nucleotide Sequence for Human Leukocyte Interferon. *Proc. Natl. Acad. Sci. USA* **1980**, *77*, 7010–7013. [CrossRef]
11. Venkat, S.K.; Revathi, J.C.; Venkata, S.A.; Narayana, P.K.S.; Venkata, R. A Process for the Production of Human Interferon Alpha from Genetically Engineered Yeast. European Patent EP1272624, 20 September 2001.
12. Digan, M.E.; Lair, S.V.; Brierley, R.A.; Siegel, R.S.; Williams, M.E.; Ellis, S.B.; Kellaris, P.A.; Provow, S.A.; Craig, W.S.; Velicelebi, G.; et al. Continuous Production of a Novel Lysozyme Via Secretion from the Yeast, *Pichia pastoris*. *Bio/Technology* **1989**, *7*, 160–164. [CrossRef]
13. Yang, W. Bulking up Fuzeon. *BioCentury* **2002**, *23*, 10–18.
14. Sahu, P.K.; Patel, T.S.; Sahu, P.; Singh, S.; Tirkey, P.; Sharma, D. Molecular Farming: A Biotechnological Approach in Agriculture for Production of Useful Metabolites. *Int. J. Biotechnol. Biochem.* **2014**, *4*, 23–30.
15. Tarafdar, A.; Kamle, M.; Prakash, A.B.; Padaria, J.C. Transgenic Plants: Issues and Future Prospects. *Biotechnol. Adv.* **2014**, *2*, 1–47.
16. Wu, H.Y.; Liu, K.H.; Wang, Y.C.; Wu, J.F.; Chiu, W.L.; Chen, C.Y.; Lai, E.M. Agrobest: An Efficient Agrobacterium-Mediated Transient Expression Method for Versatile Gene Function Analyses in Arabidopsis Seedlings. *Plant Methods* **2014**, *10*, 19. [CrossRef]
17. Mehrizadeh, V.; Dorani, E.; Mohammadi, S.A.; Ghareyazie, B. Expression of Recombinant Human Ifn-Γ Protein in Soybean (*Glycine max* L.). *Plant Cell Tissue Organ Cult. (PCTOC)* **2021**, *146*, 127–136. [CrossRef]
18. Ren, Q.; Wang, C.Y.; Song, Z.M.; Liu, D.; Yu, H.P.; Sheng, J. Expression of Human Interferon a 2b Gene in Ginseng Cells. *Chem. Res. Chin. Univ.* **2010**, *26*, 420–426.
19. Razaghi, A.; Owens, L.; Heimann, K. Review of the Recombinant Human Interferon Gamma as an Immunotherapeutic: Impacts of Production Platforms and Glycosylation. *J. Biotechnol.* **2016**, *240*, 48–60. [CrossRef]
20. Kebeish, R.; Niessen, M.; Thiruveedhi, K.; Bari, R.; Hirsch, H.J.; Rosenkranz, R.; Stäbler, N.; Schönfeld, B.; Kreuzaler, F.; Peterhänsel, C. Chloroplastic Photorespiratory Bypass Increases Photosynthesis and Biomass Production in *Arabidopsis Thaliana*. *Nat. Biotechnol.* **2007**, *25*, 593–599. [CrossRef]
21. Herouet, C.; Esdaile, D.J.; Mallyon, B.A.; Debruyne, E.; Schulz, A.; Currier, T.; Hendrickx, K.; van der Klis, R.J.; Rouan, D. Safety Evaluation of the Phosphinothricin Acetyltransferase Proteins Encoded by the Pat and Bar Sequences That Confer Tolerance to Glufosinate-Ammonium Herbicide in Transgenic Plants. *Regul. Toxicol. Pharmacol.* **2005**, *41*, 134–149. [CrossRef]
22. Banihani, S.A. Radish (*Raphanus sativus*) and Diabetes. *Nutrients* **2017**, *9*, 1014. [CrossRef] [PubMed]
23. Adolf, G.R.; Kalsner, I.; Ahorn, H.; Maurer-Fogy, I.; Cantell, K. Natural Human Interferon-A2 Is O-Glycosylated. *Biochem. J.* **1991**, *276*, 511–518. [CrossRef] [PubMed]
24. Antonelli, G. Biological Basis for a Proper Clinical Application of Alpha Interferons. *New Microbiol.* **2008**, *31*, 305–318. [PubMed]

25. Curtis, I.S.; Nam, H.G. Transgenic Radish (*Raphanus sativus* L. Longipinnatus Bailey) by Floral-Dip Method-Plant Development and Surfactant Are Important in Optimizing Transformation Efficiency. *Transgenic Res.* **2001**, *10*, 363–371. [CrossRef] [PubMed]
26. Kebeish, R.; Aboelmy, M.; El-Naggar, A.; El-Ayouty, Y.; Peterhansel, C. Simultaneous Overexpression of Cyanidase and Formate Dehydrogenase in *Arabidopsis thaliana* Chloroplasts Enhanced Cyanide Metabolism and Cyanide Tolerance. *Environ. Exp. Bot.* **2015**, *110*, 19–26. [CrossRef]
27. Chomczynski, P.; Mackey, K. Substitution of Chloroform by Bromo-Chloropropane in the Single-Step Method of Rna Isolation. *Anal. Biochem.* **1995**, *225*, 163–164. [CrossRef]
28. Niessen, M.; Thiruveedhi, K.; Rosenkranz, R.; Kebeish, R.; Hirsch, H.J.; Kreuzaler, F.; Peterhänsel, C. Mitochondrial Glycolate Oxidation Contributes to Photorespiration in Higher Plants. *J. Exp. Bot.* **2007**, *58*, 2709–2715. [CrossRef]
29. Bradford, M.M. A Rapid and Sensitive Method for the Quantitation of Microgram Quantities of Protein Utilizing the Principle of Protein-Dye Binding. *Anal. Biochem.* **1976**, *72*, 248–254. [CrossRef]
30. Bussereau, F.; Flamand, A.; Pese-Part, D. Reproducible Plaquing System for Rabies Virus in Cer Cells. *J. Virol. Methods* **1982**, *4*, 277–282. [CrossRef]
31. Lansky, E.P.; Newman, R.A. Punica Granatum (Pomegranate) and Its Potential for Prevention and Treatment of Inflammation and Cancer. *J. Ethnopharmacol.* **2007**, *109*, 177–206. [CrossRef]
32. Mosmann, T. Rapid Colorimetric Assay for Cellular Growth and Survival: Application to Proliferation and Cytotoxicity Assays. *J. Immunol. Methods* **1983**, *65*, 55–63. [CrossRef]
33. Ozgur, O.; Karti, S.; Sonmez, M.; Yilmaz, M.; Karti, D.; Ozdemir, F.; Ovali, E. Effects of Interferon-Alfa-2a on Human Hepatoma Hepg2 Cells. *Exp. Oncol.* **2003**, *25*, 105–107.
34. Barta, A.; Sommergruber, K.; Thompson, D.; Hartmuth, K.; Matzke, M.A.; Matzke, A.J. The Expression of a Nopaline Synthase—Human Growth Hormone Chimaeric Gene in Transformed Tobacco and Sunflower Callus Tissue. *Plant Mol. Biol.* **1986**, *6*, 347–357. [CrossRef] [PubMed]
35. Hiatt, A.; Caffferkey, R.; Bowdish, K. Production of Antibodies in Transgenic Plants. *Nature* **1989**, *342*, 76–78. [CrossRef] [PubMed]
36. Burnett, M.J.; Burnett, A.C. Therapeutic Recombinant Protein Production in Plants: Challenges and Opportunities. *Plants People Planet* **2020**, *2*, 121–132. [CrossRef]
37. Obembe, O.O.; Popoola, J.O.; Leelavathi, S.; Reddy, S.V. Advances in Plant Molecular Farming. *Biotechnol. Adv.* **2011**, *29*, 210–222. [CrossRef]
38. Curtis, I.S. The Noble Radish: Past, Present and Future. *Trends Plant Sci.* **2003**, *8*, 305–307. [CrossRef]
39. El-Ayouty, Y.; El-Manawy, I.; Nasih, S.; Hamdy, E.; Kebeish, R. Engineering *Chlamydomonas Reinhardtii* for Expression of Functionally Active Human Interferon-A. *Mol. Biotechnol.* **2019**, *61*, 134–144. [CrossRef]
40. Babaeipour, V.; Shojaosadati, S.A.; Khalilzadeh, R.; Maghsoudi, N.; Farnoud, A.M. Enhancement of Human Γ-Interferon Production in Recombinant *E. Coli* Using Batch Cultivation. *Appl. Biochem. Biotechnol.* **2010**, *160*, 2366–2376. [CrossRef]
41. Mizukami, T.; Komatsu, Y.; Hosoi, N.; Itoh, S.; Oka, T. Production of Active Human Interferon-A in *E. coli* I. Preferential Production by Lower Culture Temperature. *Biotechnol. Lett.* **1986**, *8*, 605–610. [CrossRef]
42. Xu, J.; Dolan, M.C.; Medrano, G.; Cramer, C.L.; Weathers, P.J. Green Factory: Plants as Bioproduction Platforms for Recombinant Proteins. *Biotechnol. Adv.* **2012**, *30*, 1171–1184. [CrossRef] [PubMed]
43. Karg, S.R.; Kallio, P.T. The Production of Biopharmaceuticals in Plant Systems. *Biotechnol. Adv.* **2009**, *27*, 879–894. [CrossRef] [PubMed]
44. Pestka, S.; Langer, J.A.; Zoon, K.C.; Samuel, C.E. Interferons and Their Actions. *Annu. Rev. Biochem.* **1987**, *56*, 727–777. [CrossRef] [PubMed]
45. Nishiguchi, S.; Kuroki, T.; Nakatani, S.; Morimoto, H.; Takeda, T.; Nakajima, S.; Shiomi, S.; Seki, S.; Kobayashi, K.; Otani, S. Randomised Trial of Effects of Interferon-α on Incidence of Hepatocellular Carcinoma in Chronic Active Hepatitis C with Cirrhosis. *Lancet* **1995**, *346*, 1051–1055. [CrossRef]
46. Masumura, T.; Morita, S.; Miki, Y.; Kurita, A.; Morita, S.; Shirono, H.; Koga, J.; Tanaka, K. Production of Biologically Active Human Interferon-A2 in Transgenic Rice. *Plant Biotechnol.* **2006**, *23*, 91–97. [CrossRef]
47. Gladilina, Y.A.; Sokolov, N.N.; Krasotkina, J.V. Cloning, Expression, and Purification of *Helicobacter Pylori* L-Asparaginase. *Biochem. (Mosc.) Suppl. Ser. B Biomed. Chem.* **2009**, *3*, 89–91. [CrossRef]
48. Kebeish, R.; El-Sayed, A.; Fahmy, H.; Abdel-Ghany, A. Molecular Cloning, Biochemical Characterization, and Antitumor Properties of a Novel L-Asparaginase from *Synechococcus elongatus* Pcc6803. *Biochemistry* **2016**, *81*, 1173–1181. [CrossRef]
49. Elshafei, A.M.; Hassan, M.M.; Abouzeid, M.A.E.; Mahmoud, D.A.; Elghonemy, D.H. Purification, Characterization and Antitumor Activity of L-Asparaginase from *Penicillium brevicompactum* 2 Nrc 829 3. *Br. Microbiol. Res. J.* **2011**, *2*, 158–174. [CrossRef]
50. Enomoto, H.; Tao, L.; Eguchi, R.; Sato, A.; Honda, M.; Kaneko, S.; Iwata, Y.; Nishikawa, H.; Imanishi, H.; Iijima, H.; et al. The in Vivo Antitumor Effects of Type I-Interferon against Hepatocellular Carcinoma: The Suppression of Tumor Cell Growth and Angiogenesis. *Sci. Rep.* **2017**, *7*, 12189. [CrossRef]
51. Melén, K.; Keskinen, P.; Lehtonen, A.; Julkunen, I. Interferon-Induced Gene Expression and Signaling in Human Hepatoma Cell Lines. *J. Hepatol.* **2000**, *33*, 764–772. [CrossRef]

52. Hindi, N.N.; Saleh, M.I. Patient Characteristics associated with Peglyated Interferon Alfa-2a Induced Neutropenia in Chronic Hepatitis C Patients. *Clin. Exp. Pharmacol. Physiol.* **2018**, *45*, 636–642. [CrossRef] [PubMed]
53. Liu, C.J.; Chuang, W.L.; Lee, C.M.; Yu, M.L.; Lu, S.N.; Wu, S.S.; Liao, L.Y.; Chen, C.L.; Kuo, H.T.; Chao, Y.C.; et al. Peginterferon Alfa-2a Plus Ribavirin for the Treatment of Dual Chronic Infection with Hepatitis B and C Viruses. *Gastroenterology* **2009**, *136*, 496–504.e3. [CrossRef] [PubMed]
54. Ryff, J.-C. To Treat or Not to Treat? The Judicious Use of Interferon-A-2a for the Treatment of Chronic Hepatitis B. *J. Hepatol.* **1993**, *17*, S42–S46. [CrossRef]

bioengineering

Article

Neuroprotective Effect of *Abelmoschus manihot* Flower Extracts against the H₂O₂-Induced Cytotoxicity, Oxidative Stress and Inflammation in PC12 Cells

Shih-Wei Wang [1,2,†], Chi-Chang Chang [3,4,†], Chin-Feng Hsuan [1,5], Tzu-Hsien Chang [4], Ya-Ling Chen [4], Yun-Ya Wang [6], Teng-Hung Yu [1,5], Cheng-Ching Wu [1,5] and Jer-Yiing Houng [7,8,*]

1   School of Medicine, College of Medicine, I-Shou University, Kaohsiung 82445, Taiwan
2   Division of Allergy, Immunology, and Rheumatology, Department of Internal Medicine, E-Da Hospital, Kaohsiung 82445, Taiwan
3   School of Medicine for International Students, College of Medicine, I-Shou University, Kaohsiung 82445, Taiwan
4   Department of Obstetrics & Gynecology, E-Da Hospital/E-Da Dachang Hospital, Kaohsiung 82445, Taiwan
5   Division of Cardiology, Department of Internal Medicine, E-Da Hospital/E-Da Dachang Hospital/E-Da Cancer Hospital, Kaohsiung 82445, Taiwan
6   School of Chinese Medicine for Post-Baccalaureate, College of Medicine, I-Shou University, Kaohsiung 82445, Taiwan
7   Department of Nutrition, I-Shou University, Kaohsiung 82445, Taiwan
8   Department of Chemical Engineering, I-Shou University, Kaohsiung 82445, Taiwan
*   Correspondence: jyhoung@isu.edu.tw; Tel.: +886-7-6151100 (ext. 7915)
†   These authors contributed equally to this work.

Citation: Wang, S.-W.; Chang, C.-C.; Hsuan, C.-F.; Chang, T.-H.; Chen, Y.-L.; Wang, Y.-Y.; Yu, T.-H.; Wu, C.-C.; Houng, J.-Y. Neuroprotective Effect of *Abelmoschus manihot* Flower Extracts against the H₂O₂-Induced Cytotoxicity, Oxidative Stress and Inflammation in PC12 Cells. *Bioengineering* 2022, 9, 596. https://doi.org/10.3390/bioengineering9100596

Academic Editors: Minaxi Sharma, Kandi Sridhar and Zeba Usmani

Received: 15 September 2022
Accepted: 20 October 2022
Published: 21 October 2022

Publisher's Note: MDPI stays neutral with regard to jurisdictional claims in published maps and institutional affiliations.

Abstract: The progression of neurodegenerative diseases is associated with oxidative stress and inflammatory responses. *Abelmoschus manihot* L. flower (AMf) has been shown to possess excellent antioxidant and anti-inflammatory activities. This study investigated the protective effect of ethanolic extract (AME), water extract (AMW) and supercritical extract (AMS) of AMf on PC12 neuronal cells under hydrogen peroxide (H₂O₂) stimulation. This study also explored the molecular mechanism underlying the protective effect of AME, which was the best among the three extracts. The experimental results showed that even at a concentration of 500 µg/mL, neither AME nor AMW showed toxic effects on PC12 cells, while AMS caused about 10% cell death. AME has the most protective effect on apoptosis of PC12 cells stimulated with 0.5 mM H₂O₂. This is evident by the finding when PC12 cells were treated with 500 µg/mL AME; the viability was restored from 58.7% to 80.6% in the Treatment mode ($p < 0.001$) and from 59.1% to 98.1% in the Prevention mode ($p < 0.001$). Under the stimulation of H₂O₂, AME significantly up-regulated the expression of antioxidant enzymes, such as catalase, glutathione peroxidase and superoxide dismutase; promoted the production of the intracellular antioxidant; reduced glutathione; and reduced ROS generation in PC12 cells. When the acute inflammation was induced under the H₂O₂ stimulation, AME significantly down-regulated the pro-inflammatory cytokines and mediators (e.g., TNF-α, IL-1β, IL-6, COX-2 and iNOS). AME pretreatment could also greatly promote the production of nucleotide excision repair (NER)-related proteins, which were down-regulated by H₂O₂. This finding indicates that AME could repair DNA damage caused by oxidative stress. Results from this study demonstrate that AME has the potential to delay the onset and progression of oxidative stress-induced neurodegenerative diseases.

Keywords: *Abelmoschus manihot*; neurodegenerative disease; oxidative stress; proliferation; inflammation; nucleotide excision repair

1. Introduction

Oxidative stress and inflammation have been implicated in the development of neurodegenerative diseases, including stroke, post-stroke cerebral ischemia–reperfusion, Huntington's disease, Alzheimer's disease and Parkinson's disease [1–3]. Oxidative stress leads to excessive production and progressive accumulation of reactive oxygen species (ROS) such as hydrogen peroxide (H_2O_2), hydroxy free radicals and superoxide anion, as well as reactive nitrogen species (RNS) such as nitric cation, nitrogen dioxide and peroxynitrite anion, in various pathological conditions. Oxidative stress, an important cause of neuronal degeneration and injury, induces cell apoptosis or necrosis through cellular oxidative damage such as DNA damage, lipid peroxidation, protein oxidation, expression of inflammatory and apoptotic genes and decrease in nitric oxide bioactivity [4–6]. Inflammation is known to be essential in regulating the central nervous system. However, this function may run out of control when microglial cells are over-activated and pro-inflammatory cytokines or mediators are over-produced, ultimately leading to neuronal cell damage and dysfunction [3,7].

Hydrogen peroxide is produced from oxidative energy metabolism by the reaction with oxidant-generating agents in vivo. It has the ability to cross lipid bilayers and react with membrane or protein-bound metal ions to form hydroxyl radicals, making H_2O_2 an extremely dangerous reactive oxygen [8]. It causes oxidative damage to cells and various lesions in DNA, which causes DNA helixes to twist, hindering base pairing, transcription and replication. Unrepaired oxidative damage can easily cause cancer and shorten telomeres, leading to premature aging and senescence. Mammals have a DNA repair mechanism that can repair individual base damages or genomes by recognizing specific DNA spiral distortions. The nucleotide excision repair (NER) mechanism is the common function that combines with a variety of proteins in a spatially and temporally specific manner to repair major lesions. It repairs bulky DNA adducts and helix-twist damages and is also a synergistic system of base excision repair that can be used to repair oxidative stress-induced DNA damage [9,10].

Many studies have shown that the application of antioxidative and anti-inflammatory strategies to alleviate oxidative stress-stimulated neuronal damage by scavenging free radicals and inhibiting amyloid deposition in neuronal cells is promising in the treatment of neurodegenerative diseases [11,12]. Numerous studies have also reported that some plant extracts or active plant ingredients have a protective effect on neuronal cells and can attenuate the progression of neurological diseases [13–15].

Abelmoschus manihot (L.) Medic, a member of the Malvaceae family, is widely distributed in valleys and grasslands from Asia to Europe. *A. manihot* flowers (AMf) have been used to treat malignant skin ulcers, burns and cellulitis [16–21]. AMf extracts and their bioactive components possess many biochemical activities, including antioxidant, anti-inflammatory, antiviral, anti-diabetic nephropathy, anti-lipogenic, antidepressant, antiplatelet, analgesic, anticonvulsant, cardioprotective, hepatoprotective, immunomodulatory; they are also effective against the cerebral infarction, diarrhea, bone loss, preventing menorrhagia, relieving labor, stimulating lactation and enhancing sexual arousal and reproduction [18,22–24]. In China, AMf extracts have been applied in clinical practice for the treatment of diabetic nephropathy and chronic glomerulonephritis [25,26]. In terms of neuroprotective activity, AMf extract has been shown to significantly reduce the incidence of cerebral edema in rats with acute incomplete cerebral ischemia and alleviate the pathological changes in brain tissues [27]. In addition, AMf extract could protect rat hippocampal neurons by inhibiting NMDA (*N*-methyl-D-aspartate) receptor responses [28]. However, the molecular mechanism involved in the neuroprotective activity of AMf extracts remains to be elucidated.

In this study, differentiated rat adrenal pheochromocytoma PC12 neuronal cells, a cell model often used to study neurodegenerative diseases [29,30], were applied to evaluate the neuroprotective effect of AMf extracts on the growth of neuronal cells stimulated by H_2O_2, and effects on H_2O_2-induced oxidative stress and acute inflammatory response in PC12 cells. The effect of AMf extract on the repair of DNA damage through the NER pathway was also examined.

2. Materials and Methods

2.1. Preparation of AMf Extracts

The AMf raw materials were purchased from Kangmei Chinese Medicine Store (Bozhou, Anhui, China), and its nucleotide sequence was 99.43% identical to that of Gen-Bank ID: KY218782.1 [24]. In order to prepare AMf ethanol (AME) extract, 2.6 kg of dried flowers were ground to powder and then extracted at room temperature with 16 L of 95% ethanol for 24 h. After filtering and collecting the extract solution, the residue was extracted two more times with 16 L of 95% ethanol each. The extraction solutions were pooled, the solvent was removed using an evaporator (Panchum Scientific, Kaohsiung City, Taiwan), and the residue was dried with a freeze-dryer (Panchum Scientific); finally, the dried AME samples were obtained.

In order to prepare AMf water (AMW) extract, 1.0 kg of dried flower powder was heated in 6 L of water to boiling point and then extracted at 90 °C for 3 h. Then the extract was cooled down to room temperature and filtered. The filtrate was evaporated with an evaporator and finally freeze-dried the residue to obtain the AMW extract

The AMf supercritical-CO_2 fluid (AMS) extract was prepared by extracting 1.0 kg of flower powder in a supercritical fluid extractor (5 L/1000 bar R&D unit, Natex, Ternitz, Austria). The extraction procedure was as follows: the tank temperature was raised to 40 °C and then kept at this temperature; then, the tank pressure was raised to 150 bar in 20 min, 250 bar in 20 min, 300 bar in 10 min, 350 bar in 10 min and stayed at 350 bar for 2 h. After the extraction, freeze-dried the residue to obtain AMS extract.

These three dried extracts were all kept in a −20 °C freezer until use. The total polyphenol content (TPC) and total flavonoid content (TFC) of the various extracts were analyzed using the method of Tsai et al. [31] and expressed as gallic acid equivalents and catechin equivalents, respectively. The contents of five flavonoids were analyzed by HPLC, according to the method of Chang et al. [24].

2.2. Cell Culture and Analyses

The rat adrenal pheochromocytoma PC12 cell line was obtained from Bioresource Collection and Research Center (Hsinchu, Taiwan). The cells were cultivated in Dulbecco's Modified Eagle Medium supplemented with 10% fetal bovine serum (Gibco, Grand Island, NY, USA), 100 units/mL penicillin—100 µg/mL streptomycin, 1% L-glutamine, 0.02% $NaHCO_3$, pH 7.2–7.4. The cells were cultured in an incubator at 37 °C, with an atmosphere of 5% CO_2 and 95% air.

The general protocol of cell culture in this study was to plate 5×10^4 cells in each well of 96-well plates and incubate for 24 h. In the Treatment mode, PC12 cells were first stimulated with H_2O_2 for 4 h and then treated with different concentrations of the extracted sample for 24 h. In the Prevention mode, PC12 cells were first treated with different concentrations of the extracted sample for 24 h and then stimulated by H_2O_2 for 4 h.

Cell viability was analyzed using the MTT (3-(4,5-Dimethylthiazol-2-yl)-2,5-diphenyltetrazolium bromide) assay kit (Sigma-Aldrich Chemicals, St. Louis, MO, USA). The medium was removed from the 96-well dish, and after washing the cells with PBS (phosphate buffered saline), 100 µL of 5 mg/mL MTT were added and incubated for 2–4 h. Subsequently, after replacing the MTT reagent with 100 µL of DMSO (dimethyl sulfoxide), the dish was shaken to dissolve the crystals. The amount of PC12 cells was then measured using an ELISA reader at 570 nm wavelength.

Morphological changes in cell nuclei were detected using Hoechst staining. Cells were first fixed with 3.7% paraformaldehyde for 30 min, washed with PBS and stained with Hoechst 33,258 (Sigma-Aldrich) for 30 min at 37 °C. The morphological changes were observed and photographed with a fluorescence microscope.

2.3. Assay of Cellular Oxidative System

2.3.1. Intracellular ROS Content of H_2O_2-Stimulated PC12 Cells

Intracellular ROS content was analyzed by the Fluorometric Intracellular ROS Kit (Sigma-Aldrich), which detected ROS with 2',7'-dichlorodihydrofluorescein diacetate (DCF-DA). The cells treated with AME for 24 h and stimulated by H_2O_2 for 4 h were washed with PBS and then reacted with 5 µM DCF-DA at room temperature for 30 min. The ROS content was then determined by fluorescence at 502 nm excitation and 524 nm emission in a multi-detection microplate reader (Synergy^{TM} 2, BioTek, Winooski, VT, USA).

2.3.2. Antioxidant Enzymes Activity and Glutathione Content

Cell extracts were prepared from PC12 cells, which were pretreated with AME for 24 h and then stimulated by H_2O_2 for 4 h, according to the manufacturer's instructions. The activities of catalase, glutathione peroxidase (GPx) and superoxide dismutase (SOD) and the content of reduced glutathione (GSH) were analyzed by Sigma-Aldrich assay kits with catalog numbers STA-341, K762-100, STA-340 and STA-312, respectively. The protein content in cell extract was determined by the Bicinchoninic Acid Protein Assay Kit (Sigma-Aldrich).

2.4. Western Blot Assay

Western blot was used to detect the protein expression of antioxidant enzymes, inflammatory cytokines, NER-related proteins and β-actin internal standard. This assay mainly followed the procedures described in our previous paper [32]. Briefly, PBS-washed PC12 cells were lysed with a modified RIPA buffer (Sigma-Aldrich). A certain amount of protein was denatured by heating at 95 °C for 5 min. After cooling, it was placed in the sample tank for electrophoresis. The proteins were separated on an SDS-PAGE gel, then transferred to a PVDF membrane, and a primary antibody was added. Subsequently, the separated protein was detected with ECL Plus Western Blotting Detection Reagents (Sigma-Aldrich) after treatment with a secondary antibody conjugated to horseradish peroxidase (HRP). All antibodies used were purchased from Sigma-Aldrich (Table 1). The ChemiDoc XRS+ system (Bio-Rad, Hercules, CA, USA) was used to detect protein expression, and the Quantity One$^{®}$ software (version 5.2.1, Bio-Rad) was used for quantification.

Table 1. The sources of the antibodies used in this study.

1° Ab	2° Ab	Molecular Weight (kDa)
Catalase	Rabbit	60
GPx 1/2	Rabbit	24
SOD-1	Mouse	16
p-NF-κB	Rabbit	65
iNOS	Rabbit	130
COX-2	Rabbit	75
TNF-α	Rabbit	25
IL-1β	Rabbit	17
XPE	Rabbit	127
XPC	Rabbit	106
RPA	Rabbit	37
XPA	Rabbit	31
XPF	Rabbit	104
ERCC	Rabbit	36
XPG	Rabbit	133
PCNA	Rabbit	36
β-actin	Mouse	48

2.5. Statistical Analysis

Each experiment was repeated at least three times, and the data were presented as arithmetic means ± standard deviation (SD). For the chemical composition analysis, the significance of differences between mean values was determined by one-way analysis of variance (ANOVA) and Duncan's multiple range test when the $p < 0.05$. The Student's *t*-test was used to examine the statistical difference of each parameter in other experiments. These tests were all performed with SPSS 25.0 software (SPSS Inc., Chicago, IL, USA). The significance of the data difference compared with the vehicle group were defined as * $p < 0.05$, ** $p < 0.01$ and *** $p < 0.001$.

3. Results

3.1. Extraction Yield and Chemical Composition Analysis of AMf Extracts

AMf was extracted by 95% ethanol, hot water and supercritical-CO_2 fluid, and yields were 25.2%, 16.4% and 1.0%, respectively; the extracts were denoted as AME, AMW and AMS, respectively. Previous studies reported that the main active ingredients in AMf extracts are polyphenols and flavonoids [18,33,34]. Therefore, the TPC and TFC in these three extracts were analyzed. Table 2 shows that the richest content of TPC (80.6 mg/g) was found in AME, followed by AMW (61.5 mg/g) and AMS (47.3 mg/g). In terms of TFC, AME was still the highest (38.0 mg/g), followed by AMS (30.9 mg/g) and AMW (28.9 mg/g). The chromatograms of the five major flavonoid ingredients in AMf extracts analyzed by high-performance liquid chromatography (HPLC) are shown in Figure 1, and the content of each ingredient is also shown in Table 2. In these three extracts, hyperoside had the highest content, followed by myricetin, isoquercitrin, rutin and quercetin. Compared with the three extracts, the content of hypericin and quercetin in AMS was higher than that in AME, and the content of other components was still less than that of AME. The content of active ingredients of AMW was generally less than that of AME and AMS.

Table 2. Extraction yield and chemical composition of the three AMf extracts *.

Extract	AME	AMW	AMS
Extraction yield (*w/w* %)	25.2	16.4	1.0
TPC (mg/g extract)	120.8 ± 0.8 [A]	61.5 ± 1.4 [B]	47.3 ± 0.8 [C]
TFC (mg/g extract)	57.0 ± 1.0 [A]	28.9 ± 1.7 [B]	30.9 ± 1.4 [B]
Rutin (1)	10.0 ± 0.4 [Ac]	6.8 ± 0.2 [Bc]	5.0 ± 0.2 [Cc]
Hyperoside (2)	37.8 ± 2.1 [Ba]	18.9 ± 1.2 [Ca]	42.4 ± 2.2 [Aa]
Isoquercitrin (3)	20.6 ± 1.7 [Ab]	10.9 ± 1.3 [Bb]	17.9 ± 1.0 [Ab]
Myricetin (4)	36.9 ± 1.4 [Aa]	17.5 ± 0.5 [Ba]	19.2 ± 1.1 [Bb]
Quercetin (5)	2.6 ± 0.1 [Bd]	1.1 ± 0.1 [Cd]	4.1 ± 0.3 [Ac]

* The significance of differences between mean values was determined by one-way ANOVA and Duncan's multiple range test when the $p < 0.05$. [A, B, C] Different letters in the same row indicate significant differences ($p < 0.05$). [a, b, c] Different letters in the same column indicate significant differences ($p < 0.05$).

3.2. Cytotoxicity of AMf Extracts on PC12 Cells

The cytotoxicity of the three extracts on PC12 cells was demonstrated by the cell viability. As shown in Figure 2, AME increased cell viability only when concentration reached 500 µg/mL (Figure 2A), AMW did not affect cell viability over the range of concentrations tested (Figure 2B), while AMS exhibited a dose-dependent cytotoxic effect on PC12 (Figure 2C). However, the viability of PC12 cells remained higher than 90% even when AMS concentration was up to 500 µg/mL.

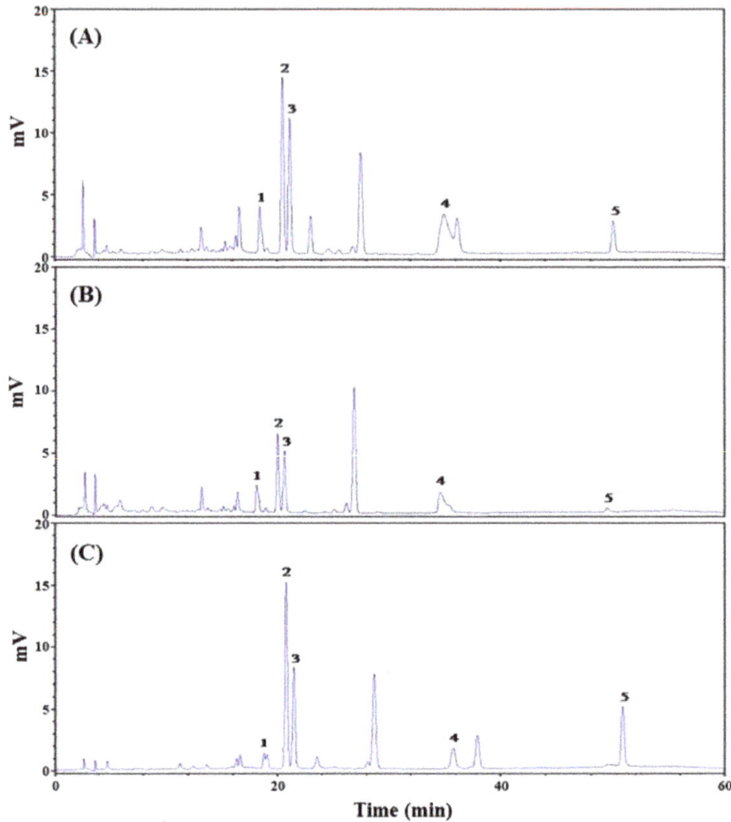

Figure 1. HPLC chromatograms of AMf extracts ((**A**) AME, (**B**) AMW and (**C**) AMS). Identified compounds: rutin (<u>1</u>), hyperoside (<u>2</u>), isoquercitrin (<u>3</u>), myricetin (<u>4</u>) and quercetin (<u>5</u>).

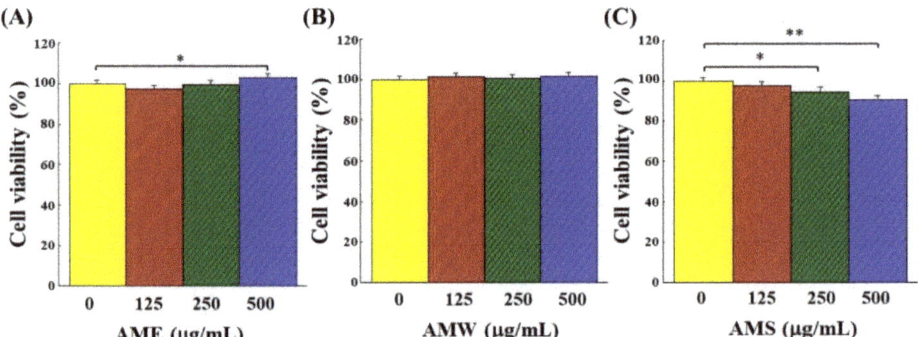

Figure 2. Cytotoxicity of the three AMf extracts on PC12 cells: (**A**) AME, (**B**) AMW and (**C**) AMS. PC12 cells were incubated for 24 h and then treated with specific extract for additional 24 h, and the viability was analyzed with an MTT kit. The vehicle group (group 0) was the PC12 cells grown only with the treatment by the sample solvent DMSO. The cell number of this group was set as 100%. This experiment was repeated five times. The significance of the difference in data compared to the vehicle is labeled as * $p < 0.05$ and ** $p < 0.01$.

3.3. Effects of AMf Extracts on Proliferation of H_2O_2-Stimulated PC12 Cells

Cell proliferation (expressed as cell viability) in PC12 cells stimulated with different concentrations of H_2O_2 is shown in Figure 3. With the H_2O_2 concentration increased, the extent of inhibition on cell proliferation was also increased. When the concentration of H_2O_2 was at 0.5 mM, the viability of PC12 cells dropped to 59.9%. In order to observe the protective effect of AMf extracts on the H_2O_2-stimulated PC12 cells, the H_2O_2 concentration was set at 0.5 mM in the subsequent experiments.

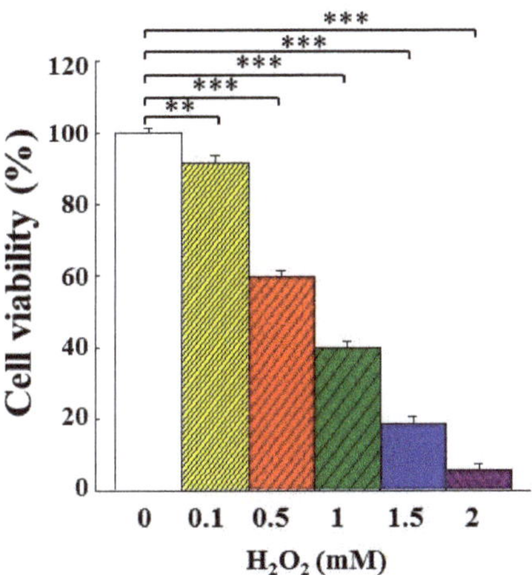

Figure 3. Inhibitory effect of H_2O_2 concentration on the growth of PC12 cells. The cells were first incubated for 24 h, then stimulated with the indicated concentration of H_2O_2 for 24 h, and cell viability was analyzed with an MTT kit. The vehicle group (group 0) was the PC12 cells grown on the treatment with water. The cell number of this group was set as 100%. This experiment was repeated five times. The significance of the difference in data compared to the vehicle is marked as ** $p < 0.01$ and *** $p < 0.001$.

The protective effects of AMf extract on the growth of the H_2O_2-challenged PC12 cells were examined in two modes: (1) in the Treatment mode, the cells were stimulated by H_2O_2 for 4 h and then treated with three different extract samples, respectively, for additional 24 h; (2) in the Prevention mode, the cells were treated with three different extract samples, respectively, for 24 h and then stimulated by H_2O_2 for another 4 h. Figure 4 shows the experimental results under the Treatment mode operation. AME exhibited the most significant cytoprotective effect, and as the treatment dose increased, the cell viability rose remarkably (Figure 4A). The cytoprotective effect of AMW was not obvious (Figure 4B), while AMS showed a protective effect only when it reached 500 µg/mL (Figure 4C).

Under the Prevention mode operation, compared to the Treatment mode, AME and AMS had a more profound protective effect on PC12 cells (Figure 5A,C), while AMW had no protective effect (Figure 5B). When 500 µg/mL AME was introduced, the cell viability in the Treatment mode recovered from 58.7% to 80.6%, while that in the Prevention mode recovered from 59.1% to 98.1%. These findings clearly demonstrated that AME had the best protective effect on PC12 cells, and more so in the Prevention mode. Therefore, all subsequent experiments were conducted using AME in the Prevention mode.

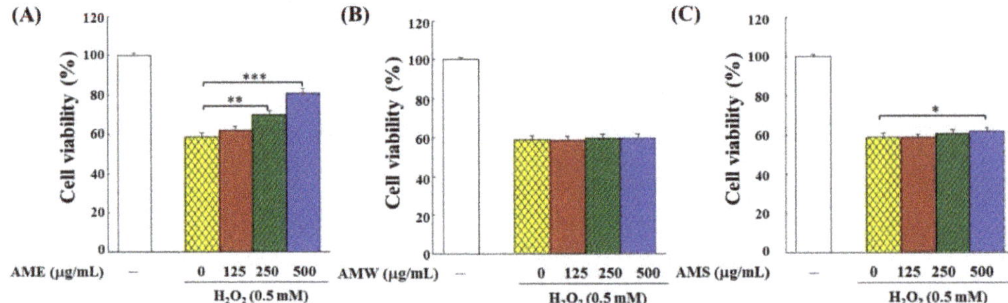

Figure 4. Protective effects of AMf extracts ((**A**) AME, (**B**) AMW and (**C**) AMS) on growth of PC12 cells under the H_2O_2 stimulation in Treatment mode. PC12 cells were first cultivated for 24 h, stimulated with 0.5 mM H_2O_2 for 4 h, then treated with specific extract for 24 h, and the viability was analyzed with an MTT kit. The control group was the cells grown without H_2O_2 stimulation and extract sample treatment, and the cell number of this group was set as 100%. This experiment was repeated five times. The significance of the difference in data compared to the vehicle are labeled as * $p < 0.05$, ** $p < 0.01$ and *** $p < 0.001$.

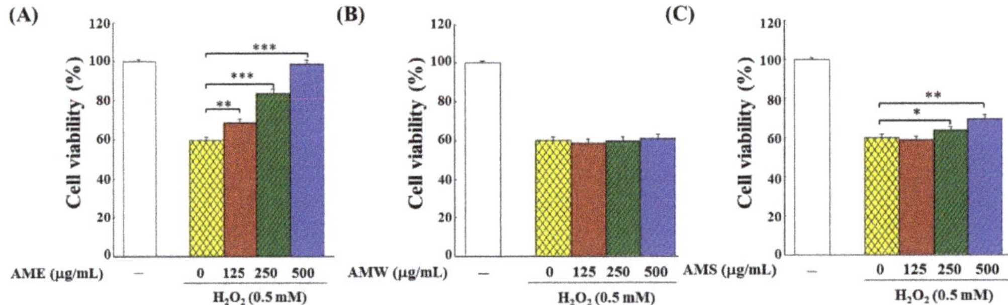

Figure 5. Protective effects of the three AMf extracts on the growth of PC12 cells under the H_2O_2 stimulation in Prevention mode: (**A**) AME, (**B**) AMW and (**C**) AMS. The cells were first cultivated for 24 h, treated with specific extract at indicated concentration for another 24 h, then stimulated with 0.5 mM H_2O_2 for 4 h. Cell viability was analyzed with an MTT kit. The control group was the PC12 cells grown without treatment with extract sample and H_2O_2 stimulation. The cell number of this group was set as 100%. This experiment was repeated five times. The significance of the difference in data compared to the vehicle (group 0) are labeled as * $p < 0.05$, ** $p < 0.01$ and *** $p < 0.001$.

3.4. Effect of AME Pretreatment on Cell Morphology and Nuclear Staining of H_2O_2-Stimulated PC12 Cells

The cell nucleus morphology changes were detected by Hoechst staining and photographed with a fluorescent microscope. Figure 6A shows that the number of PC12 cells significantly decreased when treated with 0.5 mM H_2O_2. The cells originally grew in suspension stacking and had become thin and loose after treatment. In Hoechst nucleus staining (Figure 6B), brighter blue fluorescent nuclei marked by the red arrows indicated nuclei with severe DNA damage. When cells were treated with H_2O_2, there was a dramatic increase in the number of cells with DNA damage.

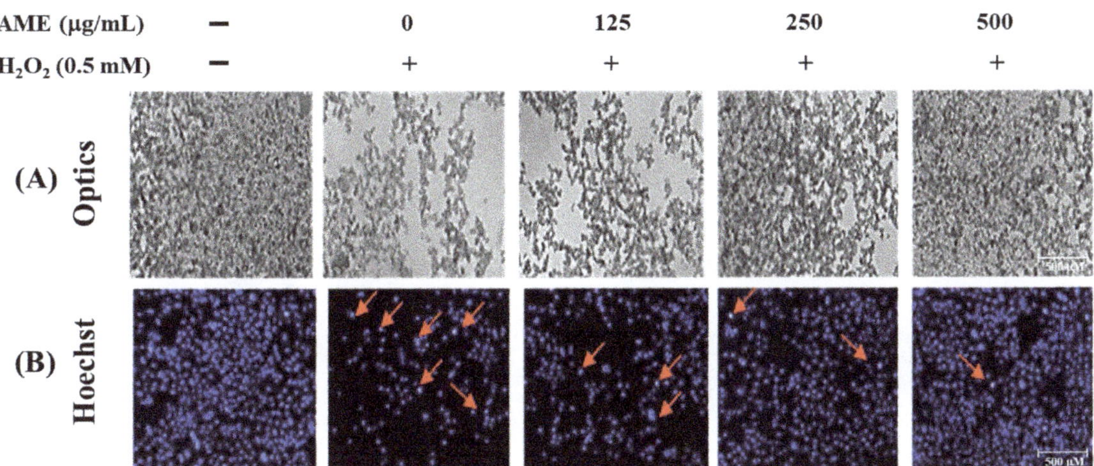

Figure 6. Changes in nuclear morphology of H_2O_2-stimulated PC12 cells under the AME pretreatment. The cells were first cultivated for 24 h, treated with AME for another 24 h, then stimulated with 0.5 mM H_2O_2 for 4 h. (**A**) The cell morphology was observed by a phase-contrast microscope. (**B**) Nuclear changes were examined by Hoechst staining and observed with a fluorescence microscope. The arrows indicate nuclei with severe DNA damage.

When PC12 cells were pretreated with AME, the number of cells increased as the concentration of AME increased; the patterns of cell growth were also returning to a dense and stacked state (Figure 6A). As shown by the Hoechst staining, the number of DNA-damaged cells decreased dramatically when the AME concentration increased (Figure 6B). Therefore, these findings further confirmed that AME possessed a protective effect on the H_2O_2-stimulated PC12 cells.

3.5. Effect of AME Pretreatment on the Regulation of NER

NER is a repair mechanism for oxidative stress-induced DNA damage [9,10]. In order to examine whether oxidative stress and AME pretreatment affect the NER system of PC12 cells, this study used Western blot assay to analyze the expression of several NER-related proteins. As shown in Figure 7, when PC12 cells were stimulated with H_2O_2 alone, all other proteins were down-regulated (20–86% to the control), with the exception of XPF, which was up-regulated (around 110% to the control). Pretreatment of cells with AME enhanced the production of all tested proteins, implying that AME could substantially raise the NER capacity to repair oxidative stress-induced DNA damage.

3.6. Effect of AME Pretreatment on Antioxidant System

An increase in the amount of ROS would modulate the intracellular antioxidant system. In this study, the levels of antioxidant enzymes and reduced glutathione (GSH) were used to reflect the changes in the intracellular antioxidant system. Figure 8 indicates that the H_2O_2-stimulation significantly decreased (about 64–73% of the control group) the activities of the three intracellular antioxidant enzymes, i.e., catalase, glutathione peroxidase (GPx) and superoxide dismutase (SOD). The H_2O_2 stimulation also significantly reduced the GSH content (73.5% of the control group). This finding implies that the H_2O_2 stimulation reduced the performance of the antioxidant system in PC12 cells. When cells were pretreated with AME for 24 h and then challenged by 0.5 mM H_2O_2, the activities of these three suppressed intracellular antioxidant enzymes and GSH generation were all recovered. The higher the AME concentration, the higher the recovery of these enzymes in cells. At 500 µg/mL, all these four indexes returned (GPx, SOD and GSH) to or exceeded (catalase) those in the control, indicating that AME could protect the intracellular antioxidant system when PC12 cells were subjected to oxidative stress.

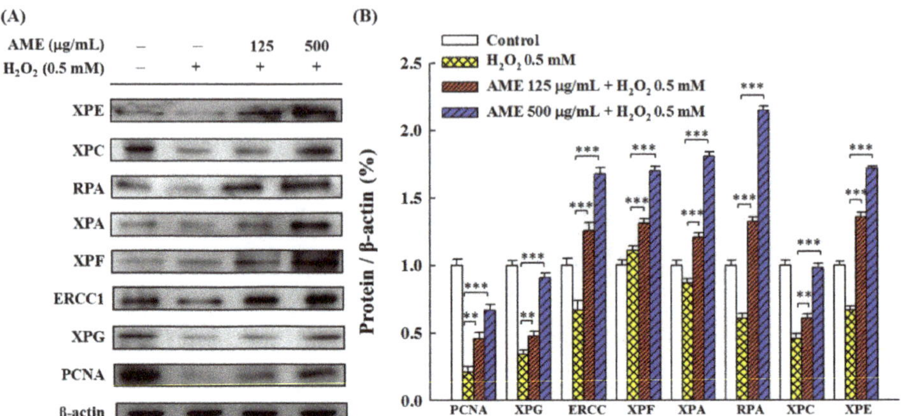

Figure 7. Effect of AME pretreatment on the expressions of NER-related proteins in H_2O_2-stimulated PC12 cells. (**A**) Changes in protein expression. (**B**) Densitometric quantitation. The data were estimated from triplicate independent experiments. The control group was the cells grown without treatment with extract sample and H_2O_2 stimulation. The protein expressed by this group was set as 1.0. The significance of the difference in data compared to the vehicle group is labeled as ** $p < 0.01$ and *** $p < 0.001$.

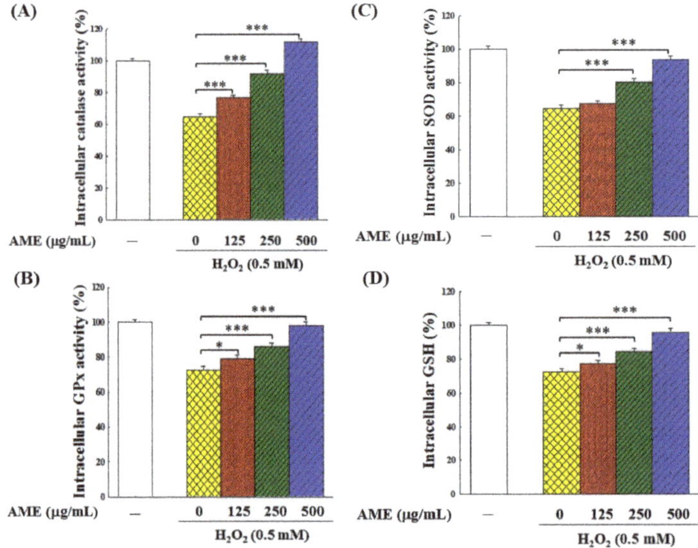

Figure 8. Effect of AME pretreatment on the antioxidant enzymes' activities and the GSH content in H_2O_2-stimulated PC12 cells: (**A**) catalase, (**B**) GPx, (**C**) SOD and (**D**) GSH. The cells were first cultivated for 24 h, treated with AME at the indicated concentration for another 24 h, then stimulated with 0.5 mM H_2O_2 for 4 h. The activities of catalase, GPx, SOD and the content of GSH were analyzed. This experiment was repeated five times. The significance of the difference in data compared to the vehicle (group 0) are labeled as * $p < 0.05$ and *** $p < 0.001$.

In order to examine whether the increase in activity of these antioxidant enzymes was due to increased production of enzyme proteins in the cells, Western blot analysis was carried out. Figure 9 shows that when cells were treated with H_2O_2 only, the production of these antioxidant enzymes was significantly inhibited. However, when cells were pretreated with AME, the H_2O_2-induced inhibitory effect was suppressed, and the production of all three enzyme proteins was up-regulated in a dose-dependent fashion. When the AME concentration reached 500 µg/mL, the production of these enzyme proteins was comparable to the group (SOD) or exceeded (GPx and catalase) that of the control. This result is in good agreement with the data shown in Figure 8; it shows that the protective effect of AME on the intracellular antioxidant system was achieved by up-regulating the production of antioxidant enzymes.

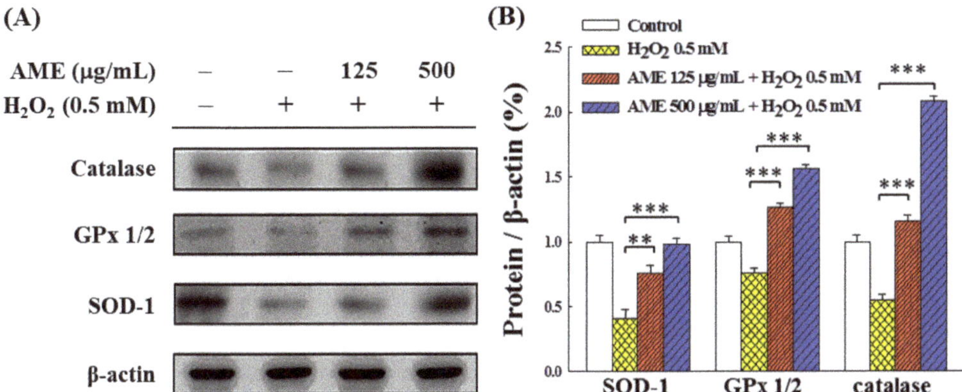

Figure 9. Effect of AME pretreatment on the production of antioxidant enzymes in H_2O_2-stimulated PC12 cells. (**A**) Changes in protein formation. (**B**) Densitometric quantitation. The data were estimated from triplicate independent experiments. The control group was the cells grown without treatment with extract sample and H_2O_2 stimulation. The protein expressed by this group was set as 1.0. The significance of the difference in data compared to the vehicle group is labeled as ** $p < 0.01$ and *** $p < 0.001$.

When PC12 cells were challenged with 0.5 mM H_2O_2, the intracellular ROS production, detected by fluorescence microscope, increased significantly by 1.43 times compared with the control group (Figure 10A,B). However, when PC12 cells were pretreated with AME and then stimulated with H_2O_2, the amount of intracellular ROS decreased in direct relationship with the rise of AME concentration. As the AME concentration reached 500 µg/mL, the intracellular ROS content dropped down to the same level as the control group (no H_2O_2 induction).

3.7. Effect of AME Pretreatment on Inflammatory Responses

Oxidative stress can stimulate the inflammatory response. Figure 11 shows that the H_2O_2 induction substantially increased the expression of pro-inflammatory cytokines and mediators (e.g., IL-1β, TNF-α, COX-2 and iNOS) and phosphorylation of NF-κB in PC12 cells. Pretreatment with AME, even at a low AME dose (125 µg/mL), significantly reduced the expression of IL-1β, TNF-α and COX-2. However, effectively down-regulating the iNOS expression and NF-κB phosphorylation required AME concentration at 500 µg/mL, indicating that AME pretreatment was less sensitive in regulating the iNOS expression and NF-κB phosphorylation.

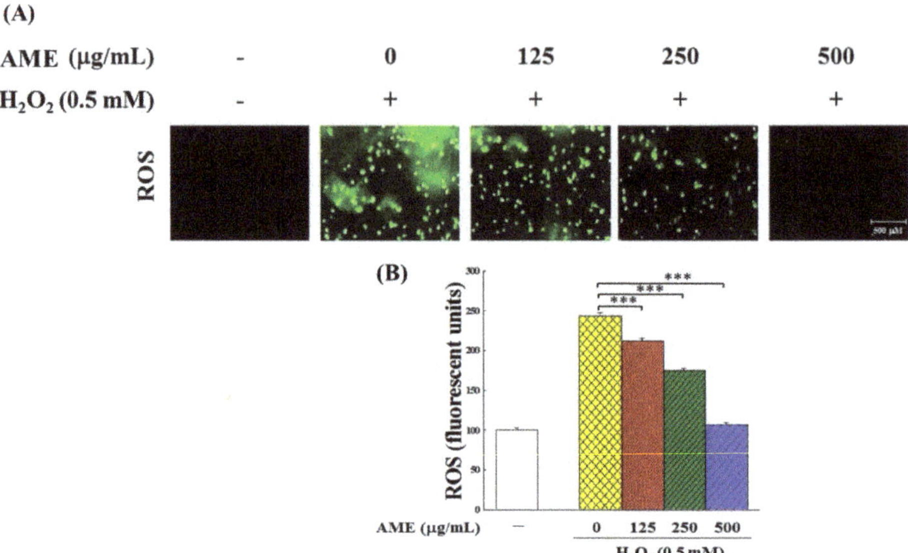

Figure 10. Effect of AME pretreatment on the ROS generation in H_2O_2-stimulated PC12 cells. (**A**) The intracellular ROS levels indicated by DCF-DA fluorescence intensity. (**B**) Fluorescent quantitation. This experiment was repeated three times. The significance of the difference in data compared to the vehicle (group 0) is labeled as *** $p < 0.001$.

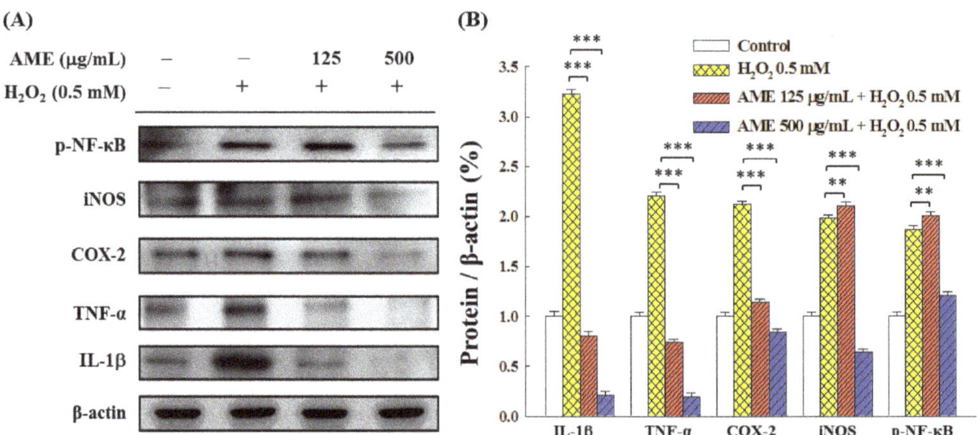

Figure 11. Effect of AME pretreatment on the expressions of pro-inflammatory cytokines and mediators in H_2O_2-stimulated PC12 cells. (**A**) Changes in protein expression. (**B**) Densitometric quantitation. The data were estimated from triplicate independent experiments. The control group was the PC12 cells grown without treatment with extract sample and H_2O_2 stimulation. The protein expressed by this group was set as 1.0. The significance of the difference in data compared to the vehicle group is labeled as ** $p < 0.01$ and *** $p < 0.001$.

4. Discussion

The present study investigated the protective effects of three AMf extracts on the growth of PC12 neuronal cells stimulated by H_2O_2. This study also examined the inhibitory effects of AME pretreatment on intracellular ROS production and acute inflammatory response and enhancing effects on intracellular antioxidant system and NER capacity. It should be noted here that the immortalized PC12 cell line is a model of neuronal cells but does not fully represent neurons.

To date, there are more than 128 chemical constituents identified mostly from the flowers but also from seeds, stems and leaves of *A. manihot*. These components can be classified as polyphenols, flavonoids, amino acids, nucleosides, steroids, organic acids, volatile oils and polysaccharides [18]. Polyphenols and flavonoids are recognized and widely studied as the main active substances of AMf [20,33,34]. Among them, hyperoside, myricetin, isoquercetin, rutin, quercetin, hibifolin and quercetin-3-O-robinobioside are regarded as the main active ingredients [33,34]. In this study, the content of five flavonoids in three extracts of AMf was analyzed by HPLC, as shown in Figure 1 and Table 2. Among the three extracts, AME has the highest TPC and TFC values. In terms of active ingredient content, only AMS has a higher content of hypericin and quercetin than AME, while the content of other components of AMS and all components of AMW was lower than that of AME.

Polyphenols and flavonoids can scavenge reactive oxygen species and nitrogen species and activate redox transcription factors to regulate gene expression, so they often possess antioxidant, anti-inflammatory, anticancer, antiviral and neurological and cardiac protection activities. Thus, these cost-effective medicinal ingredients have significant biological activity. Depending on the type, mode of action and bioavailability, polyphenols and flavonoids have been proven to be effective in managing a variety of diseases [35–37].

When the extraction is carried out with different solvents, types of ingredients can be extracted depending on the characteristics of the solvent. In this study, hot water, 95% ethanol and supercritical CO_2, the polarity ranks from high to low, were used to prepare AMW, AME and AMS. By comparing the extraction yield and the TPC and TFC in the extract (Table 2), it can be seen that ethanol had the highest extraction efficiency. The results from this study have also shown that AME exhibited the best protective effect on the H_2O_2-stimulated PC12 cells, which may be related to the high content of TPC and TFC in this extract.

Incubation with AME and AMW did not affect the growth of PC12 cells, but AMS showed slight cytotoxicity (Figure 2). AME exhibited a significant protective effect on the proliferation of the H_2O_2-induced PC12 cells, while AMS had little effect, and AMW had no effect (Figures 4 and 5). This shows that the components contained in the three extracts should be different, and it is worth further analysis of the differences in the components contained in these three extracts.

Oxidative stress induced by H_2O_2 stimulation generally leads to apoptosis of most human cells [38,39]. Previously, numerous studies reported that AMf extracts could inhibit oxidative stress-induced apoptosis through regulation of caspase-3/8/9, Bax and Bcl-2 in cells such as glomerular podocyte [25,40], human umbilical vein endothelial cells (HUVECs) [41], rat kidney epithelial NRK-52E cells [42] and human hepatoma HepG2 cells [43]. This information prompted us to examine whether AME could also protect the growth of PC12 cells by inhibiting apoptosis. Furthermore, oxidative stress can lead to DNA damage. The Hoechst staining experiment shown in Figure 6B demonstrated that AME had the effect of protecting intracellular DNA.

The NER mechanism, a common DNA repair system, can recognize lesions and bind to a variety of proteins in a spatially and temporally specific manner to remove damaged, short single-stranded DNA fragments and synthesize short complementary double strands using undamaged single-stranded DNA as a template, to repair oxidative stress-induced DNA damage [10,44,45].

The NER involves two major sub-pathways: global genomic NER (GG-NER) and transcription-coupled NER (TC-NER). These two sub-pathways differ in how they recognize DNA lesions but function similarly in cleavage, repair and ligation of damaged genes. The former is responsible for repairing damages throughout the genome, while the latter is responsible for lesions in active gene transcription [46].

NER in mammalian cells involves nine main proteins. The deficiency of specific proteins can lead to specific diseases; for example, XPA, XPB, XPC, XPD, XPE, XPF and XPG are derived from xeroderma pigmentosum and belong to the GG-NER sub-pathway, while CSA and CSB proteins are associated with Cockayne syndrome, which belongs to the TC-NER sub-pathway. In addition, ERCC1, RPA and PCNA are involved in the excision and repair of nucleotides.

It was reported that H_2O_2 stimulation reduces the expression of most related proteins in the NER system [9,47]. This study illustrated that the challenge of the H_2O_2-induced oxidative stress reduced the expression of most NER-related proteins tested to 20–86% while slightly increasing the expression of XPF to 110% (Figure 7). Both XPF and XPG are endonucleases that specifically incise damaged strands at 50 and 30 from lesions, respectively, and facilitate the release of lesions containing 22–32 nt long oligomers [10]. ERCC1 expression is considered to be a major key target for redox-regulated NER capabilities [47]. It was shown that the incision activity of ERCC1 requires heterodimerization with XPF (ERCC4) [48]; however, the expression of ERCC1 and XPF are not correlated [47,49]. This might explain why H_2O_2 significantly retarded the expression of XPG, whereas it enhanced the formation of XPF in PC12 cells, as shown in this study. Further studies are needed to clarify this difference. Nonetheless, the present study found that treatment with AME up-regulated all tested NER-related proteins in the H_2O_2-stimulated cells, showing that AME could protect the NER system to maintain the DNA repair function in cells under oxidative stress.

The results in this study show that AME pretreatment could effectively decrease intracellular ROS, enhance GSH content and up-regulate the production of antioxidant enzymes in the H_2O_2-stimulated PC12 cells (Figures 8–10). AME could also attenuate the inflammatory responses which were induced by H_2O_2 stimulation (Figure 11). Our findings are consistent with those reported in the literature. Deng et al. reported that Huangkui capsule, a flavonoid extract of AMf, could suppress inflammation and oxidative stress in the lipopolysaccharide-induced RAW 264.7 cells and mice [50]. Huangkui capsule could also attenuate renal fibrosis in diabetic nephropathic rats by modulating oxidative stress [51]. Liu et al. reported that ethyl acetate extract of AMf could exert significant suppressive effects on oxidative stress in the H_2O_2-stimulated HepG2 cells and D-galactose-stimulated aging mice [43]. The histological analysis demonstrated that this extract effectively reduced the brain and liver injury of these mice. Qiu et al. also reported similar results [52]. Moreover, AMf extract was shown to protect mice against CCl_4-induced liver injury through anti-oxidative stress and anti-inflammatory effects [53]. AMf extract also exhibited a hepatoprotective effect on the α-naphthyl isothiocyanate-induced cholestatic liver damage through anti-inflammation, anti-oxidative injury and regulation of liver transporters expression [54].

5. Conclusions

This study demonstrates that the administration of AME can protect the survival of cells, upregulate the expression of intracellular antioxidant enzymes, promote the generation of intracellular GSH and substantially decrease ROS in the H_2O_2-stimulated PC12 neuronal cells. In addition, AME has significant anti-inflammatory activity on the cells and the repairing effect on the NER system to reduce gene damage caused by oxidative stress. Therefore, AME has the potential to be useful in the prevention and treatment of neurological diseases caused by oxidative stress and inflammation.

Author Contributions: Conceptualization, S.-W.W. and J.-Y.H.; methodology, S.-W.W., C.-C.C. and C.-F.H.; investigation, T.-H.C., Y.-L.C., Y.-Y.W., T.-H.Y. and C.-C.W.; formal analysis, C.-C.C. and C.-F.H.; resources, C.-C.C. and J.-Y.H.; writing—original draft preparation, S.-W.W. and Y.-Y.W.; writing—review and editing, C.-C.C. and J.-Y.H.; project administration, C.-C.C. and J.-Y.H.; funding acquisition, S.-W.W., C.-C.C. and J.-Y.H. All authors have read and agreed to the published version of the manuscript.

Funding: This research was funded by E-Da Hospital (EDAHP106019 and EDAHI111004) and the Ministry of Science and Technology of Taiwan (MOST 110-2221-E-214-002). The APC was funded by E-Da Hospital.

Institutional Review Board Statement: Not applicable.

Informed Consent Statement: Not applicable.

Data Availability Statement: Data are contained within the article.

Conflicts of Interest: The authors declare no conflict of interest.

References

1. Gaki, G.S.; Papavassiliou, A.G. Oxidative stress-induced signaling pathways implicated in the pathogenesis of Parkinson's disease. *Neuromolecular Med.* **2014**, *16*, 217–230. [CrossRef] [PubMed]
2. Turunc Bayrakdar, E.; Uyanikgil, Y.; Kanit, L.; Koylu, E.; Yalcin, A. Nicotinamide treatment reduces the levels of oxidative stress, apoptosis, and PARP-1 activity in $A\beta$(1-42)-induced rat model of Alzheimer's disease. *Free Radic. Res.* **2014**, *48*, 146–158. [CrossRef]
3. Xu, Y.; Tang, D.; Wang, J.; Wei, H.; Gao, J. Neuroprotection of andrographolide against microglia-mediated inflammatory injury and oxidative damage in PC12 neurons. *Neurochem. Res.* **2019**, *44*, 2619–2630. [CrossRef] [PubMed]
4. Gao, Y.; Dong, C.; Yin, J.; Shen, J.; Tian, J.; Li, C. Neuroprotective effect of fucoidan on H_2O_2-induced apoptosis in PC12 cells via activation of PI3K/Akt pathway. *Cell. Mol. Neurobiol.* **2012**, *32*, 523–529. [CrossRef]
5. Li, J.; Wuliji, O.; Li, W.; Jiang, Z.G.; Ghanbari, H.A. Oxidative stress and neurodegenerative disorders. *Int. J. Mol. Sci.* **2013**, *14*, 24438–24475. [CrossRef] [PubMed]
6. Pei, J.P.; Fan, L.H.; Nan, K.; Li, J.; Dang, X.Q.; Wang, K.Z. HSYA alleviates secondary neuronal death through attenuating oxidative stress, inflammatory response, and neural apoptosis in SD rat spinal cord compression injury. *J. Neuroinflamm.* **2017**, *14*, 97. [CrossRef]
7. Ning, X.; Guo, Y.; Ma, X.; Zhu, R.; Tian, C.; Wang, X.; Ma, Z.; Zhang, Z.; Liu, J. Synthesis and neuroprotective effect of E-3,4-dihydroxy styryl aralkyl ketones derivatives against oxidative stress and inflammation. *Bioorg. Med. Chem. Lett.* **2013**, *23*, 3700–3703. [CrossRef]
8. Hoffschir, F.; Daya-Grosjean, L.; Petit, P.X.; Nocentini, S.; Dutrillaux, B.; Sarasin, A.; Vuillaume, M. Low catalase activity in xeroderma pigmentosum fibroblasts and SV40-transformed human cell lines is directly related to decreased intracellular levels of the cofactor, NADPH. *Free Radic. Biol. Med.* **1998**, *24*, 809–816. [CrossRef]
9. Ting, A.P.; Low, G.K.; Gopalakrishnan, K.; Hande, M.P. Telomere attrition and genomic instability in xeroderma pigmentosum type-b deficient fibroblasts under oxidative stress. *J. Cell. Mol. Med.* **2010**, *14*, 403–416. [CrossRef] [PubMed]
10. Lee, T.H.; Kang, T.H. DNA oxidation and excision repair pathways. *Int. J. Mol. Sci.* **2019**, *20*, 6092. [CrossRef]
11. Ataie, A.; Shadifar, M.; Ataee, R. Polyphenolic antioxidants and neuronal regeneration. *Basic Clin. Neurosci.* **2016**, *7*, 81–90. [CrossRef]
12. Kaur, N.; Chugh, H.; Sakharkar, M.K.; Dhawan, U.; Chidambaram, S.B.; Chandra, R. Neuroinflammation mechanisms and phytotherapeutic intervention: A systematic review. *ACS Chem. Neurosci.* **2020**, *11*, 3707–3731. [CrossRef] [PubMed]
13. Hu, M.; Liu, Y.; He, L.; Yuan, X.; Peng, W.; Wu, C. Antiepileptic effects of protein-rich extract from *Bombyx batryticatus* on mice and its protective effects against H_2O_2-induced oxidative damage in PC12 cells via regulating PI3K/Akt signaling pathways. *Oxid. Med. Cell. Longev.* **2019**, *2019*, 7897584. [CrossRef] [PubMed]
14. Sharma, P.; Srivastava, P.; Seth, A.; Tripathi, P.N.; Banerjee, A.G.; Shrivastava, S.K. Comprehensive review of mechanisms of pathogenesis involved in Alzheimer's disease and potential therapeutic strategies. *Prog. Neurobiol.* **2019**, *174*, 53–89. [PubMed]
15. Yang, F.; Ma, Q.; Matsabisa, M.G.; Chabalala, H.; Braga, F.C.; Tang, M. *Panax notoginseng* for cerebral ischemia: A systematic review. *Am. J. Chin. Med.* **2020**, *48*, 1331–1351. [CrossRef] [PubMed]
16. Rubiang-Yalambing, L.; Arcot, J.; Greenfield, H.; Holford, P. Aibika (*Abelmoschus manihot* L.): Genetic variation, morphology and relationships to micronutrient composition. *Food Chem.* **2016**, *193*, 62–68. [CrossRef] [PubMed]
17. Kim, H.; Dusabimana, T.; Kim, S.R.; Je, J.; Jeong, K.; Kang, M.C.; Cho, K.M.; Kim, H.J.; Park, S.W. Supplementation of *Abelmoschus manihot* ameliorates diabetic nephropathy and hepatic steatosis by activating autophagy in mice. *Nutrients* **2018**, *10*, 1703. [CrossRef]
18. Luan, F.; Wu, Q.; Yang, Y.; Lv, H.; Liu, D.; Gan, Z.; Zeng, N. Traditional uses, chemical constituents, biological properties, clinical settings, and toxicities of *Abelmoschus manihot* L.: A comprehensive review. *Front. Pharmacol.* **2020**, *11*, 1068. [CrossRef]
19. Tu, Y.; Sun, W.; Wan, Y.G.; Che, X.Y.; Pu, H.P.; Yin, X.J.; Chen, H.L.; Meng, X.J.; Huang, Y.R.; Shi, X.M. Huangkui capsule, an extract from *Abelmoschus manihot* (L.) medic, ameliorates adriamycin-induced renal inflammation and glomerular injury via inhibiting p38MAPK signaling pathway activity in rats. *J. Ethnopharmacol.* **2013**, *147*, 311–320. [CrossRef]
20. Zhang, Y.; He, W.; Li, C.; Chen, Q.; Han, L.; Liu, E.; Wang, T. Antioxidative flavonol glycosides from the flowers of *Abelmouschus manihot*. *J. Nat. Med.* **2013**, *67*, 78–85. [CrossRef] [PubMed]

21. Hou, J.; Qian, J.; Li, Z.; Gong, A.; Zhong, S.; Qiao, L.; Qian, S.; Zhang, Y.; Dou, R.; Li, R.; et al. Bioactive compounds from *Abelmoschus manihot* L. alleviate the progression of multiple myeloma in mouse model and improve bone marrow microenvironment. *Onco. Targets Ther.* **2020**, *13*, 959–973. [CrossRef]
22. Bourdy, G.; Walter, A. Maternity and medicinal plants in Vanuatu. I. The cycle of reproduction. *J. Ethnopharmacol.* **1992**, *37*, 179–196. [CrossRef]
23. Puel, C.; Mathey, J.; Kati-Coulibaly, S.; Davicco, M.J.; Lebecque, P.; Chanteranne, B.; Horcajada, M.N.; Coxam, V. Preventive effect of *Abelmoschus manihot* (L.) Medik. on bone loss in the ovariectomised rats. *J. Ethnopharmacol.* **2005**, *99*, 55–60. [CrossRef]
24. Chang, C.C.; Houng, J.Y.; Peng, W.H.; Yeh, T.W.; Wang, Y.Y.; Chen, Y.L.; Chang, T.H.; Hung, W.C.; Yu, T.H. Effects of *Abelmoschus manihot* flower extract on enhancing sexual arousal and reproductive performance in zebrafish. *Molecules* **2022**, *27*, 2218. [CrossRef] [PubMed]
25. Chen, Y.; Cai, G.; Sun, X.; Chen, X. Treatment of chronic kidney disease using a traditional Chinese medicine, Flos *Abelmoschus manihot* (Linnaeus) Medicus (Malvaceae). *Clin. Exp. Pharmacol. Physiol.* **2016**, *43*, 145–148. [CrossRef] [PubMed]
26. Li, N.; Tang, H.; Wu, L.; Ge, H.; Wang, Y.; Yu, H.; Zhang, X.; Ma, J.; Gu, H.F. Chemical constituents, clinical efficacy and molecular mechanisms of the ethanol extract of *Abelmoschus manihot* flowers in treatment of kidney diseases. *Phytother. Res.* **2021**, *35*, 198–206. [CrossRef]
27. Gao, S.; Fan, L.; Dong, L.Y.; Cen, D.Y.; Jiang, Q.; Fang, M. Protection effects of the total flavone from *Abelmoschus manihot* L. medic on acute incompletely cerebral ischemia in rats. *Chin. J. Clin. Pharmacol. Ther.* **2003**, *8*, 167–169.
28. Cheng, X.P.; Qin, S.; Dong, L.Y.; Zhou, J.N. Inhibitory effect of total flavone of *Abelmoschus manihot* L. Medic on NMDA receptor-mediated current in cultured rat hippocampal neurons. *Neurosci. Res.* **2006**, *55*, 142–145.
29. Clementi, M.E.; Lazzarino, G.; Sampaolese, B.; Brancato, A.; Tringali, G. DHA protects PC12 cells against oxidative stress and apoptotic signals through the activation of the NFE2L2/HO-1 axis. *Int. J. Mol. Med.* **2019**, *43*, 2523–2531. [CrossRef] [PubMed]
30. Wu, Y.; Qin, D.; Yang, H.; Wang, W.; Xiao, J.; Zhou, L.; Fu, H. Neuroprotective effects of deuterium-depleted water (DDW) against H_2O_2-induced oxidative stress in differentiated PC12 cells through the PI3K/Akt signaling pathway. *Neurochem. Res.* **2020**, *45*, 1034–1044. [CrossRef]
31. Tsai, Y.D.; Hsu, H.F.; Chen, Z.H.; Wang, Y.T.; Huang, S.H.; Chen, H.J.; Wang, C.P.; Wang, S.W.; Chang, C.C.; Houng, J.Y. Antioxidant, anti-inflammatory, and anti-proliferative activities of extracts from different parts of farmed and wild *Glossogyne tenuifolia*. *Ind. Crops Prod.* **2014**, *57*, 98–105. [CrossRef]
32. Chen, T.H.; Chang, C.C.; Houng, J.Y.; Chang, T.H.; Chen, Y.L.; Hsu, C.C.; Chang, L.S. Suppressive effects of *Siegesbeckia orientalis* ethanolic extract on proliferation and migration of hepatocellular carcinoma cells through promoting oxidative stress, apoptosis and inflammatory responses. *Pharmaceuticals* **2022**, *15*, 826. [CrossRef]
33. Lai, X.; Liang, H.; Zhao, Y.; Wang, B. Simultaneous determination of seven active flavonols in the flowers of *Abelmoschus manihot* by HPLC. *J. Chromatogr. Sci.* **2009**, *47*, 206–210. [CrossRef] [PubMed]
34. Wan, Y.; Wang, M.; Zhang, K.; Fu, Q.; Wang, L.; Gao, M.; Xia, Z.; Gao, D. Extraction and determination of bioactive flavonoids from *Abelmoschus manihot* (Linn.) Medicus flowers using deep eutectic solvents coupled with high-performance liquid chromatography. *J. Sep. Sci.* **2019**, *42*, 2044–2052. [CrossRef]
35. Gao, H.; Yuan, X.; Wang, Z.; Gao, Q.; Yang, J. Profiles and neuroprotective effects of *Lycium ruthenicum* polyphenols against oxidative stress-induced cytotoxicity in PC12 cells. *J. Food Biochem.* **2020**, *44*, e13112. [CrossRef]
36. Li, H.; Liu, M.W.; Yang, W.; Wan, L.J.; Yan, H.L.; Li, J.C.; Tang, S.Y.; Wang, Y.Q. Naringenin induces neuroprotection against homocysteine-induced PC12 cells via the upregulation of superoxide dismutase 1 expression by decreasing miR-224-3p expression. *J. Biol. Regul. Homeost. Agents* **2020**, *34*, 421–433. [PubMed]
37. Ullah, A.; Munir, S.; Badshah, S.L.; Khan, N.; Ghani, L.; Poulson, B.G.; Emwas, A.H.; Jaremko, M. Important flavonoids and their role as a therapeutic agent. *Molecules* **2020**, *25*, 5243. [CrossRef]
38. Du, G.T.; Ke, X.; Meng, G.L.; Liu, G.J.; Wu, H.Y.; Gong, J.H.; Qian, X.D.; Cheng, J.L.; Hong, H. Telmisartan attenuates hydrogen peroxide-induced apoptosis in differentiated PC12 cells. *Metab. Brain Dis.* **2018**, *33*, 1327–1334. [CrossRef]
39. Guo, Z.; Yuan, Y.; Guo, Y.; Wang, H.; Song, C.; Huang, M. Nischarin attenuates apoptosis induced by oxidative stress in PC12 cells. *Exp. Ther. Med.* **2019**, *17*, 663–670. [CrossRef]
40. Zhou, L.; An, X.F.; Teng, S.C.; Liu, J.S.; Shang, W.B.; Zhang, A.H.; Yuan, Y.G.; Yu, J.Y. Pretreatment with the total flavone glycosides of Flos *Abelmoschus manihot* and hyperoside prevents glomerular podocyte apoptosis in streptozotocin-induced diabetic nephropathy. *J. Med. Food* **2012**, *15*, 461–468. [CrossRef]
41. Tang, L.; Pan, W.; Zhu, G.; Liu, Z.; Lv, D.; Jiang, M. Total flavones of *Abelmoschus manihot* enhances angiogenic ability both in vitro and in vivo. *Oncotarget* **2017**, *8*, 69768–69778. [CrossRef] [PubMed]
42. Li, W.; He, W.; Xia, P.; Sun, W.; Shi, M.; Zhou, Y.; Zhu, W.; Zhang, L.; Liu, B.; Zhu, J.; et al. Total extracts of *Abelmoschus manihot* L. attenuates adriamycin-induced renal tubule injury via suppression of ROS-ERK1/2-mediated NLRP3 inflammasome activation. *Front. Pharmacol.* **2019**, *10*, 567. [CrossRef]
43. Liu, C.; Jia, Y.; Qiu, Y. Ethyl acetate fraction of *Abelmoschus manihot* (L.) Medic flowers exerts inhibitory effects against oxidative stress in H_2O_2-induced HepG2 cells and D-galactose-induced aging mice. *J. Med. Food* **2021**, *24*, 997–1009. [CrossRef]
44. Lans, H.; Hoeijmakers, J.H.J.; Vermeulen, W.; Marteijn, J.A. The DNA damage response to transcription stress. *Nat. Rev. Mol. Cell Biol.* **2019**, *20*, 766–784. [CrossRef]

45. Kusakabe, M.; Onishi, Y.; Tada, H.; Kurihara, F.; Kusao, K.; Furukawa, M.; Iwai, S.; Yokoi, M.; Sakai, W.; Sugasawa, K. Mechanism and regulation of DNA damage recognition in nucleotide excision repair. *Genes Environ.* **2019**, *41*, 2. [CrossRef]
46. Yamamoto, A.; Nakamura, Y.; Kobayashi, N.; Iwamoto, T.; Yoshioka, A.; Kuniyasu, H.; Kishimoto, T.; Mori, T. Neurons and astrocytes exhibit lower activities of global genome nucleotide excision repair than do fibroblasts. *DNA Repair* **2007**, *6*, 649–657. [CrossRef]
47. Langie, S.A.; Knaapen, A.M.; Houben, J.M.; van Kempen, F.C.; de Hoon, J.P.; Gottschalk, R.W.; Godschalk, R.W.; van Schooten, F.J. The role of glutathione in the regulation of nucleotide excision repair during oxidative stress. *Toxicol. Lett.* **2007**, *168*, 302–309. [CrossRef]
48. Wood, R.D.; Araújo, S.J.; Ariza, R.R.; Batty, D.P.; Biggerstaff, M.; Evans, E.; Gaillard, P.H.; Gunz, D.; Köberle, B.; Kuraoka, I.; et al. DNA damage recognition and nucleotide excision repair in mammalian cells. *Cold Spring Harb. Symp. Quant. Biol.* **2000**, *65*, 173–182. [CrossRef]
49. Vogel, U.; Dybdahl, M.; Frentz, G.; Nexo, B.A. DNA repair capacity: Inconsistency between effect of over-expression of five NER genes and the correlation to mRNA levels in primary lymphocytes. *Mutat. Res.* **2000**, *461*, 197–210. [CrossRef]
50. Deng, J.; He, Z.; Li, X.; Chen, W.; Yu, Z.; Qi, T.; Xu, S.; Xu, Z.; Fang, L. Huangkui capsule attenuates lipopolysaccharide-induced acute lung injury and macrophage sctivation by suppressing inflammation and oxidative stress in mice. *Evid. Based Complement. Alternat. Med.* **2021**, *2021*, 6626483. [CrossRef]
51. Mao, Z.M.; Shen, S.M.; Wan, Y.G.; Sun, W.; Chen, H.L.; Huang, M.M.; Yang, J.J.; Wu, W.; Tang, H.T.; Tang, R.M. Huangkui capsule attenuates renal fibrosis in diabetic nephropathy rats through regulating oxidative stress and p38MAPK/Akt pathways, compared to α-lipoic acid. *J. Ethnopharmacol.* **2015**, *173*, 256–265. [CrossRef]
52. Qiu, Y.; Ai, P.F.; Song, J.J.; Liu, C.; Li, Z.W. Total flavonoid extract from *Abelmoschus manihot* (L.) Medic flowers attenuates d-galactose-induced oxidative stress in mouse liver through the Nrf2 pathway. *J. Med. Food* **2017**, *20*, 557–567. [CrossRef] [PubMed]
53. Ai, G.; Liu, Q.; Hua, W.; Huang, Z.; Wang, D. Hepatoprotective evaluation of the total flavonoids extracted from flowers of *Abelmoschus manihot* (L.) Medic: In vitro and in vivo studies. *J. Ethnopharmacol.* **2013**, *146*, 794–802. [CrossRef]
54. Yan, J.Y.; Ai, G.; Zhang, X.J.; Xu, H.J.; Huang, Z.M. Investigations of the total flavonoids extracted from flowers of *Abelmoschus manihot* (L.) Medic against α-naphthylisothiocyanate-induced cholestatic liver injury in rats. *J. Ethnopharmacol.* **2015**, *172*, 202–213. [CrossRef]

bioengineering

Article

Global Regulator AdpA_1075 Regulates Morphological Differentiation and Ansamitocin Production in *Actinosynnema pretiosum* subsp. *auranticum*

Siyu Guo [1], Tingting Leng [1], Xueyuan Sun [1], Jiawei Zheng [1], Ruihua Li [1], Jun Chen [1], Fengxian Hu [1], Feng Liu [1,*] and Qiang Hua [1,2,*]

[1] State Key Laboratory of Bioreactor Engineering, East China University of Science and Technology, 130 Meilong Road, Shanghai 200237, China
[2] Shanghai Collaborative Innovation Center for Biomanufacturing Technology, 130 Meilong Road, Shanghai 200237, China
* Correspondence: fengliu@ecust.edu.cn (F.L.); qhua@ecust.edu.cn (Q.H.); Tel./Fax: +86-21-64250972 (Q.H.)

Abstract: *Actinosynnema pretiosum* is a well-known producer of maytansinoid antibiotic ansamitocin P-3 (AP-3). Growth of *A. pretiosum* in submerged culture was characterized by the formation of complex mycelial particles strongly affecting AP-3 production. However, the genetic determinants involved in mycelial morphology are poorly understood in this genus. Herein a continuum of morphological types of a morphologically stable variant was observed during submerged cultures. Expression analysis revealed that the *ssgA_6663* and *ftsZ_5883* genes are involved in mycelial aggregation and entanglement. Combing morphology observation and morphology engineering, *ssgA_6663* was identified to be responsible for the mycelial intertwining during liquid culture. However, down-regulation of *ssgA_6663* transcription was caused by inactivation of *adpA_1075*, gene coding for an AdpA-like protein. Additionally, the overexpression of *adpA_1075* led to an 85% increase in AP-3 production. Electrophoretic mobility shift assays (EMSA) revealed that AdpA_1075 may bind the promoter regions of *asm28* gene in *asm* gene cluster as well as the promoter regions of *ssgA_6663*. These results confirm that *adpA_1075* plays a positive role in AP-3 biosynthesis and morphological differentiation.

Keywords: *Actinosynnema pretiosum*; ansamitocin P-3; *ssgA*; AdpA; morphological differentiation

Citation: Guo, S.; Leng, T.; Sun, X.; Zheng, J.; Li, R.; Chen, J.; Hu, F.; Liu, F.; Hua, Q. Global Regulator AdpA_1075 Regulates Morphological Differentiation and Ansamitocin Production in *Actinosynnema pretiosum* subsp. *auranticum*. *Bioengineering* **2022**, *9*, 719. https://doi.org/10.3390/bioengineering9110719

Academic Editors: Minaxi Sharma, Kandi Sridhar and Zeba Usmani

Received: 30 October 2022
Accepted: 17 November 2022
Published: 21 November 2022

Publisher's Note: MDPI stays neutral with regard to jurisdictional claims in published maps and institutional affiliations.

1. Introduction

Ansamitocin P-3 (AP-3) exhibits antitumor activity against various cancer cell lines [1–3]. Its derivatives are commonly used as the 'warhead' molecule in antibody-conjugated drug (ADC) for the treatment of various solid tumors [4]. AP-3 is a member of various ansamitocin congeners produced by *A. pretiosum*. Ansamitocins are of limited industrial applicability because of their low production yields. In recent decades, considerable efforts have been made to further enhance AP-3 yield to satisfy the industrial demands with medium optimiztion and genetic modifications because of its great pharmaceutical value [5–12].

From an industrial point of view, liquid culture is favorable for large scale production of antibiotics. Actinomycetes are usually subjected to submerged fermentation. Unlike other bacteria, *Actinobacteria* remarkably exhibits complex morphology during submerged cultivation. In liquid culture, their mycelium shows filamentous growth, exhibiting dispersed mycelial form or compact mycelial network [13]. It is well studied that different morphological forms lead to various degrees of nutrient and oxygen transfer during the fermentation process [14]. Three types of morphologies-freely dispersed mycelia, open mycelial networks, and compact mycelial network are generally distinguished in submerged cultures [15]. The morphological phenotypes are genetically determined and differ considerably between strains. The formation of unfavorable morphology during liquid-grown cultures may be a major bottleneck hindering the industrial production of antibiotics [16]. Submerged culture studies for different antibiotic production have made efforts to select strains with better growth characteristics. Forming dispersed mycelium helps productivity in the bioreactor for *Streptomyces hygroscopicusin* producing rapamycin [17]. A similar situation was found in previous studies of tylosin and nystatin production in both *Streptomyces fradiae* and *Streptomyces noursei* [18,19]. However, for production of nikkomycin and erythromycin pelleted growth is preferred [20,21]. Additionally, forming dense pellets also contributed to the productivity optimization of *Streptomyces lividans* TK21 for a hybrid antibiotic production as well [22].

Therefore, in order to optimize mycelial morphology in a more targeted and flexible manner, several genetic determinants have been identified that play roles in the control of morphogenesis [23]. The genetic factors include cell-matrix proteins and extracellular polymers. Morphoproteins with specific roles in liquid-culture morphogenesis for apical growth and hypha branching include the cell wall remodeling protein SsgA-like proteins (SALPs) [24], and the cellulose synthase-like protein CslA [25]. Members of the family of SALPs are required to activate cell division in both solid and liquid culture sporulation [26–28]. SsgB is the archetypal SALP and functions by colocalizing with SsgA for recruiting FtsZ during aerial hyphae early division stage [29–31]. Morphology engineering strategies for *ssgA* gene modification were employed to obtain desirable morphologies and fast growth [32,33].

Interestingly, morphological differentiation caused by *ssgA* transcriptional variations are generally controlled by AdpA [28,34,35]. AdpA is universally present in Actinomycetes, as a global regulator of morphological differentiation and secondary metabolism [36,37]. In *S. griseus*, AdpA positively controls the expression of genes involved in spore formation and aerial mycelium formation, as well as activates the transcription of various genes related to secondary metabolism [38,39]. Whereas, overexpression of the *adpAsx* gene in *S. xiamenensis* 318 had negative effects on cell division genes, such as putative *ssgA*, *ftsZ*, *ftsH*, and *whiB*. Besides, it functions as a bidirectional regulator for the biosynthesis of xiamenmycin and PTMs [40]. To date, AdpA has been proven to contain a C-terminal domain with two helix-turn-helix (HTH) DNA binding motifs [41]. As a global transcription factor, the regulatory relationship between AdpA and target genes has been investigated in *Streptomyces*. AdpA and its orthologs activate or down-regulate genes, including repression of its own transcription, by directly binding to operator regions containing a consensus sequence [38,42,43].

Although *A. pretiosum* is being developed as a sustainable industrial production platform, the genes involved in cell division and morphological development are still poorly investigated. Gene *APASM_4178* was identified as a subtilisin-like serine peptidase encoding gene responsible for mycelial fragmentation [44]. FtsZ protein from *A. pretiosum* as the analogue of β-tubulin, was demonstrated to be the AP-3 binding target. Overexpression of *APASM_5716* gene that encodes FtsZ resulted in AP-3 resistance and overproduction in *A. pretiosum* ATCC 31280 [45]. A two-component signal transduction system, PhoPR homolog was identified in the genome of the *A. pretiosum* X47 strain. PhoP is the response regulator, negatively affecting morphological development and excluding its regulation on the biosynthesis of AP-3 in X47 strain [46]. However, for filamentous microorganisms, the importance of understanding the relationship between mycelial morphological development and antibiotic biosynthesis is nontrivial. In this study, a variant of *A. pretiosum* has been observed to form dense pellets, while the control strain formed loose clumps. In addition, excessive mycelial fragmentation of the control strain was observed early in the fermentation. We investigated several putative genes that may contribute to cell division and pellet architecture. Gene *ssgA_6663* was identified as a key genetic determinant of compact mycelial network formation during solid and liquid cultures. We also characterized the roles of AdpA_1075 in controlling the morphological differentiation and AP-3 production. AdpA_1075 was determined to positively control the biosynthesis of ansamitocin by directly regulating the expression of *asm28*.

2. Materials and Methods

2.1. Bacterial Strains, Plasmids and Culture Conditions

All plasmids and strains used in this study are listed in Supplementary Table S1. *A. pretiosum* subsp. *auranticum* L40 was derived from *A. pretiosum* subsp. *auranticum* ATCC 31565 by atmospheric and room temperature plasma (ARTP) mutagenesis [12]. Strain L40 and its derivatives in this study were cultivated as described previously [47]. YMG agar plates (for solid culture) and TSBY broth (for liquid culture) were employed for strain culture. For fermentation experiments, strains were cultured in shake flasks at 28 °C for 8 days.

The stability of strain MD15 was tested following the method described by former study [10] with some modifications. In brief, strain MD15 was transferred at 24 h intervals for a total of twenty-five passages in YMG liquid medium. The original strain, fifteenth and twenty-fifth passages were selected to test the fermentation performances stability of strain MD15 in liquid fermentation.

2.2. Construction of Recombinant Strains

CRISPR-Cas9 mediated gene inactivation. Mutants with gene *ssgA_6663, adpA_1075,* or *asm28* disruption were performed by pCRISPR-Cas9apre with a unified construction process [47]. As an example, the construction of mutant with gene *ssgA_6663* deletion was described briefly. Two homologous arms (upstream 1.2 kb, downstream 1.3 kb) for *ssgA_6663* deletion were amplified and together cloned to *Stu*I-digested plasmid pCRISPR-Cas9apre by NEB DNA Assembly Master Mix (New England Biolabs, Ipswich, MA, USA) to give the pCRISPR-Cas9apreΔ*ssgA*. The ApE software (a plasmid editor, version 2.0.50. https://jorgensen.biology.utah.edu/wayned/ape/, accessed on 5 October 2016) was used to search N20 targeting sequences of sgRNAs. The sgRNA cassettes were cloned into the *Xma*JI/*Sna*BI site of pCRISPR-Cas9apreΔ*ssgA*. The amplification primers used to construct pCRISPR-Cas9apre series gene knockout plasmids are shown in Supplementary Table S2. *E. coli* ET12567 (pUZ8002) was employed to introduce the resulting plasmid into L40. According to protocol described elsewhere [47], the conjugants were induced and screened for the correct constructs by colony PCR and Sanger sequencing (Supplementary Figure S1).

Construction of plasmids for gene overexpression. pSETK derived from pSET152 was used to prepare overexpression plasmids of *ssgA_6663, adpA_1075, ftsZ_5883* or *asm28*. More specifically, the *kasOp*-rbs* fragment was introduced to *Xba*I/*Eco*RV cloning site of pSET152. Aforementioned genes were amplified from *A. pretiosum* L40 chromosome. The amplicons were cloned into *Nde*I/*Eco*RV site of pSETK, respectively. The obtained recombinant plasmids pSETKssgA, pSETKftsZ, pSETKadpA, pSETKasm28, pSETKftsZ:ssgA, and control plasmid pSETK were individually transferred into *E. coli* ET12567(pUZ8002) and then integrated into the *attB* site of strain MD02 by intergeneric conjugation. The verification of these recombinant strains was performed by PCR (Supplementary Figure S2A–C). To construct the pSETKftsZ:ssgA plasmid, the fragment containing *kasOp*-rbs-ssgA_6663* expression cassette cloned from pSETKssgA plasmid was inserted into pSETKftsZ digested by *Eco*RV to generate pSETKftsZ:ssgA. The verification of gene co-expression was double checked by PCR using two primer pairs 152yz-F/R and ftsZchk-f/ssgAchk-r (Supplementary Figure S2D–F).

2.3. RNA Isolation, cDNA Synthesis and Quantitative Real-Time PCR (qRT-PCR)

Total RNA was extracted using a bacterial RNA extraction kit (Jiangsu Cowin Biotech Co., Ltd., Taizhou, China). Isolated RNA was treated by DNase I before being reverse transcribed with cDNA Synthesis Kit (Jiangsu Cowin Biotech Co., Ltd., Taizhou, China). The cDNA templates were amplified in triplicate for each transcription analysis using MagicSYBR Mixture (Jiangsu Cowin Biotech Co., Ltd., Taizhou, China) with primers listed in Supplemental Table S2. The transcription of target genes was determined by RT-PCR using a CFX96 Real-Time System (Bio-Rad, Richmond, CA, USA). *16S rRNA* gene was used for internal normalization. Relative transcript level of genes was quantified by the $2^{-\Delta\Delta Ct}$ method [48].

2.4. Determination of AP-3 Production

AP-3 was extracted from the culture supernatant using a previously described method [12]. HPLC analysis of AP-3 was operated on Agilent series 1260 (Agilent Technologies, Inc., Santa Clara, CA, USA) equipped with a SinoChrom ODS-BP C18 column (4.6 mm × 250 mm, 5 μm, Elite, Dalian, China) coupled to UV detector at 254 nm. The column was eluted with 85% methanol and 15% water at 28 °C.

2.5. Mycelial Morphology Observation

Mycelial morphology was observed using an optical microscope (Olympus CX 31, Olympus Corporation, Tokyo, Japan). Culture broth (10 μL) was pipetted onto a standard glass slide (25 × 75 mm), dyed with crystal violet. Images were captured under an oil immersion lens (magnification, 100×).

2.6. Scanning Electron Microscope (SEM)

Mycelium was harvested by centrifugation, and washed with 0.1 M PBS. The mycelium was resuspended in 2.5% glutaraldehyde solution for 3 h. The fixed samples were then washed twice with 0.1 M PBS. Samples were subjected to gradient dehydration with ethanol solution (50%, 70%, 95% and 100%). Finally, the dehydrated samples analyses were carried out on a S3400-N scanning electron microscopy (Hitachi, Tokyo, Japan).

2.7. Heterologous Overexpression of AdpA-1075

Gene *adpA_1075* was amplified with primers 28a1075-F/R. The *adpA_1075* cassette was cloned in *Hind*III/*Nde*I-digested pET28a (+), generating plasmid pET-28a-adpA_1075. The plasmid was transformed into *E. coli* BL21(DE3) for protein overexpression. The generated strain BL21(DE3)/pET-28a-adpA_1075 was cultivated at 37 °C for 2–3 h in 100 mL LB medium containing 50 μg/mL kanamycin until OD600 reached about 0.6–0.8. Isopropyl-β-D-thiogalactoside (IPTG, 0.1 mM) was added after 30 min of cooling at 4 °C and further incubated overnight at 16 °C for AdpA_1075 expression. The cells were harvested and resuspended in 50 mM phosphate buffer solution (pH 7.5). His-tagged AdpA_1075 protein was released from cells by homogenization. Ni SepharoseTM 6 Fast Flow (GE Healthcare Life Sciences, Marlborough, MA, USA) was applied to proteins purification with elute buffer (50 mM phosphate buffer solution, 250 mM imidazole, 500 mM NaCl). The purified His-tagged AdpA-1075 was analyzed by 12.5% sodium dodecyl sulfate-polyacrylamide gel electrophoresis (SDS-PAGE).

2.8. Electrophoretic Mobility Shift Assays (EMSA)

EMSA was performed using a Chemiluminescent EMSA Kit (Beyotime Biotechnology, Shanghai, China) according to the manufacturer's instructions. The two complementary oligonucleotides were annealed, labeled with biotin, and incubated with recombinant proteins in the absence or presence of excess amounts of unlabeled wildtype oligonucleotides. The protein-DNA complexes were separated on 5% polyacrylamide gels and the signals were captured with a Chemiluminescence Imaging System (BG-gdsAUTO 720, Baygene Biotechnol Co., Ltd., Shanghai, China). As a control, for each target gene, excessive unlabeled specific DNA fragments were added to the reactions, resulting in the appearance of free un-shifted probe and demonstrating that the binding was specific.

3. Results

3.1. Identification of ssgA in A. pretiosum subsp. auranticum

The previously constructed mutant MD15 with tandem deletion of two gene clusters (cluster T1PKS-15 and T1PKS/NRPS-5) [47] showed excellent fermentation stability (Figure 1A). To better understand the improvement of fermentation performance, the mycelia from culture broth were collected at 16, 24 and 72 h, and aerial hyphae were examined on YMG medium. Scanning electron microscopy (SEM) was used to demonstrate in detail the morphological differences between the mutant strain MD15 and the control strain L40. During the first 24 h of fermentation in both control strain L40 and mutant strain MD15, pellets were formed by aggregative hyphae. Mycelial fragmentation occurred in the control strain L40 at around 72 h, while in contrast, the mycelia of MD15 remained dense pellets with well-developed mycelial boundaries (Figure 1B). Compared with the loosely interwoven mycelial morphological characteristics of L40, the aerial mycelia of MD15 formed a tight mycelial network (Figure 1C). We speculate that gene expression may be altered in strain MD15. SsgA, FtsZ, SsgB, and CslA encoding genes were found, namely *ssgA_6663*, *ftsZ_5883*, *ssgB_2072*, and *cslA_0512*, respectively, in *A. pretiosum* subsp. *auranticum* ATCC 31565 genome. To better understand the role of these genes in mutant MD15 morphological development, the transcription levels of these target genes were measured by qRT-PCR on the third day of fermentation. In mutant MD15, transcription levels of *ssgA_6663* and *ftsZ_5883* were 124% and 160% higher than those in strain L40, respectively (Figure 1D). The transcription of gene *ssgB_2072* was not detected, suggesting that the strain was defective in the initiation of sporulation, as the protein complexes SsgA and SsgB cannot colocalize in aerial hyphae. It has been supported by phenotypic observations. In strain L40 and its derivative, smooth and uncoiled aerial hyphae formed, but no spores (Figure 1C). Negligible changes were observed in *cslA_0512*, which encodes a cellulose synthase-like protein homologue essential for pellet formation [49]. This fact excludes the involvement of *cslA_0512* in the formation of compact pellets in *A. pretiosum*.

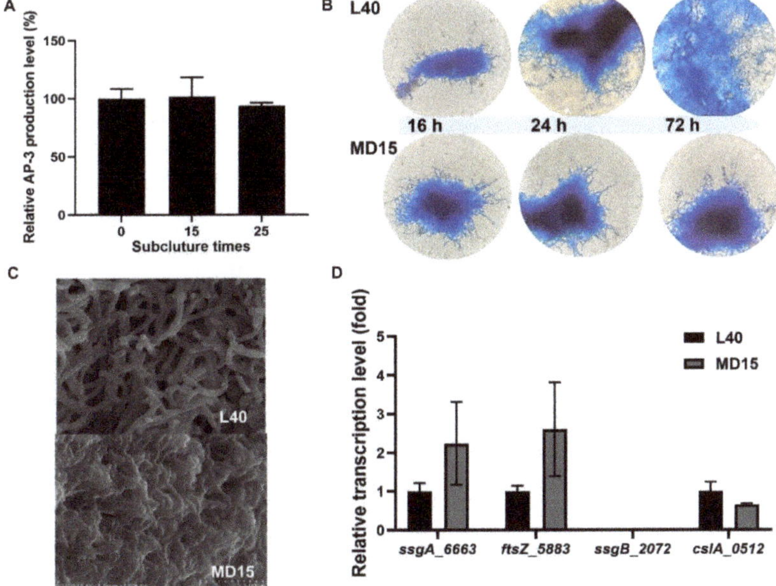

Figure 1. Fermentation stability and morphological variation of mutant MD15. (**A**) Fermentation stability of strain MD15. The strains from original, fifteenth and twenty-fifth subculture were collected to perform liquid fermentation. Fermentation experiments were carried out in three biological replicates. The average AP-3 production of the original MD15 strain was set to 100% as the standard, and the values are means ± SD (standard deviations) of three independent experiments. (**B**) Mycelium morphology changes of strain L40 and mutant MD15 during fermentation. Magnification, 100×. (**C**) Scanning electron micrographs of strain L40 and mutant MD15 on day 3 of YMG solid culture. Scale bar: 5 μm, Magnification, 7500×. (**D**) Transcription levels of the cell division genes of mutant MD15 on the third day of fermentation. The average expression value of genes in the control strain L40 was set to 1 as the standard, and the values are means ± SD of triplicate analyses. Gene expression level was determined by $2^{-\Delta\Delta Ct}$ method.

Mycelial fragmentation has previously been reported to impede AP-3 production [12,44]. Pelleted growth is preferred for *A. pretiosum* producing AP-3 in submerged cultivation [12]. The sporulation-specific gene *ssgA* directly activates cell division in *Streptomyces* [27,50]. Moreover, the expression of *ssgA_6663* also varies with strain morphology in non-spore producing *A. pretiosum* (Figure 1) [12]. We therefore assume that both *ssgA_6663* and *ftsZ_5883* play important roles in controlling the morphogenesis in *A. pretiosum*, especially in the formation of intertwined mycelial network. As reported earlier, filamentous growth and sporulation of actinobacteria require the bacterial tubulin homolog FtsZ [39,51,52]. Overexpression of *ftsZ* gene might also be used to alleviate AP-3 toxicity and improve resistance of *A. pretiosum* to AP-3. Since FtsZ was determined to be the intracellular binding target of AP-3 [45]. A high transcriptional level of *ftsZ_5883* was observed in mutant with a denser mycelial interweaving pattern (Figure 1D). Whether *ssgA_6663* and *ftsZ_5883* are involved in specific cellular process and contribute to stable fermentation remains to be investigated.

3.2. Deletion of ssgA_6663 Affected the Morphological Differentiation of A. pretiosum

We then investigated the effects of *ssgA_6663* deletion on mycelium development and AP-3 production. The length of aerial mycelia of *ssgA_6663* disruption mutant was much shorter than that of strain L40 without intertwining (Figure 2A). In submerged cultivation, mycelium of Δ*ssgA_6663* mutant developed into dispersed mycelial form (Figure 2B). The short aerial mycelium probably prevented the strain from making biomass from nutrients (Figure 2C). Further HPLC analysis revealed that AP-3 was almost undetected in the fermentation culture of the *ssgA_6663* disruption mutant, and the biomass in the fermentation was also reduced (Figure 2D). However, the introduction of multiple copies of gene *ssgA* may cause an excessive mycelial fragmentation in spore-producing *Streptomyces* [50]. To verify whether *ssgA_6663* and *ftsZ_5883* play a positive role in mycelial aggregation and entanglement, these two cell division genes were overexpressed in tandem under the drive of *kasOp**. The co-overexpression of *ftsZ_5883* and *ssgA_6663* caused the long hyphae to intertwine closely with each other and form clumps, which was more visible than that of the strain overexpressing *ftsZ_5883* alone (Figure 2B). This phenomenon may be consistent with previous reports that secondary metabolism and morphological differentiation can be regulated by global regulator, rather than morphology directly affecting secondary products biosynthesis [35]. The timing of SsgA expression in *Streptomyces* sporulation-specific cell division and morphogenesis can be regulated by global regulator AdpA [50]. As previously reported in *A. pretiosum* ATCC 31280, AdpA-like protein APASM_1021 was found [44]. Gene *adpA_1075* was identified as an AdpA orthologue from three putative AraC family protein encoding genes in *A. pretiosum* subsp. *auranticum* for the high amino acid sequence identity (97.17%) between AdpA_1075 and APASM_1021 (Supplementary Figure S3). We therefore speculate that AdpA_1075 may also play a regulatory role in the L40 strain. Furthermore, the lawns of *ssgA_6663*-null mutant on YMG plate exhibited a bald phenotype distinct from L40, whereas deletion of *adpA_1075* had little restriction on the arising of aerial hyphae (Figure 2C). Interestingly, deletion of *adpA_1075* also resulted in a significant decrease in AP-3 production without inhibition of biomass accumulation, suggesting that *adpA_1075* is essential for AP-3 biosynthesis as a pleiotropic transcriptional regulator (Figure 2D).

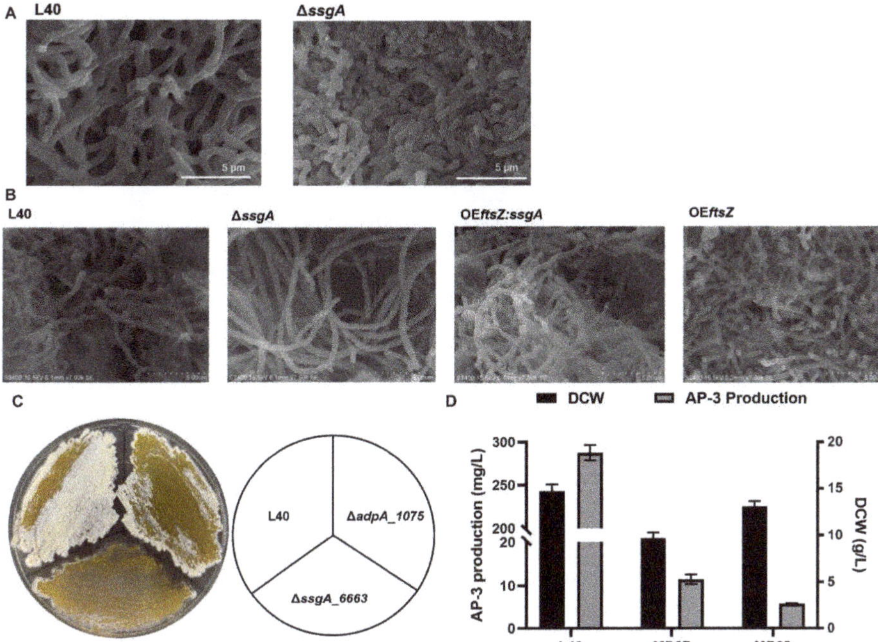

Figure 2. Differences in morphology (**A–C**) and AP-3 production (**D**) between mutant strains and L40 strain. (**A**) Scanning electron micrographs of *ssgA_6663* deletion mutant and strain L40 on day 3 of YMG plate culture. Scale bar: 5 μm. Magnification, 7500×. (**B**) Scanning electron microscopic observation of mycelium of strain L40, *ssgA_6663* deletion mutant, strain overexpressing *ssgA_6663* and *ftsZ_5883* in tandem, and mutant with *ftsZ_5883* overexpression in liquid seed culture for 16 h. Scale bar: 5 μm. Magnification, 7500×. (**C**) Phenotypes of control strain L40, *ssgA_6663* deletion mutant, and *adpA_1075* deletion mutant grown on YMG plates at 28 °C (day 3). (**D**) AP-3 production and cell growth of strains L40, MD07, and MD08. MD07, mutant with *ssgA_6663* deletion. MD08, mutant with *adpA_1075* deletion. All fermentation experiments were performed in three biological replicates.

3.3. Overexpression of adpA_1075 Increased the Production of AP-3

As reported previously, *ssgA* is essential for septum formation in aerial hyphae, a late step in morphological differentiation. Therefore, *ssgA* mutation may not affect the production of secondary metabolites [26–28]. In this study, *ssgA_6663* or *adpA_1075* was individually overexpressed under the strong promoter *kasOp** on the shuttle vector pSETK, to investigate the effects of the enhanced expression of these two proteins on ansamitocin biosynthesis. The fermentation experiments were performed for the recombinant strains overexpressing *ssgA_6663*, *ftsZ_5883*, or *adpA_1075*, and the control strain MD02::pSETK (MD02 integrated with the vector pSETK), respectively. The results showed that the overexpression of *ssgA_6663* did not improve AP-3 production, which was consistent with what was observed in *S. griseus* [26,28]. As expected, enhanced expression of *ftsZ_5883* improved strain resistance against AP-3 [45], alleviated cell toxicity, and increased AP-3 production by 45% (Figure 3A). Moreover, overexpression of *adpA_1075* increased AP-3 production by 85% without affecting dry cell weight (DCW) at the end of fermentation (Figure 3A,B). Therefore, we may conclude that overexpression of *ssgA_6663* only causes morphological changes but does not directly promote AP-3 production. Overexpression of *adpA_1075* may not only regulate ansamitocins biosynthesis, but also regulate strain morphology by controlling the expression of *ssgA_6663*.

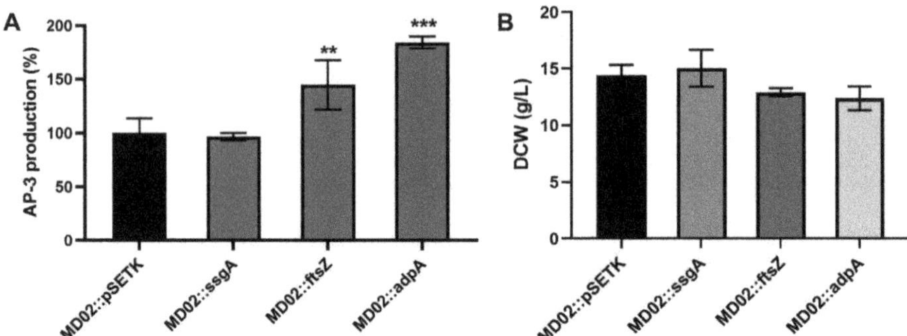

Figure 3. Effects of overexpression of cell division genes on AP-3 production (**A**) and cell growth (**B**). Three biological replicates were performed in fermentation experiments. The values are means ± SD of triplicate analyses. Differences were analyzed by one-way ANOVA, **, $p < 0.01$, ***, $p < 0.001$.

3.4. AdpA_1075 Is Involved in the Regulation of ssgA_6663 Transcription in A. pretiosum

The AraC family transcription factor known as AdpA is a global regulator of morphological differentiation and secondary metabolism [42,43,53,54]. Cell division genes are included in members of AdpA regulon, such as *ssgA* which requires AdpA to turn on transcription [28,50,55,56]. Since the fermentation morphology changes caused by *ssgA_6663* inactivation did not directly affect the production of AP-3, we hypothesized that AdpA_1075 might be involved in both *ssgA_6663* expression and ansamitocins production. Diametrically opposed transcriptional patterns of *ssgA_6663* in *adpA_1075* inactivation mutant seemed to further validate the hypothesis. In parent strain L40, the transcription of *ssgA_6663* increased significantly on day 3 of fermentation, but decreased sharply when the strain switched to the stationary growth phase (Figure 4). In contrast, in mutant MD08, *ssgA_6663* transcription was down-regulated due to the absence of AdpA_1075 and remained low from the third day of fermentation (Figure 4).

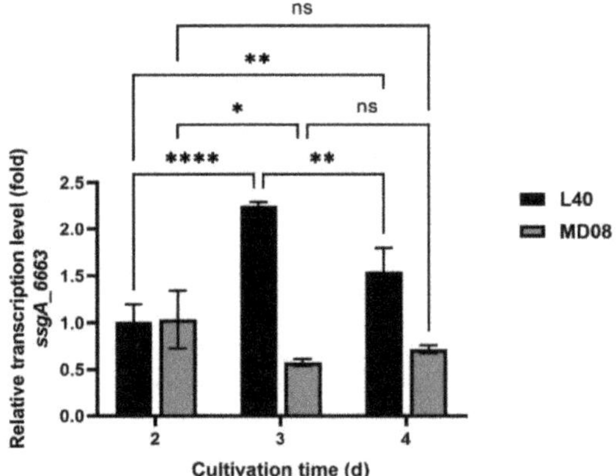

Figure 4. Relative transcription of *ssgA_6663* gene in L40 and MD08 at different culture times. RNA sample data were obtained from days 2, 3, and 4 of fermentation. Gene expression level was determined by $2^{-\Delta\Delta Ct}$ method, and the values are means ± SD of triplicate analyses. Transcription data on day 2 was were set as the baseline for comparison. *16s rRNA* was used as the internal control. Three biological replicates were performed in fermentation experiments. Significant differences were analyzed by one-way ANOVA, *, $p < 0.05$, **, $p < 0.01$, ****, $p < 0.0001$, ns, no significant. MD08, mutant with *adpA_1075* deletion.

3.5. AdpA_1075 Binds to Promoters of ssgA_6663 and asm28

The above analyses demonstrated that AdpA_1075 regulated morphological development by controlling the expression of *ssgA_6663*. To determine whether AdpA_1075 directly regulated the gene transcription of *ssgA_6663*, EMSA was performed. As previously reported, a consensus AdpA-binding sequence such as 5′-TGGCSNGWWY-3′ (S: G or C; W: A or T; Y: T or C; N: any nucleotide) was identified in *S. griseus* (Figure 5A) [57]. Consequently, the upstream region of *ssgA_6663* was analyzed by gene sequence alignment and manual correction, according to the reported AdpA-binding motifs in *Streptomyces* and *A. pretiosum* ATCC 31280 [44,57]. A conserved AdpA-binding motif, 5′-TGGCCGGAAC-3′ (in reversed orientation) was identified (Figure 5B). AdpA_1075-His$_6$ protein was then purified, and biotin labeled probes were prepared as mentioned above. EMSA results showed that the complex AdpA_1075-PssgA was formed in a protein concentration-dependent manner, confirming that AdpA_1075 bound specifically to the promoter region of *ssgA_6663* (Figure 5E). Surprisingly, we not only found an AdpA-binding sequence in the promoter region of *ssgA_6663*, but also observed a potential AdpA-binding site 5′-TGGCCCGAAC-3′ (in reversed orientation) in the upstream region of *asm28* (Figure 5C). Compared to the transcriptional changes of *ssgA_6663* and *ftsZ_5883* in *adpA_1075* overexpression mutant, transcriptional level of *asm28* in the *adpA_1075* overexpression strain was 3.5 times higher than that in the parent strain (Figure 5D). Unexpectedly, the overexpression of *adpA_1075* did not affect the expression of *ftsZ_5883*. The slight up-regulation of *ssgA_6663* indicated that another regulator may be also involved in the regulation of *ssgA_6663* transcription. Results of EMSAs showed that Pasm28 probe was shifted when incubated with 1000 nM AdpA_1075-His$_6$ (Figure 5E). The above results suggested that AdpA_1075 specifically binds to the promoters of *ssgA_6663* and *asm28*, implying that AdpA_1075 controls the transcription of these genes directly. The upstream region of *asm28* is the only potential AdpA binding site on the *asm* gene cluster, as it contains a 10-bp consensus sequence mentioned above. However, the function of *asm28* has not been reported. Therefore, we further investigated the effect of *asm28* gene on AP-3 production. Mutant with *asm28* deletion was constructed by CRISPR-Cas9-mediated vector pCRISPR-Cas9apreΔ*asm28* (Supplementary Figure S4A,B). AP-3 production of the obtained mutant MD19 was only 69% of that of parent strain MD01. However, overexpression of *asm28* promoted ansamitocin biosynthesis and resulted in an approximately 25% increase in AP-3 production, without any effect on DCW of the strain at the end of fermentation (Figure S4D,E). These results demonstrated that AdpA_1075 activated *asm28* and *ssgA_6663* on the third to fourth day of fermentation, consistent with the enhanced AP-3 production and mycelial pellet formation in strains.

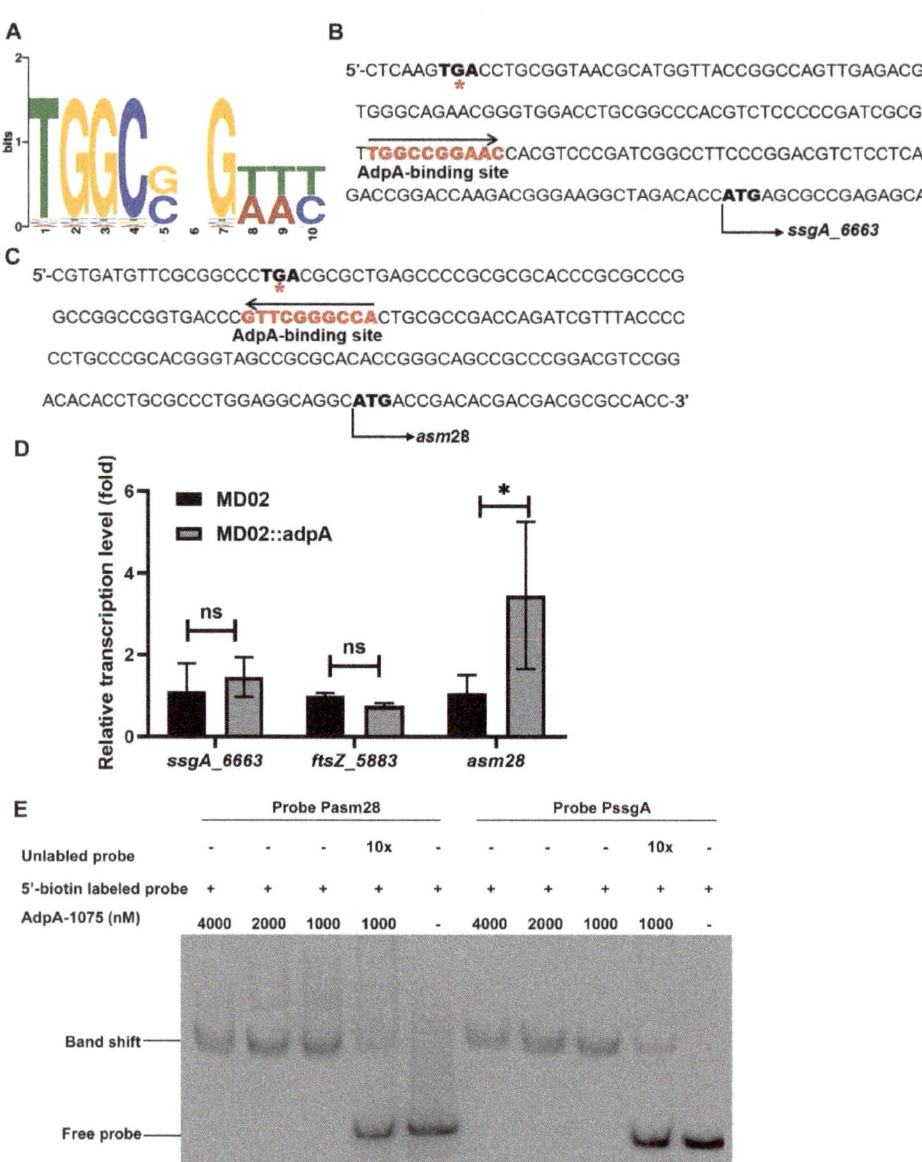

Figure 5. AdpA_1075 binds PssgA and Pasm28. (**A**) The Sequence of AdpA-binding motif. (**B**,**C**) Conserved (shown in red) AdpA-binding motifs in the upstream region of *ssgA_6663* and *asm28*. The start codon of *ssgA_6663* or *asm28* are shown in bold black. *, stop codons. (**D**) Relative transcription of gene *ssgA_6663*, *ftsZ_5883*, *asm28* in mutant with *adpA_1075* overexpression on day 3 of fermentation. Gene expression level was determined by $2^{-\Delta\Delta Ct}$ method. The average expression value of genes in the strain MD02 was set to 1 as the standard, and the values are means ± SD of triplicate analyses. Differences were analyzed by one-way ANOVA, ns, no significant, *, $p < 0.05$. (**E**) EMSAs of purified AdpA_1075-His$_6$ binding to the probes PssgA and Pasm28 labeled with biotin in the upstream region of *ssgA_6663* and *asm28*. 20 ng biotin-labeled probes were incubated for 10 min at room temperature. The experiment was repeated for three times.

4. Discussion

The mycelial fragmentation of *A. pretiosum* under submerged conditions has long attracted researchers' attention [12,44,58]. However, in our previous work, pelleted growth was found to be more conducive to the production of ansamitocin [12]. Other studies have revealed that cell-wall remodeling protein SsgA may control the development of fragmentation and promote the growth rate of spore-forming *Streptomyces* strains [32,33]. The absence of SsgB in *A. pretiosum* makes it a non-spore-forming bacterium (Figure 1C,D) [59,60]. Additionally, both mycelial fragmentation and significant biomass enhancement was observed in *ssgA_6663*-overexpressing *A. pretiosum* (Figures 2B and 3B). All these results indicate that *ssgA_6663* is not responsible for mycelial fragmentation and AP-3 yield reduction in *A. pretiosum*.

Therefore, we hypothesized that, similar to that reported in *S. limosus* [32], *A. pretiosum* strain may produce large mycelial mat structures in the early growth phase, while the enhanced expression of *ssgA_6663* leads to pellet formation during stationary phase. As expected, a significant increase in the transcriptional profile of *ssgA_6663* was observed in the parent strain L40, followed by a noticeable decrease from day 2 to day 4 of fermentation (Figure 4). However, this transcriptional pattern was disrupted when gene *adpA_1075* was inactivated (Figure 4).

AdpA as a pleiotropic transcriptional regulator can regulate both morphological differentiation and secondary metabolism [61]. We further investigated the *adpA_1075* gene, encoding an AdpA-like protein. Specific regulation of AdpA-like protein on subtilisin-family serine peptidase encoding gene resulted in a delayed fragmentation, which has been demonstrated in *A. pretiosum* ATCC 31280 [44]. Moreover, the *ssgA_6663* gene has also been identified as being primarily responsible for the formation of mycelial intertwining. Based on these findings, we hypothesized that AdpA_1075, as a global regulator, may comprehensively control the development of multicellular structures and AP-3 biosynthesis during strain fermentation.

Despite the presence of many regulatory genes in the *asm* gene cluster, few studies have been reported on the regulation of ansamitocin biosynthesis. The functions of these regulatory genes were characterized by gene inactivation, complementation, transcriptional analysis, and feeding experiments [62–65]. For example, negative regulatory gene *asm2* and *asm34*, and positive regulatory gene *asm8* and *asm18* are the putative regulatory genes on the *asm* gene cluster. Whereas, direct evidence for these regulators directly controlled *asm* structural genes was absent. Most of them probably control the expression of genes involved in resistance to xenobiotics, regulation of efflux pumps, and response to stressors [41].

AdpA homologs were well-studied, and have been proven to contain a C-terminal domain with two helix-turn-helix (HTH) DNA binding motifs [37]. In this study, the *adpA_1075* gene was identified as an ortholog of pleiotropic regulator AdpA, and its DNA-binding site is similar to the conserved sequence of the AdpA-binding motifs in many *Streptomyces* species [57]. Earlier work revealed that the expression of APASM_4178, which is responsible for mycelial fragmentation, is specifically regulated by an AdpA-like protein in *A. pretiosum* ATCC 31280 [44]. In this work, the EMSA results also demonstrated the interaction between AdpA_1075 and promoter region of the subtilisin-family serine peptidase encoding gene (data not shown). Our data showed that AdpA_1075 of the *A. pretiosum* L40 strain directly binds to the promoters of the *ssgA_6663* and *asm28*. Therefore, these findings deepen insights into the regulatory roles of the AdpA-like protein in *A. pretiosum*, revealing a high functional similarity to the AdpA homolog in *Streptomyces* reported previously [35].

In this study, an intergenic region containing AdpA-binding motif was identified in the upstream region of *asm28* (Figure 5). Our findings in vitro and in vivo experiments suggested that *asm28* may be the target of AdpA_1075 for regulating ansamitocin biosynthesis (Figures 5 and S4). Moreover, the global regulator *bldA* controls antibiotic production as well, by regulating sporulation and antibiotic production in *Streptomyces* [66]. Generally, the TTA codons are very rare in GC-rich genomes of *Actinobacteria*. TTA-containing genes are usually the cluster-situated regulators of secondary metabolite biosynthesis. In this study, we found that the *asm28* open reading frame (ORF) also contains a TTA codon, indicating that *asm28* might be an important regulatory target during ansamitocin biosynthesis. However, the function of *asm28* gene and its encoded protein remain uncharacterized at present. Follow-up studies are needed to confirm this hypothesis and to elucidate the complex regulatory network in which *asm28* is involved.

5. Conclusions

In this study, we have elucidated the function of *ssgA_6663* in mycelial development. *ssgA_6663* can dominate the mycelial intertwining and pellet formation. The silencing of the *ssgB_2072* gene resulted in the absence of SsgB, which may explain the fact that the strain developed into mycelium without sporulation septa.

Additionally, we characterized the regulatory role of AdpA_1075 in *A. pretiosum*. AdpA_1075 acts as a global regulator, affecting morphological differentiation and promoting the biosynthesis of ansamitocin. Our findings provide additional useful evidence for the regulatory mechanism of ansamitocin biosynthesis.

Supplementary Materials: The following supporting information can be downloaded at: https://www.mdpi.com/article/10.3390/bioengineering9110719/s1, Table S1: Strains and plasmids used in this study; Table S2: Primers used in this study; Figure S1: Verification of *ssgA_6663* and *adpA_1075* gene deletion mutant strains; Figure S2: Identification of recombinant strains; Figure S3: Sequence alignment of AdpA-like protein APASM_1021 with AdpA_1075; Figure S4: Effects of *asm28* gene deletion or overexpression.

Author Contributions: Conceptualization, Q.H. and F.L.; methodology, S.G. and T.L.; validation, S.G. and X.S.; investigation, S.G. and J.Z.; data curation, S.G., R.L., J.C. and F.H.; writing—original draft preparation, S.G.; writing—review and editing, Q.H. and F.L.; supervision, Q.H. and F.L.; funding acquisition, Q.H. and F.L. All authors have read and agreed to the published version of the manuscript.

Funding: This work was supported by the National Natural Science Fund for Young Scholars (32001035) and Shanghai Super postdoctoral program 2020.

Institutional Review Board Statement: Not applicable.

Informed Consent Statement: Not applicable.

Data Availability Statement: All data generated or analyzed during this study are included in this article.

Conflicts of Interest: The authors declare no competing interest.

References

1. Prelog, V.; Oppolzer, W. Ansamycins, a novel class of microbial metabolites. *Helv. Chim. Acta* **1973**, *56*, 2279–2287. [CrossRef] [PubMed]
2. Martin, K.; Müller, P.; Schreiner, J.; Prince, S.S.; Lardinois, D.; Heinzelmann-Schwarz, V.A.; Thommen, D.S.; Zippelius, A. The microtubule-depolymerizing agent ansamitocin P3 programs dendritic cells toward enhanced anti-tumor immunity. *Cancer Immunol. Immunother* **2014**, *63*, 925–938. [CrossRef] [PubMed]
3. Kashyap, A.S.; Fernandez-Rodriguez, L.; Zhao, Y.; Monaco, G.; Trefny, M.P.; Yoshida, N.; Martin, K.; Sharma, A.; Olieric, N.; Shah, P.; et al. GEF-H1 signaling upon microtubule destabilization is required for dendritic cell activation and specific anti-tumor responses. *Cell Rep.* **2019**, *28*, 3367–3380.e8. [CrossRef] [PubMed]
4. Barok, M.; Joensuu, H.; Isola, J. Trastuzumab emtansine: Mechanisms of action and drug resistance. *Breast Cancer Res.* **2014**, *16*, 209. [CrossRef] [PubMed]
5. Fan, Y.; Gao, Y.; Zhou, J.; Wei, L.; Chen, J.; Hua, Q. Process optimization with alternative carbon sources and modulation of secondary metabolism for enhanced ansamitocin P-3 production in *Actinosynnema pretiosum*. *J. Biotechnol.* **2014**, *192*, 1–10. [CrossRef]
6. Li, T.; Fan, Y.; Nambou, K.; Hu, F.; Imanaka, T.; Wei, L.; Hua, Q. Improvement of ansamitocin P-3 production by *Actinosynnema mirum* with fructose as the sole carbon source. *Appl. Biochem. Biotechnol.* **2015**, *175*, 2845–2856. [CrossRef]
7. Fan, Y.; Hu, F.; Wei, L.; Bai, L.; Hua, Q. Effects of modulation of pentose-phosphate pathway on biosynthesis of ansamitocins in *Actinosynnema pretiosum*. *J. Biotechnol.* **2016**, *230*, 3–10. [CrossRef]

8. Zhao, M.; Fan, Y.; Wei, L.; Hu, F.; Hua, Q. Effects of the methylmalonyl-CoA metabolic pathway on ansamitocin production in *Actinosynnema pretiosum*. *Appl. Biochem. Biotechnol.* **2017**, *181*, 1167–1178. [CrossRef]

9. Ning, X.; Wang, X.; Wu, Y.; Kang, Q.; Bai, L. Identification and engineering of post-PKS modification bottlenecks for ansamitocin P-3 titer improvement in *Actinosynnema pretiosum* subsp. *pretiosum* ATCC 31280. *Biotechnol. J.* **2017**, *12*, 1700484. [CrossRef]

10. Du, Z.Q.; Zhang, Y.; Qian, Z.G.; Xiao, H.; Zhong, J.J. Combination of traditional mutation and metabolic engineering to enhance ansamitocin P-3 production in *Actinosynnema pretiosum*. *Biotechnol. Bioeng.* **2017**, *114*, 2794–2806. [CrossRef] [PubMed]

11. Du, Z.Q.; Zhong, J.J. Rational approach to improve ansamitocin P-3 production by integrating pathway engineering and substrate feeding in *Actinosynnema pretiosum*. *Biotechnol. Bioeng.* **2018**, *115*, 2456–2466. [CrossRef] [PubMed]

12. Li, J.; Guo, S.; Hua, Q.; Hu, F. Improved AP-3 production through combined ARTP mutagenesis, fermentation optimization, and subsequent genome shuffling. *Biotechnol. Lett.* **2021**, *43*, 1143–1154. [CrossRef]

13. Kumar, P.; Dubey, K.K. Mycelium transformation of *Streptomyces toxytricini* into pellet: Role of culture conditions and kinetics. *Bioresour. Technol.* **2017**, *228*, 339–347. [CrossRef]

14. Celler, K.; Picioreanu, C.; van Loosdrecht, M.C.M.; van Wezel, G.P. Structured morphological modeling as a framework for rational strain design of *Streptomyces* species. *Antonie Van Leeuwenhoek* **2012**, *102*, 409–423. [CrossRef]

15. Paul, G.C.; Thomas, C.R. Characterisation of mycelial morphology using image analysis. *Adv. Biochem. Eng. Biotechnol.* **1998**, *60*, 1–59. [CrossRef] [PubMed]

16. Wang, H.; Zhao, G.; Ding, X. Morphology engineering of *Streptomyces coelicolor* M145 by sub-inhibitory concentrations of antibiotics. *Sci. Rep.* **2017**, *7*, 13226. [CrossRef] [PubMed]

17. Fang, A.; Pierson, D.L.; Mishra, S.K.; Demain, A.L. Growth of *Streptomyces hygroscopicus* in rotating-wall bioreactor under simulated microgravity inhibits rapamycin production. *Appl. Microbiol. Biotechnol.* **2000**, *54*, 33–36. [CrossRef]

18. Park, Y.; Tamura, S.; Koike, Y.; Toriyama, M.; Okabe, M. Mycelial pellet intrastructure visualization and viability prediction in a culture of *Streptomyces fradiae* using confocal scanning laser microscopy. *J. Ferment. Bioeng.* **1997**, *84*, 483–486. [CrossRef]

19. Jonsbu, E.; McIntyre, M.; Nielsen, J. The influence of carbon sources and morphology on nystatin production by *Streptomyces noursei*. *J. Biotechnol.* **2002**, *95*, 133–144. [CrossRef]

20. Vecht-Lifshitz, S.E.; Sasson, Y.; Braun, S. Nikkomycin production in pellets of *Streptomyces tendae*. *J. Appl. Bacteriol.* **1992**, *72*, 195–200. [CrossRef]

21. Wardell, J.N.; Stocks, S.M.; Thomas, C.R.; Bushell, M.E. Decreasing the hyphal branching rate of *Saccharopolyspora erythraea* NRRL 2338 leads to increased resistance to breakage and increased antibiotic production. *Biotechnol. Bioeng.* **2002**, *78*, 141–146. [CrossRef] [PubMed]

22. Sarrà, M.; Casas, C.; Poch, M.; Gòdia, F. A simple structured model for continuous production of a hybrid antibiotic by *Streptomyces lividans* pellets in a fluidized-bed bioreactor. *Appl. Biochem. Biotechnol.* **1999**, *80*, 39–50. [CrossRef]

23. van Dissel, D.; Claessen, D.; van Wezel, G.P. Chapter One-Morphogenesis of *Streptomyces* in submerged cultures. *Adv. Appl. Microbiol.* **2014**, *89*, 1–45. [PubMed]

24. Noens, E.E.; Mersinias, V.; Willemse, J.; Traag, B.A.; Laing, E.; Chater, K.F.; Smith, C.P.; Koerten, H.K.; Van Wezel, G.P. Loss of the controlled localization of growth stage-specific cell-wall synthesis pleiotropically affects developmental gene expression in an *ssgA* mutant of *Streptomyces coelicolor*. *Mol. Microbiol.* **2007**, *64*, 1244–1259. [CrossRef]

25. Nothaft, H.; Dresel, D.; Willimek, A.; Mahr, K.; Niederweis, M.; Titgemeyer, F. The phosphotransferase system of *Streptomyces coelicolor* is biased for N-acetylglucosamine metabolism. *J. Bacteriol.* **2003**, *185*, 7019–7023. [CrossRef]

26. Jiang, H.; Kendrick, K.E. Characterization of *ssfR* and *ssgA*, two genes involved in sporulation of *Streptomyces griseus*. *J. Bacteriol.* **2000**, *182*, 5521–5529. [CrossRef] [PubMed]

27. van Wezel, G.P.; van der Meulen, J.; Kawamoto, S.; Luiten, R.G.M.; Koerten, H.K.; Kraal, B. *SsgA* Is Essential for Sporulation of *Streptomyces Coelicolor* A3(2) and Affects Hyphal Development by Stimulating Septum Formation. *J. Bacteriol.* **2000**, *182*, 5653–5662. [CrossRef] [PubMed]

28. Yamazaki, H.; Ohnishi, Y.; Horinouchi, S. Transcriptional Switch on of *SsgA* by A-Factor, Which Is Essential for Spore Septum Formation in *Streptomyces Griseus*. *J. Bacteriol.* **2003**, *185*, 1273–1283. [CrossRef] [PubMed]

29. Bi, E.; Lutkenhaus, J. FtsZ Ring Structure Associated with Division in *Escherichia Coli*. *Nature* **1991**, *354*, 161–164. [CrossRef]

30. Xu, W.; Huang, J.; Lin, R.; Shi, J.; Cohen, S.N. Regulation of Morphological Differentiation in *S. Coelicolor* by RNase III (AbsB) Cleavage of MRNA Encoding the AdpA Transcription Factor: AbsB Regulates AdpA in *S. Coelicolor*. *Mol. Microbiol.* **2010**, *75*, 781–791. [CrossRef]

31. Xiao, X.; Willemse, J.; Voskamp, P.; Li, X.; Prota, A.E.; Lamers, M.; Pannu, N.; Abrahams, J.P.; van Wezel, G.P. Ectopic Positioning of the Cell Division Plane Is Associated with Single Amino Acid Substitutions in the FtsZ-Recruiting SsgB in *Streptomyces*. *Open. Biol.* **2021**, *11*, 200409. [CrossRef] [PubMed]

32. Van Wezel, G.P.; Krabben, P.; Traag, B.A.; Keijser, B.J.; Kerste, R.; Vijgenboom, E.; Heijnen, J.J.; Kraal, B. Unlocking *Streptomyces* spp. for use as sustainable industrial production platforms by morphological engineering. *Appl. Environ. Microbiol.* **2006**, *72*, 5283–5288. [CrossRef] [PubMed]

33. Nguyen, H.T.; Pham, V.T.T.; Nguyen, C.T.; Pokhrel, A.R.; Kim, T.-S.; Kim, D.; Na, K.; Yamaguchi, T.; Sohng, J.K. Exploration of cryptic organic photosensitive compound as zincphyrin IV in *Streptomyces venezuelae* ATCC 15439. *Appl. Microbiol. Biotechnol.* **2020**, *104*, 713–724. [CrossRef]

34. Traag, B.A.; Kelemen, G.H.; Van Wezel, G.P. Transcription of the sporulation gene *ssgA* is activated by the IclR-type regulator SsgR in a *whi*-independent manner in *Streptomyces coelicolor* A3(2). *Mol. Microbiol.* **2004**, *53*, 985–1000. [CrossRef]
35. Horinouchi, S.; Beppu, T. Hormonal control by A-factor of morphological development and secondary metabolism in *Streptomyces*. *Proc. Jpn. Acad. Ser. B* **2007**, *83*, 277–295. [CrossRef]
36. Ohnishi, Y.; Kameyama, S.; Onaka, H.; Horinouchi, S. The A-factor regulatory cascade leading to streptomycin biosynthesis in *Streptomyces griseus*: Identification of a target gene of the A-factor receptor. *Mol. Microbiol.* **1999**, *34*, 102–111. [CrossRef] [PubMed]
37. Ohnishi, Y.; Yamazaki, H.; Kato, J.; Tomono, A.; Horinouchi, S. AdpA, a central transcriptional regulator in the A-factor regulatory cascade that leads to morphological development and secondary metabolism in *Streptomyces griseus*. *Biosci. Biotechnol. Biochem.* **2005**, *69*, 431–439. [CrossRef]
38. Akanuma, G.; Hara, H.; Ohnishi, Y.; Horinouchi, S. Dynamic changes in the extracellular proteome caused by absence of a pleiotropic regulator AdpA in *Streptomyces griseus*. *Mol. Microbiol.* **2009**, *73*, 898–912. [CrossRef] [PubMed]
39. Bush, M.J.; Tschowri, N.; Schlimpert, S.; Flärdh, K.; Buttner, M.J. c-di-GMP signalling and the regulation of developmental transitions in streptomycetes. *Nat. Rev. Microbiol.* **2015**, *13*, 749–760. [CrossRef]
40. Bu, X.-L.; Weng, J.-Y.; He, B.-B.; Xu, M.-J.; Xu, J. A novel AdpA homologue negatively regulates morphological differentiation in *Streptomyces xiamenensis* 318. *Appl. Environ. Microbiol.* **2019**, *85*, e03107-18. [CrossRef]
41. Romero-Rodríguez, A.; Robledo-Casados, I.; Sánchez, S. An overview on transcriptional regulators in *Streptomyces*. *Biochim. Biophys. Acta* **2015**, *1849*, 1017–1039. [CrossRef] [PubMed]
42. Higo, A.; Hara, H.; Horinouchi, S.; Ohnishi, Y. Genome-wide distribution of AdpA, a global regulator for secondary metabolism and morphological differentiation in *Streptomyces*, revealed the extent and complexity of the AdpA regulatory network. *DNA Res.* **2012**, *19*, 259–273. [CrossRef]
43. Rabyk, M.; Yushchuk, O.; Rokytskyy, I.; Anisimova, M.; Ostash, B. Genomic insights into evolution of AdpA family master regulators of morphological differentiation and secondary metabolism in *Streptomyces*. *J. Mol. Evol.* **2018**, *86*, 204–215. [CrossRef] [PubMed]
44. Wu, Y.; Kang, Q.; Zhang, L.-L.; Bai, L. Subtilisin-involved morphology engineering for improved antibiotic production in actinomycetes. *Biomolecules* **2020**, *10*, 851. [CrossRef] [PubMed]
45. Wang, X.; Wang, R.; Kang, Q.; Bai, L. The antitumor agent ansamitocin P-3 binds to cell division protein FtsZ in *Actinosynnema pretiosum*. *Biomolecules* **2020**, *10*, 699. [CrossRef]
46. Zhang, P.; Zhang, K.; Liu, Y.; Fu, J.; Zong, G.; Ma, X.; Cao, G. Deletion of the response regulator PhoP accelerates the formation of aerial mycelium and spores in *Actinosynnema pretiosum*. *Front. Microbiol.* **2022**, *13*, 845620. [CrossRef] [PubMed]
47. Guo, S.; Sun, X.; Li, R.; Zhang, T.; Hu, F.; Liu, F.; Hua, Q. Two strategies to improve the supply of pks extender units for ansamitocin P-3 biosynthesis by CRISPR–Cas9. *Bioresour. Bioprocess.* **2022**, *9*, 90. [CrossRef]
48. Livak, K.J.; Schmittgen, T.D. Analysis of relative gene expression data using real-time quantitative PCR and the $2^{-\Delta\Delta CT}$ method. *Methods* **2001**, *25*, 402–408. [CrossRef]
49. Xu, H.; Chater, K.F.; Deng, Z.; Tao, M. A cellulose synthase-like protein involved in hyphal tip growth and morphological differentiation in *Streptomyces*. *J. Bacteriol.* **2008**, *190*, 4971–4978. [CrossRef]
50. Kawamoto, S.; Watanabe, H.; Hesketh, A.; Ensign, J.C.; Ochi, K. Expression analysis of the *ssgA* gene product, associated with sporulation and cell division in *Streptomyces griseus*. *Microbiology* **1997**, *143*, 1077–1086. [CrossRef]
51. McCormick, J.R.; Su, E.P.; Driks, A.; Losick, R. Growth and viability of *Streptomyces coelicolor* mutant for the cell division gene *ftsZ*. *Mol. Microbiol.* **1994**, *14*, 243–254. [CrossRef] [PubMed]
52. Santos-Beneit, F.; Roberts, D.M.; Cantlay, S.; McCormick, J.R.; Errington, J. A mechanism for FtsZ-independent proliferation in *Streptomyces*. *Nat. Commun.* **2017**, *8*, 1378. [CrossRef] [PubMed]
53. Yushchuk, O.; Ostash, I.; Vlasiuk, I.; Gren, T.; Luzhetskyy, A.; Kalinowski, J.; Fedorenko, V.; Ostash, B. Heterologous AdpA transcription factors enhance landomycin production in *Streptomyces cyanogenus* S136 under a broad range of growth conditions. *Appl. Microbiol. Biotechnol.* **2018**, *102*, 8419–8428. [CrossRef] [PubMed]
54. Kang, Y.; Wang, Y.; Hou, B.; Wang, R.; Ye, J.; Zhu, X.; Wu, H.; Zhang, H. AdpAlin, a pleiotropic transcriptional regulator, is involved in the cascade regulation of lincomycin biosynthesis in *Streptomyces lincolnensis*. *Front. Microbiol.* **2019**, *10*, 2428. [CrossRef] [PubMed]
55. Higo, A.; Horinouchi, S.; Ohnishi, Y. Strict regulation of morphological differentiation and secondary metabolism by a positive feedback loop between two global regulators AdpA and BldA in *Streptomyces griseus*. *Mol. Microbiol.* **2011**, *81*, 1607–1622. [CrossRef] [PubMed]
56. Guyet, A.; Benaroudj, N.; Proux, C.; Gominet, M.; Coppée, J.-Y.; Mazodier, P. Identified members of the *Streptomyces lividans* AdpA regulon involved in differentiation and secondary metabolism. *BMC Microbiol.* **2014**, *14*, 81. [CrossRef] [PubMed]
57. Yamazaki, H.; Tomono, A.; Ohnishi, Y.; Horinouchi, S. DNA-binding specificity of AdpA, a transcriptional activator in the A-factor regulatory cascade in *Streptomyces griseus*: DNA-binding specificity of AdpA. *Mol. Microbiol.* **2004**, *53*, 555–572. [CrossRef]
58. Watanabe, K.; Okuda, T.; Yokose, K.; Furumai, T.; Maruyama, H. *Actinosynnema mirum*, a new producer of nocardicin antibiotics. *J. Antibiot.* **1983**, *36*, 321–324. [CrossRef] [PubMed]
59. Keijser, B.J.F.; Noens, E.E.E.; Kraal, B.; Koerten, H.K.; van Wezel, G.P. The *Streptomyces coelicolor ssgB* gene is required for early stages of sporulation. *FEMS Microbiol. Lett.* **2003**, *225*, 59–67. [CrossRef]

60. Xu, Q.; Traag, B.A.; Willemse, J.; McMullan, D.; Miller, M.D.; Elsliger, M.-A.; Abdubek, P.; Astakhova, T.; Axelrod, H.L.; Bakolitsa, C.; et al. Structural and functional characterizations of SsgB, a conserved activator of developmental cell division in morphologically complex actinomycetes. *J. Biol. Chem.* **2009**, *284*, 25268–25279. [CrossRef] [PubMed]
61. Makitrynskyy, R.; Ostash, B.; Tsypik, O.; Rebets, Y.; Doud, E.; Meredith, T.; Luzhetskyy, A.; Bechthold, A.; Walker, S.; Fedorenko, V. Pleiotropic regulatory genes *bldA*, *adpA* and *absB* are implicated in production of phosphoglycolipid antibiotic moenomycin. *Open Biol.* **2013**, *3*, 130121. [CrossRef] [PubMed]
62. Bandi, S.; Kim, Y.; Chang, Y.K.; Shang, G.; Yu, T.W.; Floss, H.G. Construction of *asm2* deletion mutant of *Actinosynnema pretiosum* and medium optimization for ansamitocin P-3 production using statistical approach. *J. Microbiol. Biotechnol.* **2006**, *16*, 1338–1346.
63. Ng, D.; Chin, H.K.; Wong, V.V.T. Constitutive overexpression of *asm2* and *asm39* increases AP-3 production in the actinomycete *Actinosynnema pretiosum*. *J. Ind. Microbiol. Biotechnol.* **2009**, *36*, 1345–1351. [CrossRef] [PubMed]
64. Pan, W.; Kang, Q.; Wang, L.; Bai, L.; Deng, Z. Asm8, a specific LAL-type activator of 3-amino-5-hydroxybenzoate biosynthesis in ansamitocin production. *Sci. China Life Sci.* **2013**, *56*, 601–608. [CrossRef]
65. Li, S.; Lu, C.; Chang, X.; Shen, Y. Constitutive overexpression of *asm18* increases the production and diversity of maytansinoids in *Actinosynnema pretiosum*. *Appl. Microbiol. Biotechnol.* **2016**, *100*, 2641–2649. [CrossRef]
66. Hackl, S.; Bechthold, A. The gene *bldA*, a regulator of morphological differentiation and antibiotic production in *Streptomyces*. *Arch. Pharm.* **2015**, *348*, 455–462. [CrossRef]

Article

Storage Stability and In Vitro Bioaccessibility of Microencapsulated Tomato (*Solanum Lycopersicum* L.) Pomace Extract

Luiz C. Corrêa-Filho , Diana I. Santos , Luísa Brito , Margarida Moldão-Martins and Vítor D. Alves *

LEAF—Linking Landscape, Environment, Agriculture and Food, Associated Laboratory TERRA, Instituto Superior de Agronomia, Universidade de Lisboa, Tapada da Ajuda, 1349-017 Lisboa, Portugal; lucaalbernaz@gmail.com (L.C.C.-F.); dianaisasantos@isa.ulisboa.pt (D.I.S.); lbrito@isa.ulisboa.pt (L.B.); mmoldao@isa.ulisboa.pt (M.M.-M.)
* Correspondence: vitoralves@isa.ulisboa.pt; Tel.: +351-21-365-3195

Abstract: Tomato pomace is rich in carotenoids (mainly lycopene), which are related to important bioactive properties. In general, carotenoids are known to react easily under environmental conditions, which may create a barrier in producing stable functional components for food. This work intended to evaluate the storage stability and in vitro release of lycopene from encapsulated tomato pomace extract, and its bioaccessibility when encapsulates were incorporated in yogurt. Microencapsulation assays were carried out with tomato pomace extract as the core material and arabic gum or inulin (10 and 20 wt%) as wall materials by spray drying (160 and 200 °C). The storage stability results indicate that lycopene degradation was highly influenced by the presence of oxygen and light, even when encapsulated. In vitro release studies revealed that 63% of encapsulated lycopene was released from the arabic gum particles in simulated gastric fluid, whereas for the inulin particles, the release was only around 13%. The feed composition with 20% inulin showed the best protective ability and the one that enabled releasing the bioactives preferentially in the intestine. The bioaccessibility of the microencapsulated lycopene added to yogurt increased during simulated gastrointestinal digestion as compared to the microencapsulated lycopene alone. We anticipate a high potential for the inulin microparticles containing lycopene to be used in functional food formulations.

Keywords: agro-industry; microencapsulation; bioactive compounds; in vitro digestion; yogurt

Citation: Corrêa-Filho, L.C.; Santos, D.I.; Brito, L.; Moldão-Martins, M.; Alves, V.D. Storage Stability and In Vitro Bioaccessibility of Microencapsulated Tomato (*Solanum Lycopersicum* L.) Pomace Extract. *Bioengineering* **2022**, *9*, 311. https://doi.org/10.3390/bioengineering9070311

Academic Editors: Minaxi Sharma, Kandi Sridha, Zeba Usmani and Giorgos Markou

Received: 29 May 2022
Accepted: 12 July 2022
Published: 13 July 2022

Publisher's Note: MDPI stays neutral with regard to jurisdictional claims in published maps and institutional affiliations.

1. Introduction

Worldwide, large quantities of fruit and vegetable wastes (e.g., seeds, skins, and pomaces) are discarded annually due to processing and poor storage facilities. As these food wastes rich in organic compounds are easily susceptible to microbiological deterioration, they cannot be introduced into the food chain without some processing. However, organic by-products can be valorized by extracting the bioactive compounds present in their composition for subsequent use as functional ingredients for food, pharmaceutical and cosmetic products, as they are considered to have beneficial effects on health [1–3].

Changes in modern lifestyle and the growing awareness of the link between diet and health, as well as new processing technologies, have led to an increased interest on the development of new healthy food products. Among them, a lot of attention has been driven to the development of functional food products enriched with natural functional bioactive molecules. Many bioactive compounds, such as carotenoids, extracted from natural sources have been shown to possess several biological activities, such as antimicrobial, antibacterial, antifungal, antiviral, anti-inflammatory, anti-obesity, anticholesteremic, antihypertensive, and antioxidant functions [4].

Carotenoids such as lycopene and β-carotene are antioxidants naturally present in some food products, and may be recovered as concentrated extracts from agroindustrial by-products, such as tomato pomace [2,4,5]. Tomato and tomato-based foods are the main sources of lycopene and are considered important contributors of carotenoids to the human diet [6]. However, the same properties that make carotenoids useful in healthy tissue function create challenges in preventing the degradation of carotenoids in food products due to the presence of unsaturated bonds in molecular structure [7,8]. As such, carotenoids are compounds very instable to processing and storage and their stability is affected by pH, exposure to light, temperature, oxygen, and acid media. These factors may influence the biological activity, nutritional quality, and the development of undesired flavors and of food. Therefore, the stability is an important aspect to consider for use of carotenoids as antioxidants and colorants in foods [9–11].

Microencapsulation of bioactive compounds, such as carotenoids and anthocyanins, is a process that enables the transformation of a liquid mixture with bioactives into a powder, which facilitates their handling, transport, storage, dosage, and application [12]. Furthermore, the encapsulation process has been extensively used for the stabilization of the bioactive compounds, improving their bioaccessibility and the control of their release rate in targeted sites [12–14].

Spray drying has been widely used in the food processing field due to its ease of industrialization, continuous production, and low cost, and it produces dry particles of good quality and with low water activity. During the spray drying process, evaporation of the water from the wall material occurs rapidly which maintains the core temperature below 100 °C, despite the use of high temperatures in the process. Therefore, this is an advantageous method for the encapsulation of heat sensitive compounds [13,15].

The characteristics of the ideal protective (wall) material depends on its intended purpose. Composition of the wall material influences the efficiency of the bioactive (core) material protection and its controlled release. Generally, wall materials should produce low viscosity solutions or suspensions at high solids content, emulsifying properties (especially for hydrophobic core materials), and resistance to the gastrointestinal tract [16].

Arabic gum is a polysaccharide complex derived from acacia trees. In general, arabic gum imparts low viscosity in aqueous media, presents high solubility in water, presents good emulsifying properties and good volatile retention [15,17]. Another interesting wall material used in the food industry is inulin due to its nutritional benefits and diversified technical and functional properties. Inulin is a fructooligosaccharide (FOS) commercially obtained from chicory root and it is composed of fructose units with β (2-1) bonds with glucose at the end of the chain [18]. Inulin is a dietary fiber that improves calcium bioavailability and it has prebiotic effects as it is degraded by certain probiotic bacteria present in the colon such as bifidobacteria and lactobacilos [19].

There is an interest in designing microencapsulates to release the encapsulated bioactive compounds in the intestine, where they may be absorbed into the blood stream. The literature indicates that the various types of wall materials used for the encapsulation of bioactive compounds influence differently on the bioacessibility/bioavailability of their compounds. In vivo digestion methods generally provide more reliable results but are time consuming and expensive. Alternatively, in vitro digestion models are useful methods that have been used in the first stages of the study [16,20].

The valorization of tomato pomace with the extraction of lycopene-rich fractions, and subsequent stabilization by spray drying using various wall materials (e.g., polysaccharides and proteins) has been studied [21,22]. However, detailed studies are still needed, as many factors beyond the type of wall material influence the spray drying encapsulation process and the properties of encapsulates obtained, such as the range of air inlet temperature used and the wall material concentration. In a previous work, a comprehensive study of such factors was carried out using Arabic gum and inulin as wall materials [8]. As such, the aim of this work was to go further, and study the stability during storage of the encapsulated lycopene-rich extract from tomato pomace. Furthermore, the in vitro release rate of lycopene in simulated gastric and intestinal fluids was evaluated, both from microparticles alone and from microparticles incorporated in yogurt with probiotic bacteria.

In the latter case, beyond lycopene bioaccessibility, the effect of inulin particles on the survival of probiotic bacteria was also studied.

2. Materials and Methods

2.1. Materials

Tomato pomace (*Solanum lycopersicum* L.) was supplied by the HIT Group (Marateca, Portugal). Arabic gum (LabChem, Zelienople, PA, USA) and Inulin (Alfa Aesar, Kandel, Germany) were used to form the protective matrix. Commercial probiotic yogurt (Actimel, Danone, Paris, France) and Natural liquid yogurt (Drink, Milbona, Neckarsulm, Germany) were purchased from a local supermarket to be used as the food matrix.

2,2′-Azinobis (3-ethylbenzothiazoline-6-sulphonic acid) diammonium salt (ABTS), lycopene, β-carotene, pepsin from porcine gastric mucosa (P6887), lipase from porcine pancreas (L3126) and pancreatin from porcine pancreas (P7545) were purchased from Sigma–Aldrich (Steinheim, Germany). 6-hydroxy-2,5,7,8-tetramethylchroman-2-carboxylic acid (Trolox) was obtained from Acrós Organics (Geel, Belgium). Potassium persulfate ($K_2S_2O_8$), potassium chloride (KCl), calcium chloride ($CaCl_2$), sodium hydrogen carbonate ($NaHCO_3$) and ethanol were purchased from Panreac AppliChem. Acetonitrile, dichloromethane, ethyl acetate and sodium chloride (NaCl), were obtained from Honeywell (Seelze, Germany). MRS broth was purchased from Biokar diagnostics (Beavais, France).

2.2. Tomato Pomace Extract Preparation

Frozen tomato pomace comprised of skin and seeds was lyophilized in a Lyo Quest freeze-dryer at $-47\ °C$ and 0.100 mbar for 48 h, powdered in a cutting mill and finally classified in a system of sieves arranged in columns in order to separate particles with a diameter of less than 0.5 mm. The tomato powder was stored under vacuum and protected from light.

The extraction of bioactive compounds (mostly lycopene and β-carotene) was carried out in a soxhlet apparatus, in which 15 g of tomato pomace powder was extracted with 300 mL of ethanol until the solvent became completely colorless after contact with the powder (around 5 h). After extraction, the solvent was evaporated by rotary vacuum evaporator (Rotavapor® R II BUCHI) and the extract obtained was placed in amber glass flasks and stored at $-20\ °C$ until analysis.

2.3. Preparation of Emulsions

Arabic gum was dissolved in distilled water under stirring overnight at room temperature, while inulin was dissolved in distilled hot water (70 °C). Two wall material concentrations (10 and 20% *w/w*) were used for encapsulation of tomato pomace extract. After the full hydration of the polymer molecules, concentrated tomato pomace extract (15% *w/w*, polymer basis) was added into each polymer solution and emulsions were produced by stirring with an Ultra-Turrax T25 (IKA, Staufen im Breisgau, Germany) at 13,500 rpm for 1 min at ambient temperature. A volume of 75 mL of emulsion was prepared for each experimental condition.

2.4. Spray Drying Conditions

The resultant solutions were fed at a rate of 3.7 mL·min^{-1} to a co-current spray dryer (Lab-Plant SD-05, Huddersfield, UK) equipped with a 0.5 mm diameter nozzle. The feed solution was kept under magnetic stirring and the pressure of the compressed air set at 1.7 bar. The inlet temperature was 160 and 200 °C, for inulin and arabic gum assays, respectively. The dried powders obtained were collected and stored under vacuum protected from light.

2.5. Morphological Characterization of Microparticles

The morphology of the particles obtained by spray drying was observed by Scanning Electron Microscopy (SEM). Samples were coated with a mixture of gold (80%) and palladium (20%) in a vacuum chamber, and analyzed using a Hitachi S2400 scanning microscope (Scotia, NY, USA) operated at 10 kV with a magnification of 1000×.

2.6. Analytical Methods

To assess its antioxidant activity and the concentrations of lycopene and β-carotene, the concentrated extract was diluted into a 1:1 acetonitrile:dichloromethane (ACN:DCM) solution 0.1% (*m/v*). All analytical measurements were carried out in triplicate.

2.6.1. Carotenoids Content in the Tomato Pomace Extract

The concentration of lycopene and β-carotene in the tomato pomace extract was quantified by HPLC using an UltiMate-3000 HPLC system (Dionex, Sunnyvale, CA, USA) equipped with a column oven at 30 °C. Carotenoid separation, identification, and quantification was performed using a reversed phase C_{18} 5 μm, 120 Å (4.6 × 150 mm) column by a gradient elution of acetonitrile:water:tetraethylammonioum (900:100:1) as solvent A and ethyl acetate as solvent B. The elution started with a mixture of 75% solvent A and 25% solvent B. After 10.50 min the solvent A was decrease to 59 % and after 20.0 min to 0%. At 21.0 min the solvent A returned to the initial condition (75%), remaining constant up to 31 min. The flow rate was 1 mL.min^{-1} and the running time was 31 min. The injection volume of the samples was 20 μL. The identification of carotenoids was based on their retention time of a peak compared with the carotenoids' standards. Calibration curves were carried out using standard lycopene and β-carotene with different concentrations (1–10 mg/L) using 1:1 ACN:DCM solution as solvent. Detection was carried out with a DAD-3000 diode-array detector with a wavelength of 475 and 472 nm for lycopene and 440 nm for β-carotene. The chromatogram obtained is presented in the supplementary file (Figure S1). The amount of β-carotene and lycopene in the samples was expressed as mg/$g_{extract}$.

2.6.2. Antioxidant Activity of the Tomato Pomace Extract

The total antioxidant activity of samples was performed by radical scavenging activity assessment expressed as Trolox Equivalent Antioxidant Activity (TEAC). An ABTS stock solution was prepared by dissolving ABTS in water at a 7 mM concentration. The ABTS$^+$ solution was produced by reaction of 5 mL of ABTS stock solution and 88 μL of a 140 mM potassium persulfate ($K_2S_2O_8$) solution to give a final concentration of 2.45 mm. This solution was kept in a dark room at room temperature for 12–16 h. Before analysis, the ABTS$^+$ solution was diluted with ethanol to obtain an initial absorbance value of 0.700 ± 0.05 at 734 nm.

For the evaluation of the antioxidant activity of tomato pomace extract itself, a volume of 30 μL of diluted extract with ACN:DCM solution was mixed with 3000 μL of ABTS$^+$ solution, followed by incubation for 6 min in the dark. Then, the absorbance was measured in a spectrophotometer (Unicam, UV/Vis Spectrometer–UV4) at a wavelength of 734 nm. A calibration curve was performed using Trolox, as standard antioxidant, at the concentration range of 250–2000 μM in ethanol.

2.6.3. Loading Capacity

For the determination of the concentration of the bioactive compounds present in the microparticles a mass of 20 mg of microparticles was added to 10 mL of a mixture of ACN and DCM (1:1). The suspension was homogenized with an Ultra-Turrax T25 (IKA, Staufen, Germany) at 13,500 rpm during 3 min, in order to break the microparticles. After mixing, the suspension was placed in amber glass flasks and kept away from light for about 12 h at 5 °C. Afterwards, the suspension was filtered with a syringe filter (Nylon 25 mm diameter, pore size of 0.45 μm; Fisher Scientific, Hampton, NH, USA) into an HPLC vial. The concentration of bioactive compounds in the liquid phase was quantified by HPLC, in the same way used for the tomato pomace extract. The loading capacity (LC) of the particles was expressed as the mass of carotenoids per mass of particles.

2.6.4. Antioxidant Activity of the Encapsulated Material

For the measurement of the antioxidant activity (AA) of the encapsulated molecules, the microparticles core material were previously extracted with of a 1:1 acetonitrile:dichloromethane (ACN:DCM) as described in the previous section. Afterwards, a volume of 800 µL of supernatant was mixed with 2200 µL of $ABTS^+$ solution, followed by the steps described in Section 2.6.1 for the fresh extract.

2.7. Storage Stability of Microencapsulated Tomato Pomace Extract

Samples of microparticles loaded with tomato pomace extract, and free lycopene powder, were placed in desiccators under four conditions as follows: (1) Dark + O_2; (2) Dark + N_2; (3) Light + O_2; and (4) Light + N_2. The relative humidity and temperature were kept at 33% and 25 °C, respectively, during all the experiment. Microparticles were periodically analyzed for 27 days to access the stability of encapsulated bioactives to oxygen and light, compared to that of free lycopene.

Lycopene loss was determined as the difference between the original and the obtained content of lycopene at different storage times. The quantification of carotenoids and antioxidant activity of bioactives in microparticles was performed in triplicate, employing the same methodology described previously.

2.8. In Vitro Digestion Studies

2.8.1. Microencapsulated Tomato Pomace Extract

In vitro digestion studies of microencapsulated tomato pomace extract were performed under static conditions in simulated gastric (SGF, pH 1.7) and intestinal (SIF, pH 6.5) fluids without enzymes as described by Gomes, et al. [21]. These studies were conducted in both fluids in parallel. The compositions of the simulated gastric and intestinal fluids are shown on Table 1. The pH of the simulated digestive fluids was adjusted with 1 M HCl or 1 M NaOH solutions. The release studies were carried out in a shaking thermal water bath (type 3047, Kotterman, Germany) with controlled temperature at 37 °C. In each release essay, 50 mg of microparticles was suspended in 5 mL of simulated fluids, in 15 mL Falcon tubes, under mild stirring for 2 h.

Table 1. Salt concentrations of simulated gastrointestinal fluids.

Gastric Fluid	Intestinal Fluid
4.8 g/L NaCl	5.0 g/L NaCl
2.2 g/L KCl	0.6 g/L KCl
0.22 g/L $CaCl_2$	0.25 g/L $CaCl_2$
1.5 g/L $NaHCO_3$	

2.8.2. Yogurt with Incorporated Microparticles

In vitro digestion studies of the food matrix, yogurt, enriched with microparticles were performed using simulated gastric fluid (SGF, pH 1.7) and intestinal fluid (SIF, pH 6.5) as described in the previous section with some modifications. Enzymes were added to the simulated gastric and duodenal fluids (Table 1). Porcine pepsin and lipase from porcine pancreas were added to achieve 2000 $U \cdot mL^{-1}$ and 40 $U \cdot mL^{-1}$, respectively, in the gastric digestion mixture. Pancreatin was added to achieve 100 $U \cdot mL^{-1}$ in the intestinal digestion mixture. A mass of 2.5 g of yogurt was placed in a Falcon tube containing 50 mg of each type of microparticles. The food matrix release assay was performed sequentially. First, the mass of 2.5 mg of yogurt with microparticles was suspended in 2.5 mL of simulated gastric fluid in 15 mL Falcon tubes under mild stirring for 2 h. Subsequently, 5.0 mL of simulated duodenal fluid was added in the same tube and let under mild shaking for more 2 h.

2.8.3. Carotenoid Bioaccessibility

The bioaccessibility of carotenoids in the microparticles was determined during the in vitro digestion process in order to obtain their release profile. At regular time intervals, different tubes were taken from the thermal bath, centrifuged at 8000 rpm for 20 min at 10 °C to separate the microparticles, recovering the supernatant (micelle fraction) containing the released bioactive compounds. The bioaccessibility of carotenoids in yogurt was determined only at the end of each stage of digestion (gastric and intestinal).

For the extraction of carotenoids from both the microparticles alone and from the food matrix with microparticles, 3 mL of chloroform was added in the tubes containing the raw digesta (before centrifugation) or micelle fraction, vortexed, and then centrifuged at 4000 rpm for 10 min at 10 °C. The organic layer was collected and quantified by HPLC. The bioaccessibility of carotenoids was calculated as the ratio of the carotenoids concentration present in the micelle fraction by the carotenoids concentration in the raw digesta.

2.9. Viability of Bacteria from the Probiotic Yogurt (Actimel®)

Survival of the bacteria contained in the probiotic yogurt with and without microparticles was evaluated during the in vitro digestion studies. At regular time intervals (0, 2, 4, 6, and 8 h), different Falcon tubes were taken from thermal bath to analyze the survival of bacteria present in the yogurt. An aliquot of 0.1 mL of the sample contained in each Falcon tube was diluted in 0.9 mL of ringer solution and then 0.1 mL of that diluted sample was spread on MRS (de Man, Rogosa and Sharp) agar plate. The microbial count was made after incubation at 30 °C for 72 h. The results were expressed as $CFU \cdot mL^{-1}$ of yogurt.

MRS broth was prepared by dissolving 55.3 g MRS and 20 g agar-agar into 1 L of hot distilled water. The broth was autoclaved at 121 °C for 15 min, cooled down to 50 °C in a water bath, and then poured into the petri dishes. The media was used within 2 days.

2.10. Statistical Analysis

The experimental data were statistically evaluated using Statistica™ v.8 Software (StatSoft Inc., Tulsa, OK, USA, 2007). Statistically significant differences ($p < 0.05$) between samples were evaluated using Tukey test.

3. Results and Discussion

3.1. Characterization of Tomato Pomace Extract

The concentration of lycopene and β-carotene in the tomato pomace extract was 15.19 ± 0.42 and 0.63 ± 0.02 $mg \cdot g^{-1}$ extract, respectively, which corresponds to 6.26 mg $lycopene \cdot g^{-1}$ dry tomato pomace and 0.26 mg $\beta\text{-carotene} \cdot g^{-1}$ dry tomato pomace. In the assessment of antioxidant activity, the extract possessed 536 ± 4.75 $\mu mol\ trolox \cdot g^{-1}$ extract that is equivalent to 35.27 ± 1.25 $\mu mol\ trolox \cdot mg^{-1}$ lycopene. Lower values of lycopene content (0.143 ± 0.004 $mg \cdot g^{-1}$ tomato concentrate) and antioxidant activity (11.3 ± 0.6 $\mu mol\ Trolox \cdot g^{-1}$ tomato concentrate) in tomato concentrate obtained by reverse osmosis followed by lower diafiltration were found by Souza, et al. [23]. This difference observed in the values of the lycopene content and antioxidant activity can be explained by the different way of obtaining the extract and the source of the raw material. The lycopene content of the pomace is dependent on the type of tomato plant cultivar.

3.2. Storage Stability of Microencapsulated Tomato Pomace Extract

It is well-known that the storage conditions, i.e., the pH, temperature, light, oxygen, and water activity, are important factors for preserving sensitive materials such as bioactives, flavors, and microorganisms. The stability of microparticles is defined as a state in which their physical (particle size, shape, size diameter), chemical and microbial characteristics are maintained unchanged throughout the storage period [15]. The non-microencapsulated and microencapsulated tomato pomace extracts were stored under different environmental conditions to evaluate the effect of the wall material on the protection of lycopene and antioxidant activity.

3.2.1. Lycopene Content

Figure 1 represents the lycopene loss during storage for each environmental conditions and samples studied.

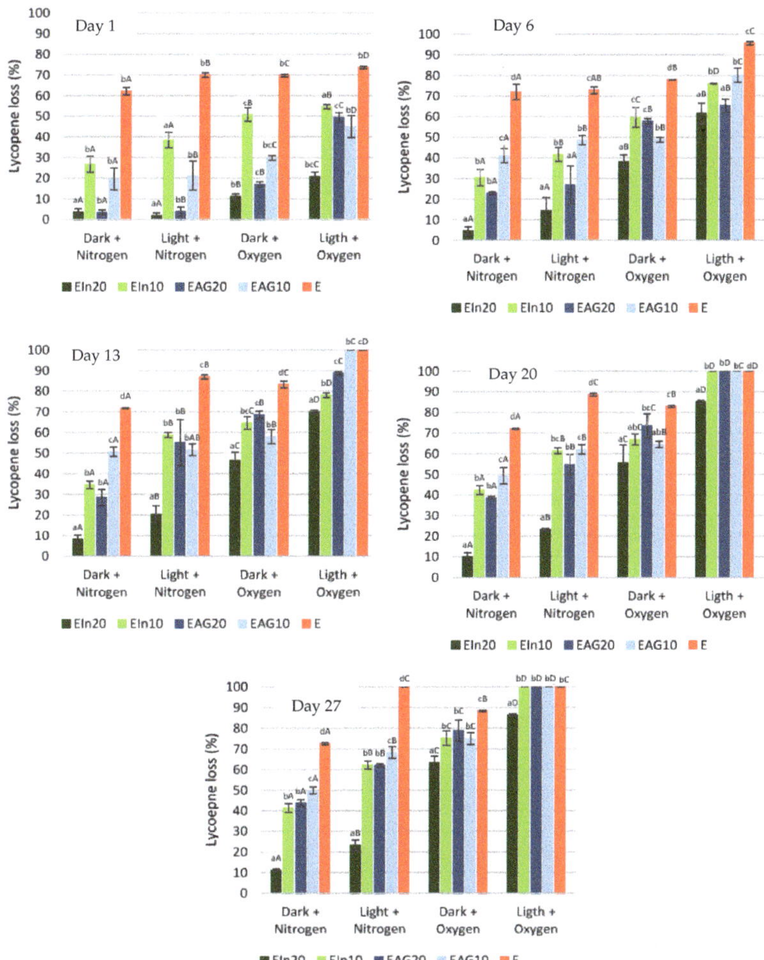

Figure 1. Lycopene loss over the storage period under different conditions. Different letters represent statistically significant differences (Tukey test $p < 0.05$): uppercase letters (effect of environmental conditions on lycopene loss); lowercase letters (effect of samples on lycopene loss). EIn20–20% Inulin; EIn10–10% Inulin; EAG20–20% Arabic gum; EAG10–10% Arabic gum; E–tomato pomace extract.

The loading capacity obtained for each wall material formulation was evaluated at the beginning of storage time (day 0). The loading capacity values obtained for 20 and 10% wall material were 1.1 and 1.3 mg lycopene·g^{-1} particles for inulin, and 0.9 and 1.5 mg lycopene·g^{-1} particles for arabic gum, respectively. The lycopene content at the start of the storage time was considered to be 100% to exclude the differences in the amount of lycopene obtained between formulations of the wall materials.

Overall, both inulin and arabic gum microcapsules with tomato pomace extract presented improved stability during storage compared to the free extract, showing the importance of microencapsulation in retarding the degradation of the bioactive compounds.

Nevertheless, the loss of lycopene and antioxidant activity increased with increasing storage time in all storage conditions.

Similar results were reported in studies on microencapsulation of carotenoids by spray drying. Souza, Hidalgo-Chávez, Pontes, Gomes, Cabral, and Tonon [23] microencapsulated carotenoids from tomato concentrate into different types of wall materials (maltodextrin, whey protein isolate and Capsul®) and Álvarez-Henao, et al. [24] studied the stability of lutein-loaded microcapsules using maltodextrin, arabic gum, and modified starch as wall materials. Çam, et al. [25] also found the same behavior in phenolic compounds when they evaluated the stability of phenolics from microencapsulated pomegranate seeds under storage at 5 °C for 3 months.

The encapsulated samples of tomato pomace extract with 20 and 10% wall material presented a lycopene degradation of 11.2 and 41.2% for inulin, and 43.6 and 49.9% for arabic gum, after 27 days of storage in dark and nitrogen. On the other hand, when stored under adverse environmental conditions (presence of light and oxygen), all the samples showed a complete degradation of the lycopene, except the particles with 20% inulin that presented a lycopene degradation of 86.4%. Overall, the results indicate that lycopene degradation was highly influenced by the presence of oxygen and light, demonstrating that the bioactive compounds should be preserved in absence of these conditions, even when encapsulated with the studied wall materials.

In terms of wall material concentration, a greater protection of lycopene was observed for the highest concentration of both wall material (20%). According to Bhandari, et al. [26], higher solids content results in greater protection of microencapsulated material and thus, greater stability. As mentioned, all samples of microencapsulated tomato pomace extract showed a better lycopene retention under less adverse environmental conditions (dark and nitrogen), presenting a degradation of less than 50%, whereas the non-microencapsulated extract was higher than 70%. Furthermore, under these conditions, the microparticles with 20% of inulin were the more efficient ones in protecting the bioactives, with a lycopene degradation of 11.2%.

Other researchers also studied carotenoid degradation in microparticles produced by spray drying using different type of wall materials. Souza, Hidalgo-Chávez, Pontes, Gomes, Cabral, and Tonon [23] stored lycopene-loaded microcapsules from tomato concentrate produced by spray drying in the dark for 28 days at 25 °C. Maltodextrin, whey protein isolate and Capsul® were used as wall materials. After the storage period, a loss of lycopene between 67 and 93% was observed. Rocha et al. [27] used Capsul® as wall material to produce lycopene-loaded microparticles. They observed that the particles stored at 25 °C in vacuum and away from the light had lycopene degradation of 41.32% after 73 days. Matioli and Rodriguez-Amaya [28] evaluated the effect of light on the degradation of lycopene from red-fleshed guava pulp microencapsulated in arabic gum during storage. The authors observed that more than 50% of encapsulated lycopene was degraded on day 19 and day 7 when stored in light and dark, respectively, both at room temperature. Álvarez-Henao, Saavedra, Medina, Jiménez Cartagena, Alzate and Londoño-Londoño [24] evaluated the storage stability of lutein-loaded microparticles produced by spray drying using arabic gum and the mixture of maltodextrin, arabic gum and modified starch (1:1:1) as wall materials. The samples were stored at 40 °C and relative humidity of 75%. At the end of 20 days of storage under the specified conditions, the microcapsules with arabic gum and the mixture of different wall materials showed a degradation in lutein content of 79 and 65%, respectively, while the degradation in non-encapsulated lutein extract was 89.5%. According to the authors, the film forming ability of the arabic gum used for producing the microparticles was not enough to provide stability over storage time, which was also observed in this work for both concentrations of arabic gum studied. However, this capacity increased with the addition of the modified starch and maltodextrin in the wall material formulation, which according to the authors can be due to the amylose present in the starch.

In terms of the different storage conditions studied, in the presence of light and oxygen, the stability of the lycopene contained within the microparticles was more negatively affected by the presence of oxygen. The best storage condition was Dark + Nitrogen, followed by Light + Nitrogen. The conditions containing oxygen, Dark + Oxygen followed by Light + Oxygen, were the two worst for storage of the microcapsules. Pelissari, et al. [29] reported that the degradation of lycopene was highly influenced by the presence of oxygen and storage temperature when evaluating the storage stability of microcapsules produced by spray chilling using shortening composed of hydrogenated and interesterified cottonseed, soy, and palm oils, and arabic gum as wall materials. According to Rodriguez-Amaya [30], a loss of carotenoids during storage occurs mainly due to non-enzymatic oxidation, since the carotenoid molecules are highly unsaturated and easily prone to oxidative degradation. This degradation depends on the availability of oxygen and the structure of the carotenoid being stimulated by the presence of light and heat.

3.2.2. Antioxidant Activity

The antioxidant activity (AA) values obtained for each wall material formulation was evaluated at the beginning of storage time (day 0). The antioxidant activity values obtained for microcapsules with 20 and 10% of wall material were 10.7 and 14.0 μmol Trolox$\cdot$mg^{-1} lycopene for inulin, and 22.2 and 11.6 μmol Trolox$\cdot$mg^{-1} lycopene for arabic gum, respectively. Additionally, the AA value of non-encapsulated extract was 35.3 μmol Trolox$\cdot$mg^{-1} lycopene. The AA at the start of the storage time was considered to be 100% to exclude the differences in the amount of antioxidant capacity obtained between formulations of the wall materials. Figure 2 represents the AA loss during storage for each environmental conditions and samples studied.

The encapsulated samples of tomato pomace extract with 20 and 10% of wall material showed a loss of antioxidant activity of 67.7 and 78.5% for inulin and 63.8 and 92.2% for arabic gum, after 27 days of storage under dark and nitrogen. On the other hand, when stored under the presence of light and oxygen, all samples had a high loss of antioxidant activity, above 94%. Still, from the 13th day of storage, 91% of the antioxidant activity had already been decreased, even in microencapsulated bioactive compounds. As was observed in lycopene degradation, the loss of antioxidant activity was influenced by the presence of oxygen and light as well as the type and concentration of the wall material used. Both inulin and arabic gum at 20% were significantly more able to maintain a higher antioxidant capacity than the other samples analyzed.

Gomes, et al. [31] evaluated the stability of microcapsules of arabic gum and maltodextrin loaded with lycopene-rich watermelon juice during storage for 15 days at room temperature. The results showed that the loss of AA was proportional to the loss of the lycopene content until the fifth day of storage. From this point, a greater loss of antioxidant activity was observed compared to loss of lycopene. At the end of storage time, the amount of lycopene in the particles reduced by 41.9%, while the antioxidant capacity reduced by 91.2%. According to the authors, first, the loss of the antioxidant activity was related to the degradation of the carotenoids. Afterwards, the most pronounced loss of antioxidant capacity after the fifth day was due to the continuous formation of cis isomers and other low antioxidant compounds produced by heating during the atomization process. Ramakrishna, et al. [32] studied the effect of light and temperature on the storage stability of microcapsules loaded with tamarillo juice rich in carotenoids and anthocyanins. Maltodextrin, arabic gum and modified starch were used as the wall material during the spray drying process. At the end of 28 days of storage in the presence of light, the loss of antioxidant activity varied between 6.59 and 9.85%. Meanwhile, at the end of 84 days of storage at 25 °C, losses of antioxidant activity between 11.28 and 49.63% were observed.

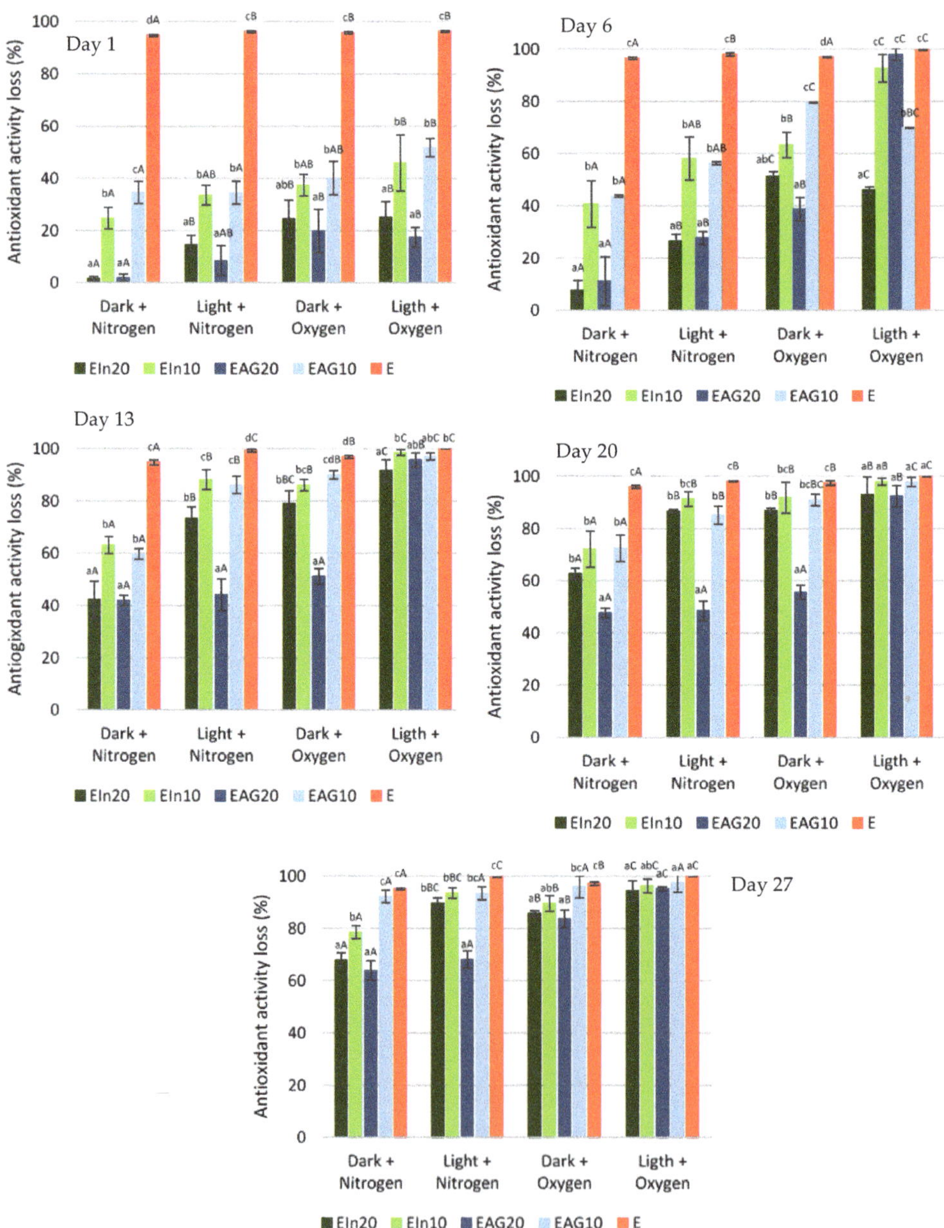

Figure 2. Antioxidant activity loss over the storage period under different conditions. Different letters represent statistically significant differences (Tukey test *p* < 0.05): uppercase letters (effect of environmental conditions on antioxidant activity loss); lowercase letters (effect of samples on antioxidant activity loss). EIn20–20% Inulin; EIn10–10% Inulin; EAG20–20% Arabic gum; EAG10–10% Arabic gum; E-tomato pomace extract.

3.3. Carotenoids Bioaccessibility

It is important to assess the release behavior of lycopene from microcapsules over the gastrointestinal tract after ingestion in order to evaluate the potential of wall materials for oral delivery in food or as a supplement for controlled/targeted release in specific zones of the gastrointestinal tract. In vitro release studies of microcapsules loaded with tomato pomace extract were performed under pH and temperature values that simulate the conditions inside the gastrointestinal tract. First, the release profile of the bioactives from the particle samples in each digestive fluid was analyzed separately. Then, the particles produced with inulin were selected to be incorporated into the natural yogurt, since they demonstrated a greater protection in the SGF, and the release rate in simulated gastrointestinal tract was also analyzed.

3.3.1. Microparticles Loaded with Tomato Pomace Extract

A sudden release of lycopene from the microparticles of both wall materials (arabic gum and inulin) occurred within the first 20 min in both simulated digestive fluids. After that, the bioactive release was quite slow and a plateau was reached for most of the cases (Figure 3).

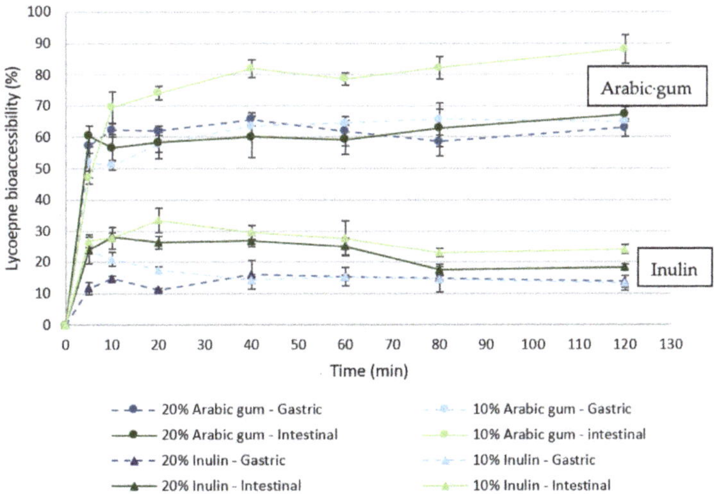

Figure 3. In vitro release profiles of lycopene from microparticles of inulin and arabic gum, under static conditions, in simulated gastric and intestinal fluids.

The carotenoid molecules that are adsorbed on the surface of the particles as well as those retained near the surface will be released more quickly, being responsible for the initial burst release. Still, the morphology of the particles can also affect the rapid release of the bioactive compounds through the cracks and the pores formed during production of the microcapsules which facilitate and accelerate the release of the core material from the polymeric matrix [33–35]. This initial burst release was also observed for the release of microcapsulated β-carotene formed by complex coacervation using casein and tragacanth gum as wall material [36] as well as in the neem seed oil extract release from arabic gum particles [37] and grape marc phenolics release from particles of whey protein and maltodextrin [38].

As shown in Figure 3, the percentage of lycopene released in simulated digestion in both gastric and intestinal fluids was higher for the arabic gum than for inulin particles. At the end of the simulated digestion period, between 63.0 and 65.1% of lycopene was released from the arabic gum particles in simulated gastric fluid, whereas for the inulin particles the release was between 13.3 and 13.8%. Furthermore, a greater release of lycopene was observed in the simulated intestinal fluid. In this fluid, the results showed a lycopene release between 67.3 and 88.1% for arabic gum particles and between 18.3 and 27.6% for inulin particles. Hence, the microencapsulation technique using inulin as the wall material appeared to be effective in protecting the bioactive compounds during the passage through the stomach (simulated gastric fluid). Rutz, et al. [39] also observed that the release profile of the encapsulated material was dependent on the type of wall material. Carotenoids from palm oil were microencapsulated in chitosan/Carboxymethylcellulose (CMC) and chitosan/sodium tripolyphosphate (TPP) by complex coacervation. They observed that the TPP particles released 53 and 67% of the carotenoids under the simulated gastric and intestinal fluids, while the CMC particles showed lower release percentages with 17 and 20% values for the simulated gastric and intestinal fluids, respectively. According to the authors, the CMC wall material promotes lipophilic interactions compared to TPP, and consequently, allowed for greater retention of carotenoids.

The fact that inulin presents lower release, being an encapsulating agent more protective than arabic gum, can be due to the solubility in aqueous medium of the studied polymers. While inulin has a low solubility in water of 2.5% (*w/v*) at 25 °C, arabic gum has a higher solubility of 50% (*w/v*) [40,41]. Another factor that may influence the release of bioactive microcapsules is their morphology. As shown in Figure 4, the type of wall material used in the production of the particles influenced its morphology. It was observed that for the particles produced with inulin, the surface was smoother while the arabic gum particles showed the formation of teeth and concavities. According to Moreno, Cocero, and Rodríguez-Rojo [38], rough and tooth-forming surfaces can accelerate the release of the microencapsulated compounds due to a larger surface area in contact with the medium as compared to the smoother surfaces.

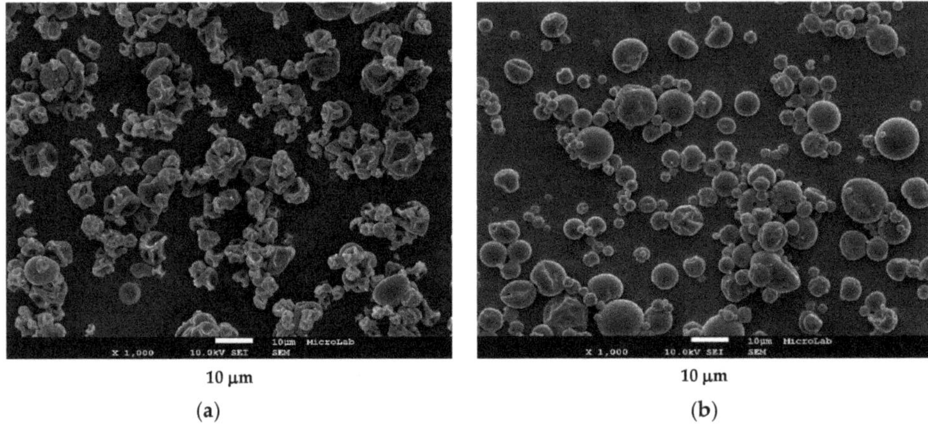

10 μm | 10 μm
(a) | (b)

Figure 4. Scanning Electron Microscopy (SEM) images (magnification ×1000) of microparticles loaded with tomato pomace extract in different wall materials. (**a**) Arabic gum particles; (**b**) Inulin particles.

Other researchers also found the release of bioactives faster on microparticles that had a rough surface. Moreno, Cocero, and Rodríguez-Rojo [38] observed that particles loaded with grape marc phenolics using whey protein isolate as wall material exhibited a wrinkled surface while the maltodextrin particles had a smooth surface. They reported that the phenolic compounds from grape marc present within the whey protein particles showed a lower release rate in the simulated gastrointestinal tract compared to that from the maltodextrin particles. In another study, the release of microencapsulated neem seed oil extract in three different wall materials (polyvinyl alcohol (PVA), arabic gum (AG), and whey protein isolate/maltodextrin) was studied by Sittipummongkol and Pechyen [37]. The authors reported that the particles with the highest release rate of neem seed extract were from the ones that showed more concavities on their surface, in the following order: WPI/MD > GA > PVA.

A significantly lower release of lycopene was observed for the particles produced with the highest wall material concentration (20%), for both arabic gum and inulin, when compared to the results of particles obtained with 10% wall material, during contact with the simulated intestinal fluid (Figure 3). However, during the passage through the simulated gastric fluid, no significant differences were observed. These results indicate that microparticles from 20% wall material demonstrate a better protection of the microencapsulated bioactive compounds in the gastrointestinal system. This difference in release rate of the bioactives found between different concentrations of the same wall material used can be attributed to the size of the particles. The microparticles obtained with the highest polymer concentration are larger when compared to that from lower concentrations, due to the increased viscosity of the solution that induces the formation of larger droplets after spraying. According to Yang, et al. [42], the microparticles produced with lower concentrations of wall material tend to have a more porous surface than microparticles with higher concentrations and consequently, facilitating the release of the encapsulated core material.

3.3.2. Yogurt Enriched with Microparticles

Microcapsulates produced with 10 and 20% inulin were chosen to be incorporated into liquid natural yogurt. Two parallel release studies were conducted: one using only microparticles and the other using yogurt enriched with microparticles. In the samples using only microparticles, 13.8 ± 0.8 and $11.1 \pm 0.5\%$ of the lycopene was released in the simulated gastric fluid from particles produced with 10 and 20% of inulin, respectively (Figure 5). Continuing to the simulated intestinal fluid, the 10 and 20% inulin particles reached the release of 18.3 ± 1.0 and $16.1 \pm 0.6\%$ at the end of this simulated stage. When using yogurt enriched with microparticles, the lycopene release increased during the simulated gastric fuid, reaching percentages of 17.4 ± 0.7 and $16.9 \pm 0.9\%$ for the concentrations of 10 and 20% of inulin, respectively. In the end of the simulated intestinal stage, a final lycopene release of 25.8 ± 0.5 and $26.1 \pm 0.2\%$ was observed for 10 and 20% of inulin, respectively.

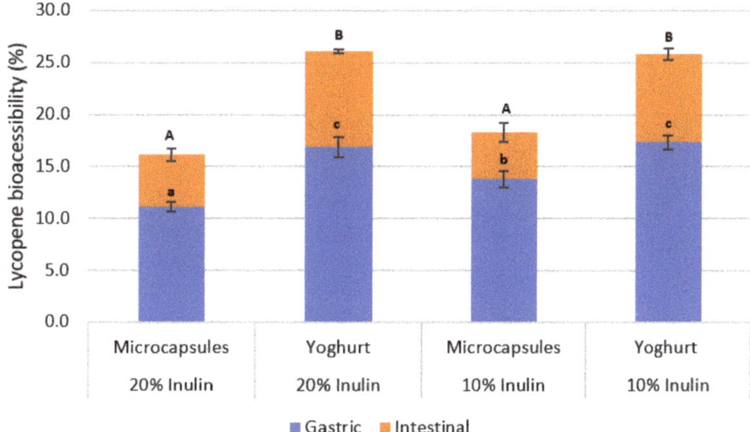

Figure 5. Lycopene release from the microcapsules alone and when incorporated into yoghurt, during a sequential static in vitro digestion: in simulated gastric fluid followed by simulated intestinal fluid. Different letters represent statistically significant differences among treatments (Tukey test $p < 0.05$): uppercase letters (during simulated intestinal fluid); lowercase letters (during simulated gastric fluid).

As such, the bioaccessibility of lycopene depends both on the digestion stage (gastric or intestinal fluid) and on the food matrix. For both inulin concentrations, a significant increase in lycopene release was observed at each step of simulated gastrointestinal digestion for yogurt with microparticles as compared to particles alone. In particular, comparing only the simulated intestinal stage, the release of lycopene in the yogurt samples was around 85% higher than that of microcapsules alone. Still, no significant differences were observed between the inulin concentrations used after the simulated gastrointestinal digestion.

The improvement in the stability of lycopene in yogurts may be related to the fact that the lycopene molecule has lipophilic characteristics, enhancing its ability to be dissolved in fats and lipids. Therefore, after its release, lycopene was dissolved by the lipid fraction of the yogurt and thus protected from the adverse conditions in the gastrointestinal tract. During simulated digestion only with particles, the lycopene released into the simulated digestive fluid, being hydrophobic, is not readily dissolved, becoming more susceptible to degradation.

Similar results were observed by Rutz, et al. [43] who analyzed the release of carotenoids in yogurt enriched with microparticles loaded with palm oil. The microparticles were produced by complex coacervation followed by lyophilization using chitosan/xanthan as a wall material. The authors observed a greater release of the carotenoids from the particles before incorporation in yogurt. However, some of these released carotenoids were degraded during simulated gastrointestinal digestion, reaching the end of the process with a release of 39.2%. After incorporation into yogurt, no degradation was observed during the simulated gastrointestinal tract, with a total release of 50.1% of microencapsulated carotenoids, thus improving its release and stabilization when added to the food matrix.

The low values in the release of lycopene found in the present work for inulin particles may be also related to its resistance to human digestive enzymes and, consequently, it is not digested in the upper gastrointestinal tract, thus promoting greater protection to the microencapsulated compounds. Inulin is able to reach the colon without being absorbed by the digestive system, and due to its prebiotic characteristics, may serve as a substrate, providing essential energy for endogenous bacteria [44].

3.4. Viability of Bacteria from Probiotic Yogurt

Based on the release results obtained for the different wall materials of the microcapsules, the particles produced with 20% inulin were chosen for the application in probiotic yogurt (Actimel®). The control yogurt (without particles) and yogurt with microencapsulated tomato pomace extract were evaluated for the survival of the probiotics contained in the yogurt during the simulated gastrointestinal tract. The survival rate of the probiotic bacteria of the yogurt with and without microparticles loaded with tomato pomace extract is shown in Figure 6.

It was observed a loss of viability of the probiotic bacteria in yogurt after the simulated gastric digestion for both cases, which was attributed to the acidic conditions present in the gastric fluid (pH = 1.7). There was a reduction of 4.07 and 4.94 log cycles of CFU in the yogurt incorporated with microparticles and in yogurt without microparticles, respectively. This behavior is in agreement with other authors who evaluated the survival of probiotic free cells during the simulated gastric treatment [45,46].

On the other hand, during the simulated intestinal digestion, the survival rate was maintained, and there was a tendency for a higher number of probiotic bacteria present in the yogurt with microparticles. In this stage, the viability of the probiotic bacteria in the yogurt with microparticles was significantly higher when compared that of yogurt without encapsulates. This difference in the viability of the bacteria may be related to the prebiotic characteristics of the inulin, thus serving as a substrate for the probiotic bacteria.

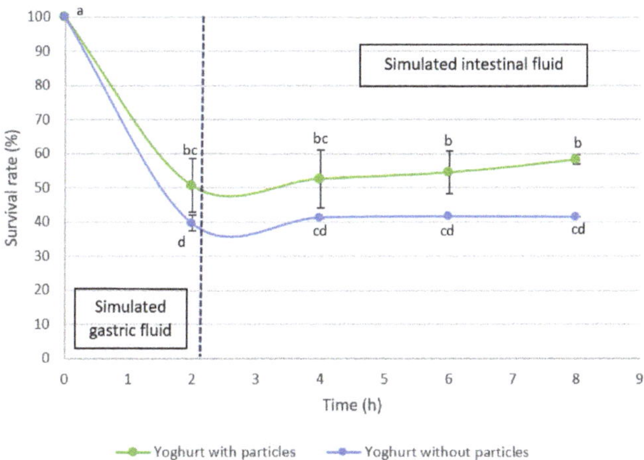

Figure 6. Viability of probiotic bacteria present in yogurt with particles and without particles during simulated gastrointestinal digestion. Different letters represent statistically significant differences between samples during simulated digestion (Tukey test $p < 0.05$).

4. Conclusions

Storage stability and in vitro release of microencapsulated ethanolic extract from tomato pomace rich in lycopene were influenced by the type and concentration of wall material. Tomato pomace extract, microencapsulated with both inulin and arabic gum as wall materials, was more stable under light and oxygen conditions than the free extract during 27 days of storage. However, between the two different types and concentrations of wall materials studied, inulin used at a concentration of 20% was the one showing the best protection ability of the core material against environmental conditions, in terms of degradation of the lycopene and antioxidant activity. Additionally, during in vitro release studies in simulated gastrointestinal fluids, it was the one that enabled reaching the intestine with a greater concentration of bioactive compounds to be released. Furthermore, after incorporation of the microparticles into yogurts, the bioaccessibility of the microencapsulated lycopene increased during simulated gastrointestinal digestion as compared to the lycopene of the microparticles studied without yogurt. Using inulin as wall material at a concentration of 20% also promoted a higher survival rate of yogurt probiotic bacteria during in vitro gastrointestinal digestion.

Supplementary Materials: The following supporting information can be downloaded at: https:// www.mdpi.com/article/10.3390/bioengineering9070311/s1, Figure S1: HPLC chromatogram obtained with a DAD-3000 diode-array detector, with a wavelength of 475 and 472 nm for lycopene and 440 nm for β-carotene, upon tomato pomace extract analysis.

Author Contributions: Conceptualization, L.C.C.-F., M.M.-M. and V.D.A.; methodology, L.C.C.-F., D.I.S., L.B., M.M.-M. and V.D.A.; formal analysis, L.C.C.-F. and D.I.S.; investigation, L.C.C.-F.; resources, L.C.C.-F., D.I.S., L.B., M.M.-M. and V.D.A.; data curation, L.C.C.-F.; writing—original draft preparation, L.C.C.-F.; writing—review and editing, L.C.C.-F., M.M.-M. and V.D.A.; supervision, M.M.-M. and V.D.A.; project administration, M.M.-M. and V.D.A.; funding acquisition, M.M.-M. and V.D.A. All authors have read and agreed to the published version of the manuscript.

Funding: The LCC-F was funded by Conselho Nacional de Desenvolvimento Científico e Tecnológico–CNPq through the Ciência sem Fronteiras program–CSF.

Institutional Review Board Statement: Not applicable.

Informed Consent Statement: Not applicable.

Data Availability Statement: Not applicable.

Acknowledgments: The authors acknowledge HIT Group, for supplying the tomato pomace used in this study.

Conflicts of Interest: The authors declare no conflict of interest.

References

1. Fierascu, R.C.; Fierascu, I.; Avramescu, S.M.; Sieniawska, E.; Barba, F.J.; Zhu, Z.; Lorenzo, J.M.; Remize, F. Recovery of Natural Antioxidants from Agro-Industrial Side Streams through Advanced Extraction Techniques. *Molecules* **2019**, *24*, 4212. [CrossRef] [PubMed]
2. Putnik, P.; Lorenzo, J.M.; Barba, F.J.; Roohinejad, S.; Jambrak, A.R.; Granato, D.; Montesano, D.; Kovačević, D.B. Novel Food Processing and Extraction Technologies of High-Added Value Compounds from Plant Materials. *Foods* **2018**, *7*, 106. [CrossRef]
3. Ojha, K.S.; Aznar, R.; O'Donnell, C.; Tiwari, B.K. Ultrasound Technology for the Extraction of Biologically Active Molecules from Plant, Animal and Marine Sources. *TrAC Trends Anal. Chem.* **2020**, *122*, 115663. [CrossRef]
4. de Andrade Lima, M.; Kestekoglou, I.; Charalampopoulos, D.; Chatzifragkou, A. Supercritical Fluid Extraction of Carotenoids from Vegetable Waste Matrices. *Molecules* **2019**, *24*, 466. [CrossRef] [PubMed]
5. Rao, A.V.; Rao, L.G. Carotenoids and Human Health. *Pharmacol. Res.* **2007**, *55*, 207–216. [CrossRef]
6. Shi, J.; Maguer, M. Le Lycopene in Tomatoes: Chemical and Physical Properties Affected by Food Processing. *Crit. Rev. Food Sci. Nutr.* **2000**, *40*, 1–42. [CrossRef]
7. Boon, C.S.; McClements, D.J.; Weiss, J.; Decker, E.A. Factors Influencing the Chemical Stability of Carotenoids in Foods. *Crit. Rev. Food Sci. Nutr.* **2010**, *50*, 515–532. [CrossRef]
8. Corrêa-Filho, L.; Lourenço, S.; Duarte, D.; Moldão-Martins, M.; Alves, V. Microencapsulation of Tomato (*Solanum lycopersicum* L.) Pomace Ethanolic Extract by Spray Drying: Optimization of Process Conditions. *Appl. Sci.* **2019**, *9*, 612. [CrossRef]
9. Shu, B.; Yu, W.; Zhao, Y.; Liu, X. Study on Microencapsulation of Lycopene by Spray-Drying. *J. Food Eng.* **2006**, *76*, 664–669. [CrossRef]
10. Provesi, J.G.; Dias, C.O.; Amante, E.R. Changes in Carotenoids during Processing and Storage of Pumpkin Puree. *Food Chem.* **2011**, *128*, 195–202. [CrossRef]
11. Trono, D. Carotenoids in Cereal Food Crops: Composition and Retention throughout Grain Storage and Food Processing. *Plants* **2019**, *8*, 551. [CrossRef] [PubMed]
12. Etzbach, L.; Meinert, M.; Faber, T.; Klein, C.; Schieber, A.; Weber, F. Effects of Carrier Agents on Powder Properties, Stability of Carotenoids, and Encapsulation Efficiency of Goldenberry (*Physalis peruviana* L.) Powder Produced by Co-Current Spray Drying. *Curr. Res. Food Sci.* **2020**, *3*, 73–81. [CrossRef] [PubMed]
13. Eun, J.B.; Maruf, A.; Das, P.R.; Nam, S.H. A Review of Encapsulation of Carotenoids Using Spray Drying and Freeze Drying. *Crit. Rev. Food Sci. Nutr.* **2019**, *60*, 3547–3572. [CrossRef] [PubMed]
14. Aizpurua-Olaizola, O.; Navarro, P.; Vallejo, A.; Olivares, M.; Etxebarria, N.; Usobiaga, A. Microencapsulation and Storage Stability of Polyphenols from Vitis Vinifera Grape Wastes. *Food Chem.* **2016**, *190*, 614–621. [CrossRef]
15. Zokti, J.A.; Baharin, B.S.; Mohammed, A.S.; Abas, F. Green Tea Leaves Extract: Microencapsulation, Physicochemical and Storage Stability Study. *Molecules* **2016**, *21*, 940. [CrossRef]
16. Calvo, P.; Lozano, M.; Espinosa-Mansilla, A.; González-Gómez, D. In-Vitro Evaluation of the Availability of ω-3 and ω-6 Fatty Acids and Tocopherols from Microencapsulated Walnut Oil. *Food Res. Int.* **2012**, *48*, 316–321. [CrossRef]
17. Troya, D.; Tupuna-Yerovi, D.S.; Ruales, J. Effects of Wall Materials and Operating Parameters on Physicochemical Properties, Process Efficiency, and Total Carotenoid Content of Microencapsulated Banana Passionfruit Pulp (*Passiflora Tripartita Var. Mollissima*) by Spray-Drying. *Food Bioprocess Technol.* **2018**, *11*, 1828–1839. [CrossRef]
18. Beirão-da-Costa, S.; Duarte, C.; Bourbon, A.I.; Pinheiro, A.C.; Januário, M.I.N.; Vicente, A.A.; Beirão-da-Costa, M.L.; Delgadillo, I. Inulin Potential for Encapsulation and Controlled Delivery of Oregano Essential Oil. *Food Hydrocoll.* **2013**, *33*, 199–206. [CrossRef]
19. Esmaeilnejad Moghadam, B.; Keivaninahr, F.; Fouladi, M.; Rezaei Mokarram, R.; Nazemi, A. Inulin Addition to Yoghurt: Prebiotic Activity, Health Effects and Sensory Properties. *Int. J. Dairy Technol.* **2019**, *72*, 183–198. [CrossRef]
20. Gonçalves, A.; Estevinho, B.N.; Rocha, F. Methodologies for Simulation of Gastrointestinal Digestion of Different Controlled Delivery Systems and Further Uptake of Encapsulated Bioactive Compounds. *Trends Food Sci. Technol.* **2021**, *114*, 510–520. [CrossRef]
21. Gomes, G.V.L.; Sola, M.R.; Marostegan, L.F.P.; Jange, C.G.; Cazado, C.P.S.; Pinheiro, A.C.; Vicente, A.A.; Pinho, S.C. Physico-Chemical Stability and in Vitro Digestibility of Beta-Carotene-Loaded Lipid Nanoparticles of Cupuacu Butter (Theobroma Grandiflorum) Produced by the Phase Inversion Temperature (PIT) Method. *J. Food Eng.* **2017**, *192*, 93–102. [CrossRef]
22. Li, J.; Pettinato, M.; Casazza, A.A.; Perego, P. A Comprehensive Optimization of Ultrasound-Assisted Extraction for Lycopene Recovery from Tomato Waste and Encapsulation by Spray Drying. *Processes* **2022**, *10*, 308. [CrossRef]
23. Souza, A.L.R.; Hidalgo-Chávez, D.W.; Pontes, S.M.; Gomes, F.S.; Cabral, L.M.C.; Tonon, R.V. Microencapsulation by Spray Drying of a Lycopene-Rich Tomato Concentrate: Characterization and Stability. *LWT* **2018**, *91*, 286–292. [CrossRef]
24. Álvarez-Henao, M.V.; Saavedra, N.; Medina, S.; Jiménez Cartagena, C.; Alzate, L.M.; Londoño-Londoño, J. Microencapsulation of Lutein by Spray-Drying: Characterization and Stability Analyses to Promote Its Use as a Functional Ingredient. *Food Chem.* **2018**, *256*, 181–187. [CrossRef] [PubMed]

25. Çam, M.; Içyer, N.C.; Erdoğan, F. Pomegranate Peel Phenolics: Microencapsulation, Storage Stability and Potential Ingredient for Functional Food Development. *LWT Food Sci. Technol.* **2014**, *55*, 117–123. [CrossRef]
26. Bhandari, B.R.; Dumoulin, E.D.; Richard, H.M.J.; Noleau, I.; Lebert, A.M. Flavor Encapsulation by Spray Drying: Application to Citral and Linalyl Acetate. *J. Food Sci.* **1992**, *57*, 217–221. [CrossRef]
27. Rocha, G.A.; Fávaro-Trindade, C.S.; Grosso, C.R.F. Microencapsulation of Lycopene by Spray Drying: Characterization, Stability and Application of Microcapsules. *Food Bioprod. Process.* **2012**, *90*, 37–42. [CrossRef]
28. Matioli, G.; Rodriguez-Amaya, D.B. Lycopene Encapsulated with Gum Arabic and Maltodextrin: Stability Study. *Braz. J. Food Technol* **2002**, *5*, 197–203.
29. Pelissari, J.R.; Souza, V.B.; Pigoso, A.A.; Tulini, F.L.; Thomazini, M.; Rodrigues, C.E.C.; Urbano, A.; Favaro-Trindade, C.S. Production of Solid Lipid Microparticles Loaded with Lycopene by Spray Chilling: Structural Characteristics of Particles and Lycopene Stability. *Food Bioprod. Process.* **2016**, *98*, 86–94. [CrossRef]
30. Rodriguez-Amaya, D.B. *Food Carotenoids: Chemistry, Biology and Technology*; John Wiley & Sons: Hoboken, NJ, USA, 2015; ISBN 1118733304.
31. Gomes, F.S.; Cabral, L.M.C.; Couri, S.; Costa, P.A.; Campos, M.B.D. Lycopene Content and Antioxidant Capacity of Watermelon Powder. *Acta Hortic.* **2014**, *1040*, 105–110. [CrossRef]
32. Ramakrishnan, Y.; Adzahan, N.M.; Yusof, Y.A.; Muhammad, K. Effect of Wall Materials on the Spray Drying Efficiency, Powder Properties and Stability of Bioactive Compounds in Tamarillo Juice Microencapsulation. *Powder Technol.* **2018**, *328*, 406–414. [CrossRef]
33. Yeo, Y.; Park, K. Control of Encapsulation Efficiency and Initial Burst in Polymeric Microparticle Systems. *Arch. Pharmacal Res.* **2004**, *27*, 1–12. [CrossRef] [PubMed]
34. Lourenço, S.C.; Torres, C.A.V.; Nunes, D.; Duarte, P.; Freitas, F.; Reis, M.A.M.; Fortunato, E.; Moldão-Martins, M.; da Costa, L.B.; Alves, V.D. Using a Bacterial Fucose-Rich Polysaccharide as Encapsulation Material of Bioactive Compounds. *Int. J. Biol. Macromol.* **2017**, *104*, 1099–1106. [CrossRef] [PubMed]
35. Flores, F.P.; Kong, F. In Vitro Release Kinetics of Microencapsulated Materials and the Effect of the Food Matrix. *Annu. Rev. Food Sci. Technol.* **2017**, *8*, 237–259. [CrossRef]
36. Jain, A.; Thakur, D.; Ghoshal, G.; Katare, O.P.; Shivhare, U.S. Characterization of Microcapsulated β-Carotene Formed by Complex Coacervation Using Casein and Gum Tragacanth. *Int. J. Biol. Macromol.* **2016**, *87*, 101–113. [CrossRef] [PubMed]
37. Sittipummongkol, K.; Pechyen, C. RETRACTED: Production, Characterization and Controlled Release Studies of Biodegradable Polymer Microcapsules Incorporating Neem Seed Oil by Spray Drying. *Food Packag. Shelf Life* **2018**, *18*, 131–139. [CrossRef]
38. Moreno, T.; Cocero, M.J.; Rodríguez-Rojo, S. Storage Stability and Simulated Gastrointestinal Release of Spray Dried Grape Marc Phenolics. *Food Bioprod. Process.* **2018**, *112*, 96–107. [CrossRef]
39. Rutz, J.K.; Borges, C.D.; Zambiazi, R.C.; da Rosa, C.G.; da Silva, M.M. Elaboration of Microparticles of Carotenoids from Natural and Synthetic Sources for Applications in Food. *Food Chem.* **2016**, *202*, 324–333. [CrossRef]
40. Kalyani Nair, K.; Kharb, S.; Thompkinson, D.K. Inulin Dietary Fiber with Functional and Health Attributes—A Review. *Food Rev. Int.* **2010**, *26*, 189–203. [CrossRef]
41. Montenegro, M.A.; Boiero, M.L.; Valle, L.; Borsarelli, C.D. Gum Arabic: More Than an Edible Emulsifier. In *Products and Applications of Biopolymers*; IntechOpen: London, UK, 2012. [CrossRef]
42. Yang, Y.Y.; Chung, T.S.; Ng, N.P. Morphology, Drug Distribution, and in Vitro Release Profiles of Biodegradable Polymeric Microspheres Containing Protein Fabricated by Double-Emulsion Solvent Extraction/Evaporation Method. *Biomaterials* **2001**, *22*, 231–241. [CrossRef]
43. Rutz, J.K.; Borges, C.D.; Zambiazi, R.C.; Crizel-Cardozo, M.M.; Kuck, L.S.; Noreña, C.P.Z. Microencapsulation of Palm Oil by Complex Coacervation for Application in Food Systems. *Food Chem.* **2017**, *220*, 59–66. [CrossRef] [PubMed]
44. Flamm, G.; Glinsmann, W.; Kritchevsky, D.; Prosky, L.; Roberfroid, M. Inulin and Oligofructose as Dietary Fiber: A Review of the Evidence. *Crit. Rev. Food Sci. Nutr.* **2001**, *41*, 353–362. [CrossRef] [PubMed]
45. Nag, A.; Han, K.S.; Singh, H. Microencapsulation of Probiotic Bacteria Using PH-Induced Gelation of Sodium Caseinate and Gellan Gum. *Int. Dairy J.* **2011**, *21*, 247–253. [CrossRef]
46. dos Santos, D.X.; Casazza, A.A.; Aliakbarian, B.; Bedani, R.; Saad, S.M.I.; Perego, P. Improved Probiotic Survival to in Vitro Gastrointestinal Stress in a Mousse Containing Lactobacillus Acidophilus La-5 Microencapsulated with Inulin by Spray Drying. *LWT* **2019**, *99*, 404–410. [CrossRef]

 *bioengineering*

Article

Development of Novel Lipid-Based Formulations for Water-Soluble Vitamin C versus Fat-Soluble Vitamin D3

Jie Chen [1], Leila Dehabadi [1], Yuan-Chun Ma [1,*] and Lee D. Wilson [2,*]

[1] Dr. Ma's Laboratories Inc., Unit 4, 8118 North Fraser Way, Burnaby, BC V5J 0E5, Canada
[2] Department of Chemistry, University of Saskatchewan, 110 Science Place, Saskatoon, SK S7N 5C9, Canada
* Correspondence: ycma@dml-labs.com (Y.-C.M.); lee.wilson@usask.ca (L.D.W.);
Tel.: +1-604-439-6089 (Y.-C.M.); +1-306-966-2961 (L.D.W.)

Abstract: The aim of this study was to develop a facile and novel lipid-based formulation of vitamin C and vitamin D3. Liposomes loaded with vitamin C and D3 were characterized using transmission electron microscopy (TEM) and zeta potential measurements for evaluating morphology, particle size and physical stability. HPLC was employed to quantify the content of vitamin C and vitamin D3 in their liposomal forms. The UHPLC analysis of the lipid-based vitamin formulation is an easy and rapid method for the characterization as well as the quantification of all components. In addition, encapsulation efficiency, vitamin loading and stability analysis were performed by the UHPLC method, in order to evaluate the reliability of the optimized lipid-based formulation. The TEM results provided key support for the core type of liposome structure in the formulations, whereas the HPLC results indicated that the liposomal vitamin C and D3 systems were homogeneous, and did not undergo phase separation. Taken together, the results demonstrate that liposomal encapsulated vitamins (vitamin C and D3) possess a unilamellar vesicle morphology with uniform particle size, despite differences in the hydrophile–lipophile profiles of the vitamins. The highly efficient encapsulation properties of such liposomal constructs are proposed to contribute to enhanced vitamin bioavailability.

Keywords: liposomes; vitamins; formulation; homogeneous systems; hydrophile–lipophile balance

Citation: Chen, J.; Dehabadi, L.; Ma, Y.-C.; Wilson, L.D. Development of Novel Lipid-Based Formulations for Water-Soluble Vitamin C versus Fat-Soluble Vitamin D3. *Bioengineering* 2022, 9, 819. https://doi.org/10.3390/bioengineering 9120819

Academic Editors: Minaxi Sharma, Kandi Sridhar, Zeba Usmani and Ali Zarrabi

Received: 5 November 2022
Accepted: 15 December 2022
Published: 19 December 2022

Publisher's Note: MDPI stays neutral with regard to jurisdictional claims in published maps and institutional affiliations.

1. Introduction

Contemporary developments in the pharmaceutical industry have seen an increased utilization of liposome carrier systems for the delivery of diverse compounds to target cells or to address drug toxicity [1–3]. Liposomes can simultaneously entrap both hydrophobic and hydrophilic substances, such as antimicrobials, antioxidants, flavor compounds, and bioactive constituents: this approach may serve to avoid decomposition prior to release of entrapped target compounds at the desired destination site [4–6]. Liposomes have drawn much attention from research and technological applications, due to their biocompatibility, biodegradability and low toxicity, including their capacity to capture both hydrophilic and lipophilic compounds [7,8]. Liposome assemblies can be achieved by amphiphilic lipids that self-assemble into bilayers driven by hydrophobic effects, which contribute to the unique structure function properties of liposomes as biomembranes and carriers for a variety of bioactive substances, due to their size, amphiphilic nature and biocompatibility [9]. Conventional liposomes are commonly classified into five types: multilamellar vesicles (MLV); small unilamellar vesicles (SUV); large unilamellar vesicles (LUV); multivesicular vesicles (MVV); and giant unilamellar liposomes (GUV) [10,11]. The size and homogeneity of the liposomes are the key properties to be considered for drug encapsulation, rather than the number of lamellar structures within the vesicles [9]. The ideal size of liposomes for drug delivery ranges between 50 and 200 nm [12,13]. Traditionally, liposomes may be prepared using a variety of techniques: reverse-phase evaporation; ether/ethanol injection;

controlled hydration; electroformation; microfluidic channels; thin-film hydration (Bangham technique); detergent depletion; heating, etc. [14–21]. On the other hand, innovative technologies such as freeze-drying double emulsions, dual asymmetric centrifugation (DAC), and supercritical fluid (SCF) treatments have been developed, in the past decade, for a liposomal-dependent drug delivery system [22]. However, a number of these methods have shortcomings, such as complex processes, a potential denaturing risk for active compounds, technically challenging operating conditions, and low drug encapsulation efficiency (EE) [23]. Moreover, the use of harmful organic solvents poses disadvantages when manufacturing dietary supplements or therapeutic medicines. For the latter case, organic solvents must be excluded from liposome-based medicinal formulations, in order to meet food and drug safety standards, and thereby conform with the compliance and regulatory requirements of the pharmaceutical industry. As a result, production costs rise, due to the additional purification processes and hazardous material management procedures that are mandated by law [24]. Notably, several emerging trends in promising liposomal drug delivery methods have been recently reviewed, including stealth liposome technology, non-PEGylated liposome techniques, lysolipid thermally sensitive liposome techniques, and depo-foam liposome techniques [25].

To analyze liposomes, a detailed and comprehensive comparison of existing imaging methods has been reported, which includes techniques such as scanning electron microscopy (SEM), transmission electron microscopy (TEM), SEM with energy dispersive X-ray analysis (SEM–EDX), and atomic force microscopy (AFM). These techniques have been widely adopted for evaluating typical properties of nanoparticles, such as size, homogeneity, and zeta potential [26,27]. The production of liposomes in the pharmaceutical industry necessitates the rigorous quality control of both chemical composition and physicochemical properties: this requires sophisticated analytical methods, such as high-performance liquid chromatography (HPLC), dynamic light scattering (DLS), and transmission electron cryomicroscopy, to provide assurance of the content and stability of the chemical components, including the structure of the nanoparticles [28,29].

In industry, numerous food and nutraceutical companies have aspired to manufacture natural health products (NHPs) with high nutritional content, through vitamin fortification, in order to enhance pharmaceutical properties and provide therapeutic effects. Vitamin C is a water-soluble vitamin, among the most commonly adopted essential nutrients and bioactive compounds, which is added to various dietary supplements, due to its potential health benefits [30–33]. As a crucial reductant and antioxidant constituent acquired from foods and health care products, vitamin C is involved in a variety of physiological functions, including the prevention of molecular oxidation, the inhibition of the enzymatic browning process, oxygen scavenging to prevent oxidative deterioration, and the suppression of nitrosamine formation. Due to vitamin C's high reactivity and sensitivity to oxidation, its degradation depends on various environmental factors, including temperature, pH, light, and oxygen, especially in the aqueous environments that occur in food and natural health products; therefore, an ongoing challenge is to maintain the stability of vitamin C during processing and preparation [30–33].

The major mechanism in aqueous media for vitamin C disintegration is oxidation to dehydroascorbic acid, which then rapidly converts to 2,3-diketogulonic acid, resulting in complete loss of the functional vitamin properties [34]. Although infusion methods via arteries or veins have higher bioavailability of vitamin C than oral administration, the disadvantages are obvious, such as strict administration criteria, risks of pathogen transmission, discomfort, and phlebitis [35–37]. The oral administration of vitamin C in the form of a crystalline powder or a liquid solution makes it prone to disintegration in the digestive system, particularly in the presence of metal ions; therefore, it is necessary to retard the breakdown of vitamin C in the stomach, and to enhance its absorption in the body, which can be achieved by encapsulation within the amphiphilic interior of the liposomes [38–41]. In addition to overcoming the oral bioavailability challenges and therapeutic issues of vitamin C, significant developments in drug optimization can improve orally administered health products, so that they reach the bloodstream by oral administration [42].

Various approaches have been employed, including microencapsulation, implementation of liposomes, and nanoparticles [43–45]. Lipid compounds that are usually regarded as harmless and biodegradable can promote transcellular delivery, by temporarily disrupting cellular lipophilic bilayers, as well as enhancing paracellular uptake of medicinal ingredients. Liposomal formulations of vitamin C not only delay its distribution, but can also increase absorption and prevent disintegration in the gut [38,39]. Generally, a healthy diet should contain lipids as a crucial element, especially phosphatidylcholines, as they have been shown to have a conducive impact on the overall wellness of patients [28,46,47]; in addition, the removal of organic solvents from the manufacturing process makes it challenging to achieve an ideal liposomal formulation; furthermore, the physical and chemical properties of phospholipids can change, upon the hydrolysis of ester linkages (between fatty acids and glycerol), the peroxidation of unsaturated acyl chains, and phospholipid degradation and oxidation, which may affect the quality and stability of the resulting liposomal forms [48,49].

Vitamin D is another important essential nutrient within a class of anti-rachitic compounds comprised of cyclopentanoperhydrophenanthrene rings, which are similar to those found in other steroids like cholesterol. In contrast to cholesterol, vitamin D is a three-ring system, and the structure of the side chains can distinguish naturally existing forms of vitamin D [50]. There are two main variants of vitamin D: vitamin D3 (active 7-dehydrocholesterol or cholecalciferol)—which is more commonly used, and is primarily transformed in dermal cells—and vitamin D2 (active ergosterol or ergocholecalciferol), which is mainly converted by irradiating ergosterol with UV light [51]. Specifically, the D3 form, or cholecalciferol, is a fat-soluble micronutrient that is required for several metabolic and immune processes. Vitamin D3 regulates about 700 genes, and consequently can control the in vivo metabolism of calcium, so as to increase intestinal calcium absorption, regulate phosphorus balance against osteoporosis, and thus promote normal bone formation and mineralization. Therefore, vitamin D3 deficiency may result in a variety of disorders, including bone disease, several cancers, type 1 diabetes, cardiovascular disease, and multiple sclerosis, especially in adults [52–54]. Humans can obtain vitamin D3 either intrinsically or extrinsically; however, an intrinsic source is only available by exposure to UVB, mostly from the sun, which raises the chance of melanoma. Based on such risks, ingestion of vitamin D3 is preferable, as the requirement of vitamin D3 from dietary intake is not typically met by the majority of accessible foods [55]. As vitamin D3 is particularly hydrophobic in nature, it is difficult to transport efficiently to the cells within the body; a change in composition, including a daily dosage of nutritional supplement, has a minor effect on the transportation of vitamin D3 and its derivatives throughout the body. Changes in vitamin formulation may have a significant impact on digestion and subsequent absorption mechanisms [56], as the biopolymer matrix found in mucus is composed of mucins, which inhibit digestion, followed by the absorption of particles containing hydrophobic substances, where particles less than 300–500 nm may only reach the epithelial cells [57]. Moreover, only competent cells may uptake the entire particle cargo of specific sizes, which allows for the endocytosis of highly hydrophobic molecules [58,59]. These features, along with the knowledge of physiologically relevant particles, have led to the design of an efficient formulation for highly hydrophobic substances in the context of liposomes, which are widely used as carriers to prevent degradation and increase the shelf life of active medicinal ingredients [60]. In addition, the low solubility of vitamin D3 prevents homogeneous physiological distribution, where liposomal vitamin D3 may further enhance the solubility profile, and regulate the release of medicinal compounds [61]. Herein, we demonstrate a robust lipid-based formulation for vitamin C and vitamin D, which is homogenous and stable without the requirement of any harmful chemicals or solvents in the formulation process.

2. Materials and Methods

Vitamin C (99.66% sodium ascorbate) and vitamin D3 (98.50% cholecalciferol) were purchased from Sigma-Aldrich (New York, NY, USA). In addition, sunflower lecithin ($\geq$40% phosphatidylcholine) was obtained from Jedwards International, Inc. (Braintree, Chicago, IL, USA). Potassium sorbate (PS) (99.47 $\pm$ 0.5%) and glycerin (99.80%) were purchased from Quadra Chemicals (Delta, BC, Canada) and EZ Chemicals Inc. (Mississauga, ON, Canada), respectively. Olive oil (100.00%) (Kirkland, WA, USA) and distilled water (ELGA LabWater (Woodridge, IL, USA)) were used for the preparation of all solutions. Methanol (99.80% HPLC grade) and acetonitrile (99.99% HPLC grade) were purchased from VWR international, LLC. Phosphoric acid (85.00% HPLC grade) was obtained from EMD Millipore Corporation (Kankakee, IL, USA). All solvents and samples were filtered through 0.2 μm wwPTFE filter (Pall Corporation (New York, NY, USA)) before injection into the HPLC.

2.1. Preparation of Liposomal Vitamin C

A solution was prepared at room temperature (25 °C), by adding 0.134 g of potassium sorbate (PS) (99.47 $\pm$ 0.5%), and dissolving in purified water (45 mL). Then, 1124 (+5%) mg of vitamin C (99.66% sodium ascorbate) was added to the aqueous solution, and stirred until completely dissolved. There was 5% overage of material input for vitamin C, according to the sample product formula. Next, 55 mL glycerin (99.80%) was added to an aqueous solution, by stirring; the solution was then placed in a blender (Hamilton Beach (Avalon, NJ, USA)), and was mixed thoroughly at a high speed (3000 rpm). Then, sunflower lecithin ($\geq$40% PC) (1.66 g) was added to the aqueous solution, and was blended at 3000 rpm to obtain liposome encapsulated vitamin C (cf. Scheme 1a).

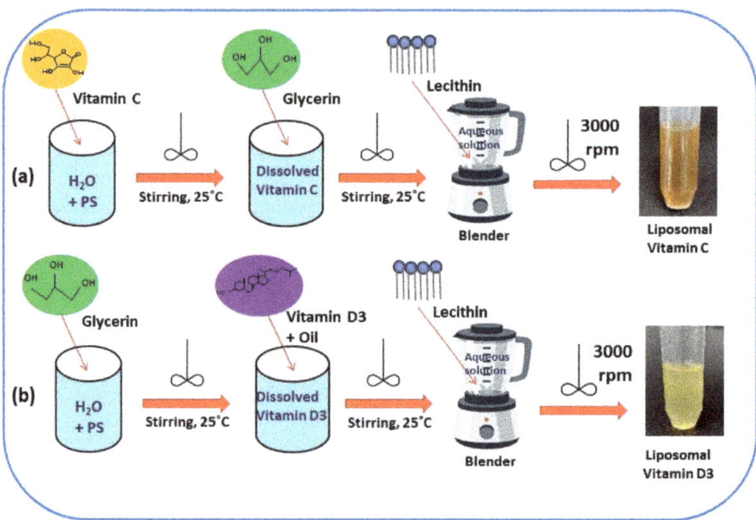

Scheme 1. Illustration of the liposome preparation method: (**a**) liposomal vitamin C, and (**b**) liposomal vitamin D3.

2.2. Preparation of Liposomal Vitamin D3

A hydrophilic solution was prepared at room temperature (25 °C), by weighing 0.134 g of PS (99.47 $\pm$ 0.5%), and dissolving this amount in 45 mL of purified water. Next, 55 mL of glycerin (99.80%) was added to the aqueous solution, by stirring; the solution was then placed in a blender container (Hamilton Beach (Avalon, NJ, USA)), and was thoroughly mixed at high speed (3000 rpm). Then, 1.25 (+10%) mg of vitamin D3 (98.50% cholecalciferol) was dissolved in 0.3 mL of olive oil (100.00%), which was added to the mixture, and blended at 3000 rpm for 30 min. There was 10% overage of material input for vitamin D3, according to the sample product formula. Next, 1.66 g of sunflower lecithin ($\geq$40% phosphatidylcholine; PC) was added to the above solution, followed by blending, to obtain liposome-encapsulated vitamin D3 (cf. Scheme 1b).

2.3. Characterization of Vitamin C and Vitamin D3 in Liposome Forms

2.3.1. Transmission Electron Microscopy (TEM) Analysis

TEM analysis was performed by negative staining. Briefly, 5 μL of liposomal sample was placed on a copper–formvar-coated TEM grid, and was allowed to settle on the grid surface for 2 min. The excess liquid was removed using an absorbent tissue. Staining of the grid was done using 0.5% phosphotungstic acid for 20 s, and excess staining was removed. Imaging was achieved using an HT 7700 TEM (Tokyo, Japan) at 80 kV.

2.3.2. Particle Size Distribution and Zeta Potential

The particle size and zeta potential of the liposomes were measured using a Malvern Zetasizer Nano ZS instrument (Malvern Instruments Ltd., Malvern, UK). The sample size distribution was obtained by measurement of scattered light ($\theta = 173°$) by particles (dynamic light scattering, DLS) illuminated with a laser beam. Zeta potential was established by an electrophoretic light scattering method based on the Doppler effect [62]. Zeta potential charge values were derived from triplicate measurements, where each consisted of a minimum of ten individual runs.

2.3.3. Encapsulation Efficiency (*EE%*) and Vitamins Loading (*VL%*)

The encapsulation efficiency (*EE%*) of the vitamins in the liposomes was calculated by Equation (1) [63–65]:

$$EE\ (\%) = \frac{Total\ vitamin\ content\ (\mathrm{mg}) - Solution\ vitamin\ content\ (\mathrm{mg})}{Total\ vitamin\ content\ (\mathrm{mg})} \times 100 \tag{1}$$

Nonencapsulated vitamins in the solution were quantified by high-performance liquid chromatography, after ultrafiltration (10 kDa) and centrifugation (Nuaire, NU-C200R, Plymouth, MA, USA) at $2500\times g$ for 20 min at 4 °C [63]. The absorbance demonstrated the nonencapsulated vitamin content; thus, the amount of trapped vitamin was calculated indirectly by Equation (1).

Vitamin loading (*VL%*) means the amount of vitamin that has been encapsulated in hydrated liposomes, and it was obtained by employing Equation (2) [64–66]:

$$VL\ (\%) = \frac{Vitamin\ content\ in\ liposomes\ (\mathrm{mg})}{Weight\ of\ liposomes\ (\mathrm{mg})} \times 100 \tag{2}$$

2.3.4. Instrumentation and Chromatographic Conditions

The UHPLC system consisted of flexible pumps, multicolumn thermostats, vial samplers, and diode array detector (Agilent 1290 Infinity II (Santa Clara, CA, USA)). The separation of vitamin C was carried out with a gradient elution program at a flow rate of 0.5 mL min^{-1} at 30.0 °C, by using a Luna C18 (100 mm × 3.0 mm, 2.5 μm) column supplied by Phenomenex (Torrance, CA, USA). The injection volume in the UHPLC system was 1 μL, where the mobile phase consisted of 0.05% (*v/v*) phosphoric acid (A) and methanol (B). The following linear gradient was applied: 0–3 min: 100% A; 3–5 min: 100–0% A, followed by re-equilibration of the column for 5 min before the next run. A wavelength of 245 nm was used for analyte detection of vitamin C. The separation of vitamin D3 was carried out with an isocratic elution program at a flow rate of 0.5 mL min^{-1} at 30.0 °C, by using a Luna C18 (100 mm × 3.0 mm, 2.5 μm) column supplied by Phenomenex, USA. The injection volume in the UHPLC system was 1 μL, and the mobile phase consisted of 75% acetonitrile (A) and 25% methanol (B). The total running time of the isocratic elution was 10 min. A wavelength of 280 nm was used for the detection of vitamin D3.

2.3.5. Sample Preparation for UHPLC Analysis

Samples of liposomal vitamin C from the top and bottom regions of the centrifuge vial were mixed with 50% methanol, at volume ratio 1:1000 (*v*/*v*), and were sonicated for 30 min before filtration and injection. A volume of 0.8 mL of liposomal vitamin D3 from the top and bottom portions of the sample tube was mixed with 100% methanol, and then the volume was brought to 10 mL in a volumetric flask. The sample was sonicated for 30 min before filtration and injection.

2.3.6. Liposomal Stability

The stability of the liposomes was assessed by determining the remainder of the vitamin in the liposomes after 3 months of storage at a refrigerated temperature of 4 °C, using Equation (3) below [63]:

$$Stability\ (\%) = \frac{Remaining\ amount\ of\ vitamin\ in\ liposomes\ (mg)}{Initial\ amount\ vitamin\ incorporated\ in\ liposomes\ (mg)} \times 100 \qquad (3)$$

3. Results and Discussion

As outlined above, a key goal of this study was to develop a robust lipid-based formulation for vitamin C and vitamin D, where the resulting system is homogenous and stable without requiring any harmful chemicals or solvents. This section describes the characterization of the lipid-based formulations that contain vitamin C and D, according to a range of methods (TEM imaging, light scattering, and HPLC). Various parameters—including particle size, zeta potential, encapsulation efficiency, and vitamin loading—which provide a measure of the stability of liposomal formulations, are discussed in the following sections.

3.1. Optimization of Formulation

Based on previous research [67], this study developed an optimized lipid-based formulation, to achieve high stability and homogeneity, where the liquid composition was critical— especially the ratio of water to glycerin, which played a key role in the sample preparation. By comparison of samples with different water–glycerin ratios, it was determined that a 45:55 (*v*/*v*) ratio should be used as the optimized water–glycerin volume ratio. For the conditions in which glycerin had a higher volume than water, the liposomes could be formed after homogenization, where the order of mixing of the respective ingredients also had a dramatic impact on the uniformity of the liposome product.

3.2. Morphology by TEM

The TEM imaging results in Figure 1 show that the liposomes were spherical in size, and that small fragments of liposomes were highly dispersed throughout them, which may indicate vesicle fracture [68,69]. Image J software (version 1.52, National Institutes of Health, Bethesda, MD, USA) was used to estimate the mean particle size of the liposomes, which was anticipated to be near 200 nm for liposomal vitamin C, and near 100 nm for liposomal vitamin D3. It is generally recognized that the size of liposomes has a significant role on tissue cell targeting: larger liposomes are usually taken by phagocytes, whereas smaller liposomes (100 to 200 nm) are more permeable to tumor cells [66,70]. The TEM images of liposomal vitamin C and D3 indicate that the liposomes were bilayer (unilamellar) vesicles, where the average sizes ranged between 100 nm to 200 nm.

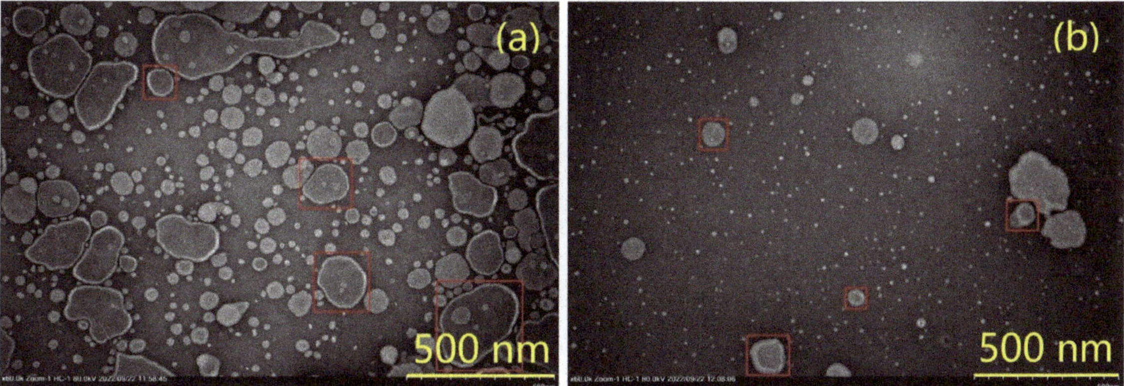

Figure 1. Transmission electron microscopy (TEM) images: (**a**) liposomal vitamin C with an average diameter near 200 nm; (**b**) liposomal vitamin D3 with an average diameter near 100 nm. Samples of bilayer (unilamellar) vesicles are indicated by a red square. The scale bars in panels (**a**,**b**) represent a 500 nm scale range (bottom right).

3.3. Size and Zeta Potential

The surface charge, electrostatic repulsion between neighboring particles, colloidal stability, and entrapment efficiency were correlated with the zeta potential (ζ) [71]. According to the ζ-value results, the liposomes that contained vitamin C had a lower ζ-value (0.8 mV) than the pristine vitamin C solution [72], which related to the enhancement effect of the carrier between the phospholipids and the active compounds. The stability of the vitamin D3 liposomes could also be characterized by the ζ-value. Based on the result, a negative ζ-value (-4.0 mV) for liposomal vitamin D3 indicated stronger particle repulsion; therefore, there was no driving force for the particles to associate with, resulting in a more stable and dispersed liposome without coagulation [63,73,74]. The PDI values were close in magnitude, and exceeded 0.2 for liposomal vitamin C (0.2) and liposomal vitamin D3 (0.53), which indicated monodisperse and polydisperse particle size distributions, respectively [75]. Liposomes may vary in size, between 200–3000 nm for liposomal vitamin C, and 100–1000 nm for liposomal vitamin D3, which is related to multiple lipid layers, fusion or aggregation phenomena [75,76].

3.4. Evaluation of Encapsulation Efficiency (EE%) and Vitamin Loading (VL%)

Table 1 summarizes the encapsulation efficiency and vitamin loading of liposomal vitamin C and liposomal vitamin D3. These parameters depended mostly on the composition of the liposomes, and may have been affected by other factors, such as liposome size and type, surface charge, bilayer rigidity, preparation method, and the properties of the vitamins. In turn, such factors may have helped to enhance the stability of the liposomes, and to have enhanced bioavailability on a physiological level without sacrificing its potency on the cellular level in the digestive system [77,78].

Table 1. Encapsulation Efficiency (EE%) and Vitamin Loading (VL%) of lipid-based vitamin C and lipid-based vitamin D3.

	Encapsulation Efficiency (*EE%*)	Vitamin Loading (*VL%*)
Vitamin C	47.15%	85.57%
Vitamin D3	>90.00%	78.45%

3.5. Determination of Vitamin C Potency in Liposomes

The potency of ascorbic acid in the lipid-based formulation of vitamin C was measured by UHPLC equipped with a DAD detector, at a wavelength of 245 nm. Table 2 indicates the concentration of vitamin C from the top and bottom regions of the sample tube (cf. inset in Figure 2a,b); the vitamin recovery values were 97.38% and 91.86%, respectively. Based on these results, the sample was homogenous, and did not show any significant phase separation; however, the concentration of the top portion was greater than the bottom region. According to Health Canada's regulation of Natural Health Products, a variation of 20% in composition is allowed for industrial production of such vitamin formulations.

Table 2. Concentrations from top and bottom layer of liposomal vitamin C.

Vitamin C	Top	Bottom
Concentration	204.50 mg/mL	192.90 mg/mL
Recovery value (%)	97.38%	91.86%

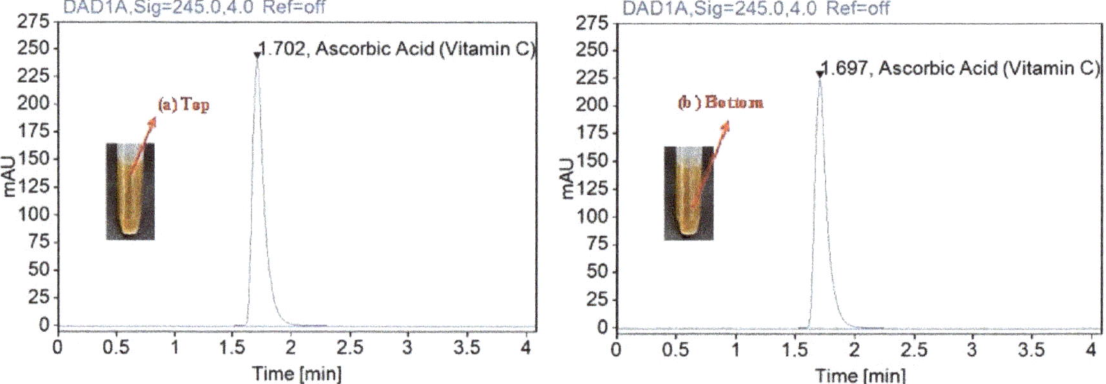

Figure 2. Representative liquid chromatograms obtained using UHPLC–DAD for liposomal vitamin C, sampled from the two separate sample regions shown in the inset: Top (**a**) and Bottom (**b**).

3.6. Determination of Vitamin D3 Potency in Liposomes

The potency of cholecalciferol in liposomal vitamin D3 was assessed by UHPLC equipped with a DAD detector, at a wavelength of 280 nm. Table 3 shows the concentration of vitamin D3 determined from the top and bottom regions of the sample tube (cf. insets in Figure 3a,b); the vitamin recovery values were 100.95% and 96.73%, respectively. Based on these results, the sample was homogenous, and did not display any notable phase separation. According to Health Canada's regulation of Natural Health Products, a variation of 20% in composition is allowed for industrial production of such vitamin formulations.

Table 3. Concentrations from top and bottom layer of liposomal vitamin D3.

Vitamin D3	Top	Bottom
Concentration	13.88 μg/mL	13.30 μg/mL
Recovery value (%)	100.95%	96.73%

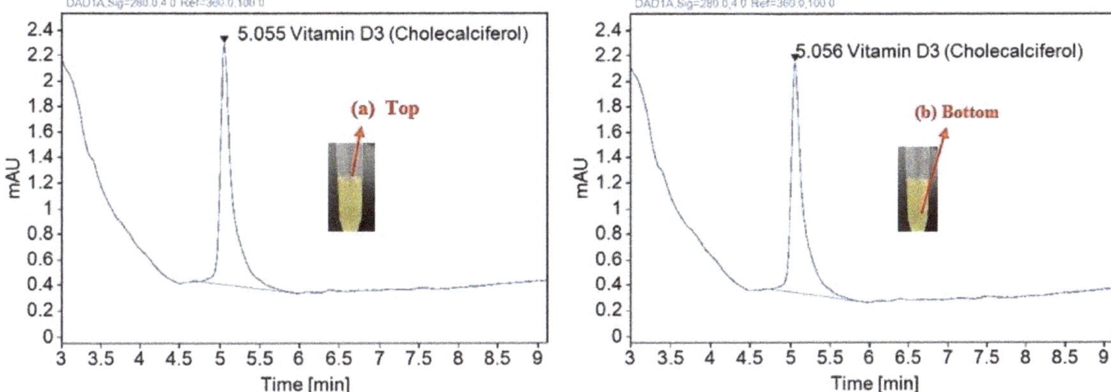

Figure 3. Representative liquid chromatograms obtained using UHPLC–DAD for liposomal vitamin D3 sampled from the two sample regions shown in the insets: Top (**a**) and Bottom (**b**).

3.7. Comparative Analysis of Vitamin Formulations

Size is a critical factor in determining the stability, biological origin, and effectiveness of formulated bioactives [79,80]. In this comparison study of liposomal vitamin C and liposomal vitamin D3, the particle size of liposomal vitamin C (~200 nm) was greater than that estimated for liposomal vitamin D3 (~100 nm), due to the higher bioactive–lipid weight ratio (B/L ratio), suggesting that the concentration of vitamin C was much greater than that of vitamin D3 [81]. While a 1:1 ratio is advantageous from a commercial standpoint, less phosphatidyl choline is required for its formulation. A 1:2 ratio, by comparison, achieves a smaller particle size (below 200 nm), which is optimal for liposomal stability: the latter is an ideal vesicle size (less than 200 nm) that is consistent with various food-based liposomes [82–84]. As a result, the upper region has a concentration of liposomal vitamin C that is higher than the bottom region of the sample. This trend indicates less homogeneity, unlike liposomal vitamin D3, where the concentration near the top region was very similar to the bottom region of the sample, indicating that a more uniform and stable solution was achieved with vitamin D3. This comparison of results from different sampling regions is consistent with the established relationship of the particle size and B/L ratio, which consequently leads to its stability [81]. We have further confirmed that at higher B/L ratios, colloids with larger size will be generated. Liposomes with reduced homogeneity concur with liposome systems that display reduced stability of the lipid-based formulation.

3.8. Stability

The results of the present study, shown in Tables 4 and 5, reveal that the liposomal entrapped vitamins leaked out only slightly over a 3-month period, for both 4 and 25 °C, as compared with the samples that were assessed immediately after the liposomal preparation.

Table 4. Stability of lipid-based vitamin C upon storage at 4 °C and 25 °C.

Vitamin C	Storage Temperature (°C)	0 Days		90 Days	
		Top	Bottom	Top	Bottom
Concentration	4	204.50 mg/mL	192.90 mg/mL	198.79 mg/mL	178.35 mg/mL
	25	204.50 mg/mL	192.90 mg/mL	175.20 mg/mL	155.54 mg/mL
Stability (%)	4	102.24%	96.42%	99.40%	89.17%
	25	102.24%	96.42%	87.60%	77.77%

Table 5. Stability of lipid-based vitamin D3 upon storage at 4 °C and 25 °C.

Vitamin D3	Storage Temperature (°C)	0 Days		90 Days	
		Top	Bottom	Top	Bottom
Concentration	4	13.88 µg/mL	13.30 µg/mL	8.8 µg/mL	9.8 µg/mL
	25	13.88 µg/mL	13.30 µg/mL	7.9 µg/mL	8.9 µg/mL
Stability (%)	4	111.00%	106.38%	70.21%	78.45%
	25	111.00%	106.38%	63.31%	70.84%

No significant differences in vitamin concentration were found between the top and bottom, for intervals ranging from 0 days to 90 days. This trend indicates that the formulations of both lipid-based vitamin C and vitamin D3 are homogenous, and stable for a minimum time of 3 months.

4. Conclusions

HPLC has been widely used for the identification and quantification of vitamins in diverse matrices, due to its relatively rapid separation capability, high sensitivity, precise quantification, and effective analytical performance; therefore, HPLC was chosen as the ideal method for assessing the potency of vitamin C and D3 in liposomal forms, and as a suitable analytical method for regular laboratory testing in the pharmaceutical industry. Based on a previous study [66], the encapsulation efficiencies of vitamin C were generally around 50%, which compares favorably to our results. A comparison of hydrophilic vitamin C to lipophilic vitamin D3 revealed a higher encapsulation efficiency (>90%) for vitamin D3, due to its high lipophilicity versus the more water-soluble (hydrophilic) vitamin C [85]. This trend was consistent with similar studies, and is related to the decreased particle size and polydispersity index, due to the role of the relative proportion of the constituents, the viscosity of the lipid phase, and the production conditions [86,87].

Our results provide support for both types of liposomal vitamin C and D3, which are homogenous colloids that remain very stable without phase separation, for a minimum of 3 months. Lipid-based formulations of this type are considered to be an effective and versatile approach to addressing the fortification of nutraceuticals and other bioactive components with variable molecular properties that cover a range of values on the hydrophile–lipophile scale. Our results concur with those of a recent study of liposomal-based vitamin C that had been readily prepared with food grade materials and simple equipment, and which provided a well-tolerated natural health product. The uptake of ingested liposomes from the gut to the mesenterial vessels and liver vessels was supported by a microbubble-enhanced ultrasound (MEU) method with real-time imaging [76]. Similarly, liposomes with a well-defined size distribution were used to achieve efficient absorption of hydrophobic vitamin D3. Clinical trials indicate that orally delivered liposomal vitamin D3 enhances the concentration of calcidiol in the serum rapidly, whereas no such function is found in oil-based vitamin D3 liquid [88].

This work focused mainly on the optimization of a lipid-based formula of water-soluble and fat-soluble vitamins, using a simple method and food-grade materials. Further studies could be extended to other applications of liposomal assemblies: for example, skin permeation studies could be carried out for both lipid-based vitamin C and lipid-based vitamin D3, in order to evaluate liposomes cellular internalization, and also to provide insight regarding the behavior of nanoparticles upon interaction with biological membranes [66,89,90]. In the case of nanocarriers, these systems have advantages for the topical and transdermal application of vitamin D3 (active 7-dehydrocholesterol or cholecalciferol), which is transformed primarily in dermal cells [51,66].

Author Contributions: Conceptualization, Y.-C.M. and L.D.W.; methodology, J.C. and L.D.; software, J.C.; validation, Y.-C.M. and L.D.W.; formal analysis, L.D. and J.C.; investigation, J.C. and L.D.; resources, Y.-C.M. and L.D.W.; data curation, J.C. and L.D.; writing—original draft preparation, J.C. and L.D.; writing—review and editing, L.D.W. and Y.-C.M.; visualization, J.C. and L.D.; supervision, Y.-C.M. and L.D.W.; and project administration, Y.-C.M. and L.D.W. All authors have read and agreed to the published version of the manuscript.

Funding: This research received no external funding.

Institutional Review Board Statement: Not applicable.

Informed Consent Statement: Not applicable.

Data Availability Statement: Not applicable.

Acknowledgments: Dr. Ma's Laboratories Inc. (DML) is acknowledged for supporting this research. All liposomal samples and HPLC analyses were carried out at Dr. Ma's Laboratories Inc. L.D acknowledges Eiko Kawamura for technical support with TEM imaging at WCV Image Centre (University of Saskatchewan). L.D.W. acknowledges Deysi J. Venegas-García for technical support with zeta potential and light scattering analysis.

Conflicts of Interest: The authors declare no conflict of interest.

References

1. Omri, A.; Suntres, Z.E.; Shek, P.N. Enhanced activity of liposomal polymyxin B against Pseudomonas aeruginosa in a rat model of lung infection. *Biochem. Pharmacol.* **2002**, *64*, 1407–1413. [CrossRef] [PubMed]
2. Schiffelers, R.M.; Storm, G.; Bakker-Woudenberg, I.A. Host factors influencing the preferential localization of sterically stabilized liposomes in Klebsiella pneumoniae-infected rat lung tissue. *Pharm. Res.* **2001**, *18*, 780–787. [CrossRef] [PubMed]
3. Stano, P.; Bufali, S.; Pisano, C.; Bucci, F.; Barbarino, M.; Santaniello, M.; Carminati, P.; Luisi, P.L. Novel camptothecin analogue (gimatecan)-containing liposomes prepared by the ethanol injection method. *J. Liposome Res.* **2004**, *14*, 87–109. [CrossRef] [PubMed]
4. Atrooz, O.M. Effects of alkylresorcinolic lipids obtained from acetonic extract of Jordanian wheat grains on liposome properties. *Int. J. Biol. Chem.* **2011**, *5*, 314–321. [CrossRef]
5. Benech, R.O.; Kheadr, E.E.; Laridi, R.; Lacroix, C.; Fliss, I. Inhibition of Listeria innocua in cheddar cheese by addition of nisin Z in liposomes or by in situ production in mixed culture. *Appl. Environ. Microbiol.* **2002**, *68*, 3683–3690. [CrossRef]
6. Shehata, T.; Ogawara, K.; Higaki, K.; Kimura, T. Prolongation of residence time of liposome by surface-modification with mixture of hydrophilic polymers. *Int. J. Pharm.* **2008**, *359*, 272–279. [CrossRef]
7. Johnston, M.J.; Semple, S.C.; Klimuk, S.K.; Ansell, S.; Maurer, N.; Cullis, P.R. Characterization of the drug retention and pharmacokinetic properties of liposomal nanoparticles containing dihydrosphingomyelin. *Biochim. Biophys. Acta* **2007**, *1768*, 1121–1127. [CrossRef]
8. Hofheinz, R.D.; Gnad-Vogt, S.U.; Beyer, U.; Hochhaus, A. Liposomal encapsulated anti-cancer drugs. *Anticancer Drugs* **2005**, *16*, 691–707. [CrossRef]
9. Lasic, D.D. Applications of liposomes. In *Handbook of Biological Physics*; Elsevier: Amsterdam, The Netherlands, 1995; pp. 491–519.
10. Elizondo, E.; Moreno, E.; Cabrera, I.; Córdoba, A.; Sala, S.; Veciana, J.; Ventosa, N. Liposomes and other vesicular systems: Structural characteristics, methods of preparation, and use in nanomedicine. *Prog. Mol. Biol. Transl. Sci.* **2011**, *104*, 1–52.
11. Stein, H.; Spindler, S.; Bonakdar, N.; Wang, C.; Sandoghdar, V. Production of isolated giant unilamellar vesicles under high salt concentration. *Front. Physiol.* **2017**, *8*, 1–16. [CrossRef]
12. Harashima, H.; Sakata, K.; Funato, K.; Kiwada, H. Enhanced hepatic uptake of liposomes through complement activation depending on the size of liposomes. *Pharm. Res.* **1994**, *11*, 402–406. [CrossRef]
13. Woodle, M.C. Sterically stabilized liposome therapeutics. *Adv. Drug Deliv. Rev.* **1995**, *16*, 249–265. [CrossRef]
14. Akbarzadeh, A.; Rezaei-Sadabady, R.; Davaran, S.; Joo, S.W.; Zarghami, N.; Hanifehpour, Y.; Samiei, M.; Kouhi, M.; Nejati-Koshki, K. Liposome: Classification, preparation, and applications. *Nanoscale Res. Lett.* **2013**, *8*, 1–9. [CrossRef]
15. Sahoo, S.K.; Labhasetwar, V. Nanotech approaches to drug delivery and imaging. *Drug Discov. Today* **2003**, *8*, 1112–1120. [CrossRef]
16. Vemuri, S.; Rhodes, C.T. Preparation and characterization of liposomes as therapeutic delivery systems: A review. *Pharm. Acta Helv.* **1995**, *70*, 95–111. [CrossRef]
17. Sharma, A.; Sharma, U.S. Liposomes in drug delivery: Progress and limitations. *Int. J. Pharm.* **1997**, *154*, 123–140. [CrossRef]
18. Jesorka, A.; Orwar, O. Liposomes: Technologies and analytical applications. *Annu. Rev. Anal. Chem.* **2008**, *1*, 801–832. [CrossRef]
19. Has, C.; Sunthar, P. A comprehensive review on recent preparation techniques of liposomes. *J. Liposome Res.* **2020**, *30*, 336–365. [CrossRef]
20. Wagner, A.; Vorauer-Uhl, K. Liposome technology for industrial purposes. *J. Drug Deliv.* **2011**, *2011*, 1–9. [CrossRef]

21. Maherani, B.; Arab-Tehrany, E.; Mozafari, M.R.; Gaiani, C.; Linder, M. Liposomes: A Review of Manufacturing Techniques and Targeting Strategies. *Curr. Nanosci.* **2011**, *7*, 436–452. [CrossRef]

22. Bulbake, U.; Doppalapudi, S.; Kommineni, N.; Khan, W. Liposomal Formulations in Clinical Use: An Updated Review. *Pharmaceutics* **2017**, *9*, 12–45. [CrossRef] [PubMed]

23. Webb, C.; Khadke, S.; Schmidt, S.T.; Roces, C.B.; Forbes, N.; Berrie, G.; Perrie, Y. The Impact of Solvent Selection: Strategies to Guide the Manufacturing of Liposomes Using Microfluidics. *Pharmaceutics* **2019**, *11*, 653–669. [CrossRef] [PubMed]

24. Beh, C.C.; Mammucari, R.; Foster, N.R. Formation of nanocarrier systems by dense gas processing. *Langmuir* **2014**, *30*, 11046–11054. [CrossRef] [PubMed]

25. Andra, V.V.S.N.L.; Pammi, S.V.N.; Bhatraju, L.V.K.P.; Ruddaraju, L.K. A Comprehensive Review on Novel Liposomal Methodologies, Commercial Formulations, Clinical Trials and Patents. *BioNanoScience* **2022**, *12*, 274–291. [CrossRef] [PubMed]

26. Robson, A.-L.; Dastoor, P.C.; Flynn, J.; Palmer, W.; Martin, A.; Smith, D.W.; Woldu, A.; Hua, S. Advantages and limitations of current imaging techniques for characterizing liposome morphology. *Front. Pharmacol.* **2018**, *9*, 80–88. [CrossRef]

27. Ruozi, B.; Belletti, D.; Tombesi, A.; Tosi, G.; Bondioli, L.; Forni, F.; Vandelli, M.A. AFM, ESEM, TEM, and CLSM in liposomal characterization: A comparative study. *Int. J. Nanomed.* **2011**, *6*, 557–563. [CrossRef]

28. Garcia, J.T.; Aguero, S.D. Phospholipids: Properties and health effects. *Nutr. Hosp.* **2015**, *31*, 76–83.

29. Van Nieuwenhuyzen, W.; Szuhaj, B.F. Effects of lecithins and proteins on the stability of emulsions. *Eur. J. Lipid Sci. Technol.* **1998**, *100*, 282–291. [CrossRef]

30. Gopi, S.; Balakrishnan, P. Evaluation and Clinical Comparison Studies on Liposomal and Non-Liposomal Ascorbic Acid (Vitamin C) and their Enhanced Bioavailability. *J. Liposome Res.* **2021**, *31*, 356–364. [CrossRef]

31. Alvim, I.D.; Stein, M.A.; Koury, I.P.; Dantas, F.B.H.; Cruz, C.C.V. Comparison between the spray drying and spray chilling microparticles contain ascorbic acid in a baked product application. *LWT–Food Sci. Technol.* **2016**, *65*, 689–694. [CrossRef]

32. Fong, S.Y.K.; Martins, S.M.; Brandl, M.; Brandl, A.B. Solid phospholipid dispersions for oral delivery of poorly soluble drugs: Investigation into celecoxib incorporation and solubility-in vitro permeability enhancement. *J. Pharm. Sci.* **2016**, *105*, 11113–11123. [CrossRef] [PubMed]

33. Matos, F.E., Jr.; Sabatino, M.D.; Passerini, N.; Favaro, T.C.S.; Albertini, B. Development and characterization of solid lipid microparticles loaded with ascorbic acid and produced by spray congealing. *Food Res. Int.* **2015**, *67*, 52–59. [CrossRef]

34. Herbig, A.L.; Renard, C.M. Factors that impact the stability of vitamin C at intermediate temperatures in a food matrix. *Food Chem.* **2017**, *220*, 444–451. [CrossRef] [PubMed]

35. Padayatty, S.J.; Sun, H.; Wang, Y.; Riordan, H.D.; Hewitt, S.M.; Katz, A.; Wesley, R.A.; Levine, M. vitamin C pharmacokinetics: Implications for oral and intravenous use. *Ann. Intern. Med.* **2004**, *140*, 533–537. [CrossRef] [PubMed]

36. Davis, J.L.; Paris, H.L.; Beals, J.W.; Binns, S.E.; Giordano, G.R.; Scalzo, R.L.; Schweder, M.M.; Blair, E.; Bell, C. Liposomal encapsulated ascorbic acid: Influence on vitamin C bioavailability and capacity to protect against ischemia-reperfusion injury. *Nutr. Metab. Insights* **2016**, *9*, 25–30. [CrossRef]

37. Parhizkar, E.; Rashedinia, M.; Karimi, M.; Alipour, S. Design and development of vitamin C-encapsulated proliposome with improved in-vitro and ex-vivo antioxidant efficacy. *J. Microencapsul.* **2018**, *35*, 301–311. [CrossRef]

38. Wechtersbach, L.; Poklar Ulrih, N.; Cigic, B. Liposomal stabilization of ascorbic acid in model systems and in food matrices. *LWT-Food Sci. Technol.* **2012**, *45*, 43–49. [CrossRef]

39. Hickey, S.; Roberts, H.J.; Miller, N.J. Pharmacokinetics of oral vitamin C. *J. Nutr. Environ. Med.* **2008**, *17*, 169–177. [CrossRef]

40. Nagle, J.F.; Tristram-Nagle, S. Structure of lipid bilayers. *Biochim. Biophys. Acta* **2000**, *1469*, 159–195. [CrossRef]

41. Pastoriza-Gallego, M.J.; Losada-Barreiro, S.; Bravo-Diiaz, C. Effects of acidity and emulsifier concentration on the distribution of vitamin C in a model food emulsion. *J. Phys. Org. Chem.* **2012**, *25*, 908–915. [CrossRef]

42. Turner, J.V.; Agatonovic-Kustrin, S. In Silico Prediction of Oral Bioavailability. In *Comprehensive Medicinal Chemistry II*, 1st ed.; Taylor, J.B., Triggle, D.J., Eds.; Elsevier: Amsterdam, The Netherlands, 2007; pp. 699–724.

43. Comunian, T.A.; Abbaspourrad, A.; Favaro, T.C.S.; Weitz, D.A. Fabrication of solid lipid microcapsules containing ascorbic acid using a microfluidic technique. *Food Chem.* **2014**, *152*, 271–275. [CrossRef]

44. Farhang, B.; Kakuda, Y.; Corredig, M. Encapsulation of ascorbic acid in liposomes prepared with milk fat globule membrane-derived phospholipids. *Dairy Sci. Technol.* **2012**, *92*, 353–366. [CrossRef]

45. Chakraborty, A.; Jana, N.R. Vitamin C-Conjugated nanoparticle protects cells from oxidative stress at low doses but induces oxidative stress and cell death at high doses. *ACS Appl. Mater. Interfaces* **2017**, *9*, 41807–41817. [CrossRef]

46. Blesso, C.N. Egg phospholipids and cardiovascular health. *Nutrients* **2015**, *7*, 2731–2747. [CrossRef]

47. Kullenberg, D.; Taylor, L.A.; Schneider, M.; Massing, U. Health effects of dietary phospholipids. *Lipids Health Dis.* **2012**, *11*, 1–16. [CrossRef]

48. Gabizon, A.; Shmeeda, H.; Barenholz, Y. Pharmacokinetics of pegylated liposomal Doxorubicin: Review of animal and human studies. *Clin. Pharmacokinet.* **2003**, *42*, 419–436. [CrossRef]

49. Gopi, S.; Amalraj, A.; Jacob, J.; Kalarikkal, N.; Thomas, S.; Guo, Q. Preparation, characterization and in vitro study of liposomal curcumin powder by cost effective nanofiber weaving technology. *New J. Chem.* **2018**, *42*, 5117–5127. [CrossRef]

50. Norman, A.W.; Henry, H.L. Vitamin D. In *Handbook of Vitamins*, 4th ed.; CRC Press: Boca Raton, FL, USA, 2007.

51. Bender, D.A. *Nutritional Biochemistry of the Vitamins*; Cambridge University Press: New York, NY, USA, 2003.

52. Carlberg, C. Nutrigenomics of vitamin D. *Nutrients* **2019**, *11*, 676–691. [CrossRef]

53. Gil, Á.; Plaza-Diaz, J.; Mesa, M.D. Vitamin D: Classic and novel actions. *Ann. Nutr. Metab.* **2018**, *72*, 87–95. [CrossRef]
54. Livney, Y.D.; Semo, E.; Danino, D.; Kesselman, E. Nanoencapsulation of Hydrophobic Nutraceutical Substances within Casein Micelles. In Proceedings of the XIVth International Workshop on Bioencapsulation, Lausanne, Switzerland, 5–7 October 2006; pp. 1–4.
55. Wilson, L.R.; Tripkovic, L.; Hart, K.H.; Lanham-New, S.A. Vitamin D deficiency as a public health issue: Using vitamin D2 or vitamin D3 in future fortification strategies. *Proc. Nutr. Soc.* **2017**, *76*, 392–399. [CrossRef]
56. Corthésy, B.; Bioley, G. Lipid-based particles: Versatile delivery systems for mucosal vaccination against infection. *Front. Immunol.* **2018**, *9*, 431–451. [CrossRef] [PubMed]
57. Cone, R.A. Barrier properties of mucus. *Adv. Drug Deliv. Rev.* **2009**, *61*, 75–85. [CrossRef] [PubMed]
58. Guo, Q.; Bellissimo, N.; Rousseau, D. The physical state of emulsified edible oil modulates its in vitro digestion. *J. Agric. Food Chem.* **2017**, *65*, 9120–9127. [CrossRef] [PubMed]
59. Gupta, R.; Behera, C.; Paudwal, G.; Rawat, N.; Baldi, A.; Gupta, P.N. Recent advances in formulation strategies for efficient delivery of vitamin D. *AAPS Pharm. Sci. Tech.* **2019**, *20*, 11–23. [CrossRef] [PubMed]
60. Goncalves Maia Campos, P.M.B.; De Camargo Junior, F.B.; De Andrade, J.P.; Gaspar, L.R. Efficacy of cosmetic formulations containing dispersion of liposome with magnesium ascorbyl phosphate, -lipoic acid and kinetin. *Photochem. Photobiol.* **2012**, *88*, 748–752. [CrossRef] [PubMed]
61. Guo, F.; Lin, M.; Gu, Y.; Zhao, X.; Hu, G. Preparation of PEG-modified proanthocyanidin liposome and its application in cosmetics. *Eur. Food Res. Technol.* **2015**, *240*, 1013–1021. [CrossRef]
62. Kirby, B.J.; Hasselbrink, E.F., Jr. Zeta Potential of Microfluidic Substrates: 1. Theory, Experimental Techniques, and Effects on Separations. *Electrophoresis* **2004**, *25*, 187–202. [CrossRef]
63. Maione-Silva, L.; De Castro, E.G.; Nascimento, T.L.; Cintra, E.R.; Moreira, L.C.; Cintra, B.A.S.; Valadares, M.C.; Lima, E.M. Ascorbic acid encapsulated into negatively charged liposomes exhibits increased skin permeation, retention and enhances collagen synthesis by fibroblasts. *Sci. Rep.* **2019**, *9*, 522–536. [CrossRef]
64. Mohammadi, M.; Ghanbarzadeh, B.; Hamishehkar, H. Formulation of Nanoliposomal Vitamin D3 for Potential Application in Beverage Fortification. *Adv. Pharm. Bull* **2014**, *4*, 569–575.
65. Afrooz, H.; Ahmadi, F.; Fallahzadeh, F.; Mousavi-Fard, H.; Alipour, S. Design and characterization of paclitaxel–verapamil co-encapsulated PLGA nanoparticles: Potential system for overcoming P-glycoprotein mediated MDR. *J. Drug Deliv. Sci. Technol.* **2017**, *41*, 174–181. [CrossRef]
66. Vakilinezhad, M.A.; Amini, A.; Akbari-Javar, H.; Baha'addini Beigi Zarandi, B.F.; Montaseri, H.; Dinarvand, R. Nicotinamide loaded functionalized solid lipid nanoparticles improves cognition in Alzheimer's disease animal model by reducing Tau hyperphosphorylation. *Daru* **2018**, *26*, 165–177. [CrossRef]
67. Jeung, J.J. Vitamin C Delivery System and Liposomal Composition Thereof. U.S. Patent 0367480 A1, 22 December 2016.
68. Cipolla, D.; Wu, H.; Gonda, I.; Eastman, S.; Redelmeier, T.; Chan, H.-K. Modifying the Release Properties of Liposomes Toward Personalized Medicine. *Pharm. Nanotechnol.* **2014**, *103*, 1851–1862. [CrossRef]
69. Bochot, A.; Fattal, E. Liposomes for intravitreal drug delivery: A state of the art. *J. Control. Release* **2012**, *161*, 628–634. [CrossRef]
70. Moghimipour, E.; Rezaei, M.; Kouchak, M.; Ramezani, Z.; Amini, M.; Angali, K.A.; Saremy, S.; Dorkoosh, F.A.; Handali, S. A mechanistic study of the effect of transferrin conjugation on cytotoxicity of targeted liposomes. *J. Microencapsul.* **2018**, *35*, 548–558. [CrossRef]
71. Raval, N.; Maheshwari, R.; Kalyane, D.; Youngren-Ortiz, S.R.; Chougule, M.B.; Tekade, R.K. *Importance of Physicochemical Characterization of Nanoparticles in Pharmaceutical Product Development Basic Fundamentals of Drug Delivery*, 1st ed.; Tekade, R.K., Ed.; Academic Press: London, UK, 2019; pp. 369–400.
72. Jacob, J.; Sukumaran, N.P.; Jude, S. Fiber-Reinforced-Phospholipid Vehicle-Based Delivery of L-Ascorbic Acid: Development, Characterization, ADMET Profiling, and Efficacy by a Randomized, Single-Dose, Crossover Oral Bioavailability Study. *ACS Omega* **2021**, *6*, 5560–5568. [CrossRef]
73. Samimi, S.; Maghsoudnia, N.; Eftekhari, R.B.; Dorkoosh, F. Lipid-Based Nanoparticles for Drug Delivery Systems. In *Characterization and Biology of Nanomaterials for Drug Delivery*, 1st ed.; Mohapatra, S., Ranjan, R., Dasgupta, N., Mishra, R.K., Thomas, S., Eds.; Elsevier: Amsterdam, The Netherlands, 2019; pp. 47–76.
74. Joseph, E.; Singhvi, G. Multifunctional Nanocrystals for Cancer Therapy: A Potential Nanocarrier. In *Nanomaterials for Drug Delivery and Therapy*, 1st ed.; Grumezescu, A.M., Ed.; Elsevier: Amsterdam, The Netherlands, 2019; pp. 91–116.
75. Soema, P.C.; Willems, G.J.; Amorij, J.-P.; Kersten, G.F. Predicting the influence of liposomal lipid composition on liposome size, zeta potential and liposome-induced dendritic cell maturation using a design of experiments approach. *Eur. J. Pharm. Biopharm.* **2015**, *94*, 427–435. [CrossRef]
76. Prantl, L.; Eigenberger, A.; Gehmert, S.; Haerteis, S.; Aung, T.; Rachel, R.; Jung, E.M.; Felthaus, O. Enhanced resorption of liposomal packed vitamin C monitored by ultrasound. *J. Clin. Med.* **2020**, *9*, 1616–1628. [CrossRef]
77. Kulkarni, S.B.; Betageri, G.V.; Singh, M. Factors affecting microencapsulation of drugs in liposomes. *J. Microencapsul.* **1995**, *12*, 229–246. [CrossRef]
78. Łukawski, M.; Dałek, P.; Borowik, T.; Foryś, A.; Langner, M.; Witkiewicz, W.; Przybyło, M. New oral liposomal vitamin C formulation: Properties and bioavailability. *J. Liposome Res.* **2019**, *30*, 227–234. [CrossRef]

79. Zhao, L.; Temelli, F.; Chen, L. Encapsulation of anthocyanin in liposomes using supercritical carbon dioxide: Effects of anthocyanin and sterol concentrations. *J. Funct. Foods* **2017**, *34*, 159–167. [CrossRef]

80. Mozafari, M.R.; Johnson, C.; Hatziantoniou, S.; Demetzos, C. Nanoliposomes and Their Applications in Food Nanotechnology. *J. Liposome Res.* **2008**, *18*, 309–327. [CrossRef] [PubMed]

81. Poudel, A.; Gachumi, G.; Wasan, K.M.; Dallal Bashi, Z.; El-Aneed, A.; Badea, I. Development and Characterization of Liposomal Formulations Containing Phytosterols Extracted from Canola Oil Deodorizer Distillate along with Tocopherols as Food Additives. *Pharmaceutics* **2019**, *11*, 185–201. [CrossRef] [PubMed]

82. Marsanasco, M.; Piotrkowski, B.; Calabró, V.; Alonso, S.; Chiaramoni, N. Bioactive constituents in liposomes incorporated in orange juice as new functional food: Thermal stability, rheological and organoleptic properties. *J. Food Sci. Technol.* **2015**, *52*, 7828–7838. [CrossRef] [PubMed]

83. Isailović, B.D.; Kostić, I.T.; Zvonar, A.; Đorđević, V.B.; Gašperlin, M.; Nedović, V.A.; Bugarski, B.M. Resveratrol loaded liposomes produced by different techniques. *Innov. Food Sci. Emerg. Technol.* **2013**, *19*, 181–189. [CrossRef]

84. Cui, H.; Zhao, C.; Lin, L. The specific antibacterial activity of liposome-encapsulated Clove oil and its application in tofu. *Food Control* **2015**, *56*, 128–134. [CrossRef]

85. Dalmoro, A.; Bochicchio, S.; Lamberti, G.; Bertoncin, P.; Janssens, B.; Barba, A.A. Micronutrients encapsulation in enhanced nanoliposomal carriers by a novel preparative technology. *RSC Adv.* **2019**, *9*, 19800–19812. [CrossRef]

86. Kiani, A.; Fathi, M.; Ghasemi, S.M. Production of novel vitamin D3 loaded lipid nanocapsules for milk fortification. *Int. J. Food Prop.* **2017**, *20*, 2466–2476. [CrossRef]

87. Zhou, H.; Yue, Y.; Liu, G.; Li, Y.; Zhang, J.; Yan, Z.; Duan, M. Characterisation and Skin Distribution of Lecithin-Based Coenzyme Q10-Loaded Lipid Nanocapsules. *Nanoscale Res. Lett.* **2010**, *5*, 1561–1569. [CrossRef]

88. Dałek, P.; Drabik, D.; Wołczańska, H.; Foryś, A.; Jagas, M.; Jędruchniewicz, N.; Przybyło, M.; Witkiewicz, W.; Langner, M. Bioavailability by design-Vitamin D3 liposomal delivery vehicles. *Nanomed. Nanotechnol. Boil. Med.* **2022**, *43*, 102552–102563. [CrossRef]

89. Bonnekoh, B.; Roding, J.; Krueger, G.R.; Ghyczy, M.; Mahrle, G. Increase of lipid fluidity and suppression of proliferation resulting from liposome uptake by human keratinocytes in vitro. *Br. J. Dermatol.* **1991**, *124*, 333–340. [CrossRef]

90. White, P.J.; Fogarty, R.D.; McKean, S.C.; Venables, D.J.; Werther, G.A.; Wraight, C.J. Oligonucleotide uptake in cultured keratinocytes: Influence of confluence, cationic liposomes, and keratinocyte cell type. *J. Infect. Dis.* **1999**, *112*, 699–705. [CrossRef]

bioengineering

Article

Pomegranate Pomace Extract with Antioxidant, Anticancer, Antimicrobial, and Antiviral Activity Enhances the Quality of Strawberry-Yogurt Smoothie

Nouf H. Alsubhi [1,*], Diana A. Al-Quwaie [1], Ghadeer I. Alrefaei [2], Mona Alharbi [3], Najat Binothman [4], Majidah Aljadani [4], Safa H. Qahl [2], Fatima A. Jaber [2], Mashael Huwaikem [5], Huda M. Sheikh [2], Jehan Alrahimi [6,7], Ahmed N. Abd Elhafez [8] and Ahmed Saad [9,*]

[1] Biological Sciences Department, College of Science & Arts, King Abdulaziz University, Rabigh 21911, Saudi Arabia
[2] Department of Biology, College of Science, University of Jeddah, Jeddah 21589, Saudi Arabia
[3] Department of Biochemistry, College of Science, King Saud University, Riyadh 11451, Saudi Arabia
[4] Department of Chemistry, College of Sciences & Arts, King Abdulaziz University, Rabigh 21911, Saudi Arabia
[5] Cinical Nutrition Department, College of Applied Medical Sciences, King Faisal University, Al Ahsa 31982, Saudi Arabia
[6] Department of Biological Sciences, Faculty of Sciences, King Abdulaziz University, Jeddah 21589, Saudi Arabia
[7] Immunology Unit, King Fahad Medical Research Centre, King Abdulaziz University, Jeddah 21589, Saudi Arabia
[8] Department of Internal Medicine, Faculty of Medicine, Zagazig University, Zagazig 44519, Egypt
[9] Biochemistry Department, Faculty of Agriculture, Zagazig University, Zagazig 44511, Egypt
* Correspondence: nhalsubhi@kau.edu.sa (N.H.A.); ahmedm4187@gmail.com (A.S.)

Citation: Alsubhi, N.H.; Al-Quwaie, D.A.; Alrefaei, G.I.; Alharbi, M.; Binothman, N.; Aljadani, M.; Qahl, S.H.; Jaber, F.A.; Huwaikem, M.; Sheikh, H.M.; et al. Pomegranate Pomace Extract with Antioxidant, Anticancer, Antimicrobial, and Antiviral Activity Enhances the Quality of Strawberry-Yogurt Smoothie. *Bioengineering* **2022**, *9*, 735. https://doi.org/10.3390/bioengineering9120735

Academic Editors: Minaxi Sharma, Kandi Sridhar, Zeba Usmani and Xiaohu Xia

Received: 16 October 2022
Accepted: 19 November 2022
Published: 28 November 2022

Publisher's Note: MDPI stays neutral with regard to jurisdictional claims in published maps and institutional affiliations.

Abstract: Valorizing the wastes of the food industry sector as additives in foods and beverages enhances human health and preserves the environment. In this study, pomegranate pomace (PP) was obtained from the company Schweppes and exposed to the production of polyphenols and fiber-enriched fractions, which were subsequently included in a strawberry-yogurt smoothie (SYS). The PP is rich in carbohydrates and fibers and has high water-absorption capacity (WAC) and oil-absorption capacity (OAC) values. The LC/MS phenolic profile of the PP extract indicated that punicalagin (199 g/L) was the main compound, followed by granatin B (60 g/L) and pedunculagin A (52 g/L). Because of the high phenolic content of PP extract, it ($p \leq 0.05$) has high antioxidant activity with SC50 of 200 µg/mL, besides scavenging 95% of DPPH radicals compared to ascorbic acid (92%); consequently, it reduced lung cancer cell lines' viability to 86%, and increased caspase-3 activity. Additionally, it inhibited the growth of pathogenic bacteria and fungi i.e., *L. monocytogenes*, *P. aeruginosa*, *K. pneumonia*, *A. niger*, and *C. glabrata*, in the 45–160 µg/mL concentration range while killing the tested isolates with 80–290 µg/mL concentrations. These isolates were selected based on the microbial count of spoiled smoothie samples and were identified at the gene level by 16S rRNA gene sequence analysis. The interaction between Spike and ACE2 was inhibited by 75.6%. The PP extract at four levels (0.4, 0.8, 1.2, and 1.4 mg/mL) was added to strawberry-yogurt smoothie formulations. During 2 months storage at 4 °C, the pH values, vitamin C, and total sugars of all SYS decreased. However, the decreases were gradually mitigated in PP-SYS because of the high phenolic content in the PP extract compared to the control. The PP-SYS3 and PP-SYS4 scored higher in flavor, color, and texture than in other samples. In contrast, acidity, fat, and total soluble solids (TSS) increased at the end of the storage period. High fat and TSS content are observed in PP-SYS because of the high fiber content in PP. The PP extract (1.2 and 1.6 mg/mL) decreases the color differences and reduces harmful microbes in PP-SYS compared to the control. Using pomegranate pomace as a source of polyphenols and fiber in functional foods enhances SYS's physiochemical and sensory qualities.

Keywords: pomegranate; pomace; extraction; biological activity; functional beverages; sensory quality

1. Introduction

Agriculture and the food industry produce a huge amount of waste that poses a threat to health and the environment, although it is rich in bioactive components. Because of the global food shortage, we must reuse the waste in various applications, such as foods and feedstuffs [1–3]. Unfortunately, separating bioactive compounds from the consumable parts of the agricultural product could have an excessive cost for the starting material. Accordingly, using agro-industrial waste (such as fruit peels) for separating such compounds to control the spread of foodborne and human pathogens is more economical [4,5]. Pomegranate produces about 1,500,000 tons of waste worldwide per annum [6] and is expected to reach 12,500,000 tons by 2022; about 60% of the fruit weight corresponds to peels, which are often discarded as agro-waste.

Pomegranate (*Punica granatum* L.) has a well-known medicinal history. It is rich in bioactive molecules, phenolic compounds, and flavonoids and has many medicinal properties [7]. The level of phenolic compounds changes according to cultivars and fruit parts [8]. Gözlekçi et al. [9] demonstrated that peel extract contains higher total phenolics, flavonoids, and anthocyanins content than pulp and juice extracts and possesses stronger biological activities. The phytochemical constituents of the ethanolic extract of pomegranate juice were determined by Sorrenti et al. [10], who identified many phytochemical compounds; phenolic compounds were the main constituents. On the other hand, Khaleel et al. [11] demonstrated the presence of 292 different phytocompounds with different retention times and chemical structures. Eleven elements were considered most significant as antimicrobials.

In this regard, Priyanka et al. [12] studied the antimicrobial activity of methanol, chloroform, and methanol chloroform (1:1) extracts of peel and juice of pomegranates in different bacterial and fungal strains. They found that peel methanolic extract showed the highest antimicrobial activity, while chloroform extract was less effective against the tested microorganisms. Other studies have also reported that pomegranate methanol peel extract showed maximum growth inhibition (85.71%) compared to all tested organic solvents against some pathogenic and drug-resistant bacterial strains [13]. Li et al. [14] determined the MICs of punicalagin, as the main active component in pomegranate peel, for ten *Salmonella* strains and found the minimum inhibitory concentrations (MICs) in the range of 0.20–1.0 mg/mL. Moreover, Zam and Khaddour [15] detected a reduction in the adhesion index of up to 80% and a reduction in *E. coli* motility in the urinary tract using aqueous pomegranate peel extract at MIC. Abou El-Nour [16] recorded a large decrease in biofilm formation of *Pseudomonas aeruginosa* and inhibition in swarming motility when treated with aqueous and alcoholic pomegranate peel extracts.

Pomegranate extracts also exhibit antioxidant activity. It has been reported by many investigators that there is a relationship between chemopreventive agents and antioxidant activity; during the oxidation of biological materials, superoxide, H_2O_2, and hydroxyl radicals (OH·) are produced. Reducing the destructive effect on human cells and tissues stops unorganized radical reactions. Mekni et al. [17] illustrated that methanol peel extracts of five pomegranate varieties have high antioxidant power (above 80%) against DPPH radicals. Ohshima et al. [18] demonstrated that oxidative stress generates toxic metabolites which can initiate and stimulate cancers. Breast cancer is the most important life disease a woman may face during her lifetime. Rocha et al. [19] found that the growth of two breast cancer cell lines was completely inhibited by a 1% concentration of pomegranate juice, which also can decrease cell migration to bones. Li et al. [14] recorded a reduction in cell viability in different human cancers by methanol PPE, and MCF-7 breast cells were the most responsive to the antitumor effect (growth inhibition of 81.0% at 5 µg/mL) compared to A549 lung cell cancer (80.0% at dose 250 µg/mL). A recent study by Mahmoudi et al. [20] suggested a decrease in the viability of MCF-7 and MDA-MB-468 breast cancer by slowing down their progression when treated with pomegranate seed oil.

Because of all the above-mentioned significant activities, pomegranate peels can be added to commercial tomato and orange juices that contain strawberries. The extract of dried pomegranate peel substantially boosted the antioxidant content of both flavored juices.

Due to the astringent flavor of pomegranate peel, the maximum extract concentration in the juices had the lowest sensory acceptability. According to the results, adding 0.5% dried pomegranate peel extract to strawberries was approved in tomato and orange juice [21]. This study investigated the antioxidant, total phenolic, tannin, and phytic acid levels of four fruit peels (pomegranates, grapes, apples, and mocha palms). The maximum antioxidant activity and phenolic content were found in pomegranate peel. It was somewhat poor in flavonoids and vitamin C. The sensory assessment of guava juice supplemented with pomegranate peel did not vary from the control. Incorporating pomegranate peel into guava juice is therefore highly well received, increases the antioxidant components, protects health, and enhances quality of life [22]. Altunkaya et al. [23] formulated apple juice supplemented with pomegranate peel extract (PPE) at concentrations ranging from 0.5% to 2%. The most pomegranate peel was found in the juice, as well as the most antioxidant activity.

In comparison, only 0.5 g PPE per 100 mL was determined to be acceptable for color, smell, and taste. The conclusion was that PPE-enriched apple juice is superior to regular apple juice. Therefore, adding PPE at concentrations between 0.5% and 1.0% proved acceptable and generated a healthier product without altering apple juice's sensory qualities or toxicological safety.

No available studies cast light on the pomegranate pomace and its application in the food industry, in which it may be more valuable than peels. Therefore, this study aimed to complete the evaluation of pomegranate pomace concerning the phenolic content by LC/MS, and antioxidant, anticancer, antibacterial, antifungal, and antiviral properties. Additionally, it encourages using natural and safe alternatives to control pathogenic microorganisms and component oxidation. Therefore, highly functional pomegranate pomace extract was added to the strawberry-yogurt smoothie to enhance its color, flavor, and texture, besides extending its shelf life and quality.

2. Materials and Methods

2.1. Preparation of Pomegranate Pomace Aqueous Extract (PP)

Pomegranate pomace was obtained from the juice company. The PP was homogenized and washed with tap water, followed by distilled water. The pomace was dried in an oven at 50 °C for 2 days, then powdered and kept dry until extraction. Ten grams of pomegranate pomace powder was stirred in 150 mL of sterilized distilled water for 12 h, then filtered by Whatman No. 1 paper. The filtered extracts were centrifuged at 4000 rpm/15 min and kept at 4 °C.

2.2. Proximate Analysis of Pomegranate Pomace Powder

Moisture, protein, fat, and ash levels were determined according to AOAC guidelines [24]. The carbohydrates were determined by differential analysis. As per Saad et al. [25], the water-absorption capacity (WAC) and oil-absorption capacity (OAC) were evaluated as follows: PP powder (1 g) was stirred in 10 mL of water or oil for 30 min, then centrifuged at 7500 rpm/30 min in weighted test tubes. The residue was weighed, and WAC and OAC were estimated as mL of water or oil retained per gram of sediment.

2.3. Phenolic Profile in PP Extract

2.3.1. Total Phenols and Total Flavonoids Content

Total phenolics were determined in PP extracts by the Folin–Ciocalteu method described by Saad et al. [26], and the data were calculated as mg of gallic acid equivalents (GAE)/g. The total flavonoid content was estimated using the $AlCl_3$ colorimetric method following Namir et al. [27]. The data were estimated as mg quercetin equivalents (QE)/g.

2.3.2. High-Resolution LC/MS Profile of Phenolic Content of PP Extract

Accela U-HPLC system (Thermo Fisher Scientific, USA) was used to identify the phenolic compounds in PP. LC/MS data were obtained by an Executive Orbitrap mass spectrometer linked with the HPLC system. The compounds were separated following

Colantuono et al. [28]. Separation column (Gemini C18-110, Phenomenex, Torrance, CA, USA), 150 × 2.0 mm, 5 μm and mobile phases were (A) aqueous Formic acid 0.1% and (B) Formic acid in acetonitrile 0.1%. The column was heated to 30 °C, and the flow rate was adjusted at 0.2 mL/min. The PP extract was dissolved in a mixture of methanol and water (50:50). Then, 10 μL of the prepared sample was injected into the column. Negative ionization methods collected MS data in the mass range of 100–1400 m/z. The capillary temperature was 275 °C, the auxiliary gas had a pressure of 15 arbitrary units, and the sheath gas had a pressure of 30 arbitrary units. Linear calibration curves of PC (1–50 μg/mL) and EA and GA (0.1–5 μg/mL) were established. With a mass tolerance of ±5 ppm, the metabolites were identified using accurate mass values to the fifth decimal place.

2.4. Pomegranate Pomace Extract Activity

2.4.1. DPPH-Radical-Scavenging Activity

The antioxidant activity was determined using the DPPH technique, as described by Abdel-Moniem et al. [29]. A total of 0.5 mL of ethanolic DPPH was added to 1 mL PP extract and incubated for 30 min in the dark; the produced color was read spectrophotometrically at a wavelength of 517 nm. The absorbance was applied as in Equation (1). The lowest concentration that scavenges 50% of DPPH radical was calculated as IC_{50} [30].

$$\% \ DPPH \ scavenging \ activity = \frac{Abs \ control - Abs \ sample}{Abs \ control} \times 100 \tag{1}$$

2.4.2. Anticancer

The lung carcinoma cell line (A-549) was assayed for cytotoxicity; the doxorubicin reference standard was used as a positive control. According to Mosmann [31] and Gomha et al. [32], the cytotoxicity evaluation was carried out using a viability assay. The apoptosis-inducing potential of PP extract was assessed by measuring its caspase-3 activity. After treating the 25 cm^2 flasks with DMSO or PP at the right doses and times, the A-549 cells were collected. For each experiment, 2×10^6 cell lysates were used and analyzed as specified. After 45 min of incubation with substrate Z-DEVD-R110 in a microplate reader, fluorescence was seen, and the excitation and emission were measured at 485/530 nm [33].

2.4.3. Antimicrobial

Twelve pathogenic microorganisms were employed for the antimicrobial screening test. *S. aureus*, *L. monocytogenes*, *B. cereus*, *P. aeruginosa*, *K. pneumonia*, *E. coli*, *A. niger*, *A. flavus*, *C. glabrata*, *C. albicans*, *C. davenportii*, and *P. expansum*. These strains were selected based on the microbial count of spoiled yogurt samples. It was found during microbial examination with a light microscope and biochemical and morphological definitions that these isolates are the most isolates that cause smoothie spoilage. These isolates were confirmed by identification at the gene level through isolating DNA and using PCR to detect genes. The bacterial isolates were identified based on 16S rRNA and the fungal isolates on 18S rRNA gene sequence analysis. Sequencing was performed via the automated DNA sequencer (ABI Prism 3130 Genetic Analyzer by Applied Biosystems Hitachi, Japan). Genomic DNA was obtained by the hexadecyltrimethylammonium bromide (CTAB) technique, and the integrity and level of purified DNA were established by agarose gel electrophoresis. The DNA level was customized to 20 ng/μL for PCR amplification. The forward primer used with the isolates is (5 AGA GTT TGA TCC TGG CTC AG 3), and the reverse is (5 GGT TAC CTT GTT ACG ACT T 3). PCR products were isolated by electrophoresis on 1.5% agarose gels stained with ethidium bromide and documented in the Alphaimager TM1200 documentation and analysis system. The obtained polymorphic differences were analyzed via the program NTSYS-PC2 by assessing the distance of isolates by Jaccard's Similarity Coefficient. All the bacterial strains were maintained by sub-culturing on nutrient agar or potato dextrose slant and stored at 4 °C. The PP extract was evaluated for its antibacterial and antifungal activity by the well-disc-diffusion method, according to Dahham et al. [34] and El-Saadony et al. [35]. Next, 50 mL of molten Mueller–Hinton agar (MHA) or potato dextrose agar (PDA) was poured into plates. A loopful of bacterial or fungal inoculum was spread over the surface of the plates. Then, 8 mm wells were punched into each plate and filled with 50 µL of PP extract concentrations (200, 400, 600, 800, and 100 µg/mL), and negative control wells were filled with water. MHA and PDA plates were incubated at 37 °C for 24–48 h (bacteria) or 28 °C for 5 days (fungi) [36]. The resultant inhibition zone diameters (mm) indicated the antimicrobial activity [37]. The MIC, MBC, and MFC values were determined by Saad et al. [25].

2.4.4. Antiviral

ACE2 was created as a recombinant protein in human cells. SARS-CoV-2 Inhibitor Screening Assay kit (Adipogen, San Diego, CA, USA) was used to inhibit the binding between the Spike and ACE2. The SARS-CoV-2 antiviral activity of the PP extract was evaluated as follows: 100 µL of Spike RBD was added to each well of a 96-well plate and kept for 16 h at 4 °C. Then, 100 µL of PP concentrations (200, 400, 600, 800, and 1000 µg/mL) was added to wells containing Spike and incubated for one hour at 37 °C in the presence of Inhibitor Mix Solution ((0.2% BSA, 0.05% Tween R 20 in PBS, and biotin-conjugated ACE2 (0.5 g/mL)). After incubation, HRP-labeled streptavidin (1:200 dilution) was added to each well and incubated for one hour at room temperature. To end the reaction, 100 µL of tetramethyl benzidine was added and incubated for 5 min. The absorbance was read at 450 nm by the microplate reader (BioTek Elx808, Winooski, VT, USA).

2.5. PP-Strawberry-Yogurt Smoothie Processing and Preservation

2.5.1. Preparation of Strawberry Smoothie

Fresh strawberry fruits were washed, cleaned, and processed for juice production using a Moulinex mixer following Saad et al. [26] with some modifications. In a blender, 50 mL of PP extract (25 g/100 mL) at different concentrations and yogurt (40 g) were mixed well with filtered strawberry juice (100 mL). The water and sugar were added as shown in Table 1. The smoothies, SYSC, PP-SYS1, PP-SYS2, PP-SYS3, and PP-SYS4, were transferred into 250 mL bottles and pasteurized in a TOMY Sx-700 autoclave (Tokyo, Japan) at 95 °C for 2 min under 50 MPa. The bottles were cooled and stored at 4 °C for two months.

Table 1. The ingredients of pomegranate pomace-enriched strawberry-yogurt smoothie.

Ingredients	SYSC	PP-SYS1	PP-SYS2	PP-SYS3	PP-SYS4
Strawberry Juice (mL)	100	100	100	100	100
PP-yogurt (g)	40	40	40	40	40
Sugar (g)	10	10	10	10	10
PP (mL)	-	50	50	50	50
Water (mL)	50	-	-	-	-

Strawberry-yogurt smoothie (SYS), control (C), smoothie supplemented with pomegranate pomace extract 0.4 mg/mL (PP-SYS1), smoothie supplemented with pomegranate pomace extract 0.8 mg/mL (PP-SYS2), smoothie supplemented with pomegranate pomace extract 1.2 mg/mL (PP-SYS3), smoothie supplemented with pomegranate pomace extract 1.6 mg/mL (PP-SYS4).

2.5.2. Physiochemical Properties

The pH values of the smoothies, SYSC, PP-SYS1, PP-SYS2, PP-SYS3, and PP-SYS4, were calculated using a pH meter (pH 211, HANNA, Nusfalau, Romania). Standard method 942.15 was used to estimate the titratable acidity, which was recorded as a percentage of citric acid. A refractometer was used to calculate the TSS. AOAC [24] was also used to assess vitamin C concentration. The Gerber technique was used to determine fat content; an increase in fat percentage from 0.4 to 4% while keeping protein constant improved texture, stability, and perceived viscosity. The total sugars were estimated by El-Saadony et al. [38].

2.5.3. Color Differences and Sensory Properties

Hunter spectrophotometer (AA Color Flex EZ) was used to determine the color parameters (L^*, a^*, and b^*) of PP-SYS (HunterLab, Murnau, Germany). The following parameters were considered: Equation (2) calculated the color difference (ΔE) between smoothie samples during the storage period [39].

$$\Delta E = \sqrt{\left(\Delta L\right)^2 + \left(\Delta a\right)^2 + \left(\Delta b\right)^2} \tag{2}$$

The sensory characteristics (color, flavor, texture, and overall acceptability) of PP-SYS were estimated using a 9-point hedonic scale. PP-SYS samples were randomly coded and served in one-use cups to 30 semi-trained panelists, 20 males and 10 females (aged 21–30), in a fluorescent-lit laboratory with an air-conditioning temperature of 23 °C. Water was used to change the mouth contact areas after each judgment to avoid interfering with the results.

2.5.4. Microbial Load

Ten PP-SYS samples were stirred in 90 mL of peptone water for 30 min to obtain a 10^{-1} suspension. Serial dilutions were generated up to 10^{-8}. The dilutions were put into disposable Petri plates with a specific medium [40]. The total viable count (TBC) was determined after 24 h incubation at 30 °C using plate count agar. The results for microorganisms were converted to logarithms (CFU/mL).

2.6. Statistical Analysis

A one-way ANOVA test was used to distinguish the significant differences between sample means at a probability level ($p \leq 0.05$). The LSD test determined significant differences between means. All statistical calculations were performed with SPSS 20 for Windows.

3. Results and Discussion
3.1. Chemical Composition

The findings revealed that pomegranate pomace is a rich source of carbohydrates and fiber with very low-fat content. Table 2 shows the proximate analysis of pomegranate pomace, which comprised 0.61% fat, 4.12% ash, 6.98% moisture, 9.11% protein, and 20.66% fiber, with total carbohydrates accounting for 58.52% of the sample. There are no studies on pomegranate pomace, but the findings are comparable with those of Ranjitha et al. [41], except for pomegranate peels that contain 0.85% fat, 4.32% ash, 7.27% moisture, 3.74% protein, 17.31% fiber, and 66.51% carbohydrates. Moreover, El-Hadary and Taha [42] found that the chemical composition of pomegranate peel was crude carbohydrates (78%), fiber (12%), protein (3.5%), ash (3.4%), and lipids (2.25%).

Table 2. Physiochemical parameters of pomegranate pomace powder.

PP Composition	Content (g/100 g)
Chemical	
Moisture	6.98 ± 0.2
Protein	9.11 ± 0.1
Fat	0.61 ± 0.01
Ash	4.12 ± 0.3
Fiber	20.66 ± 0.8
Soluble carbohydrates	58.52 ± 0.6
Physical	
WAC (mL/g)	8.21 ± 0.1
OAC (mL/g)	7.77 ± 0.1
L^*	61.22 ± 0.5
a^*	25.52 ± 0.8
b^*	15.66 ± 0.3

Data are presented as mean $\pm$ SD, lightness (L^*), redness (a^*), yellowness (b^*), water-absorption capacity (WAC), and oil-absorption capacity (OAC).

The high content of carbohydrates and fibers in PP influences the WAC and OAC values; in this study, the WAC of pp was 8.21 mL/g, compared to lower WACs in cantaloupe waste, dates, pear, and tomato pomace with values of 6.17, 5.70, 4.90, and 4.12 mL/g, respectively [43]. The OH in polysaccharide chains can build hydrogen bonds with water, enhancing fiber-rich materials' capacity to retain moisture [44]. Regarding the OAC of PP, it was 7.77 mL/g, which is higher than lemon byproducts (6.60 mL/g) and tiger nut pomace (6.90 mL/g) [45]. The OAC is crucial since lipids serve as taste preservers and enhance the texture of food. The OAC is also a technological property related to the chemical structure of the plant polysaccharides. It depends on surface properties, overall charge density, thickness, and hydrophobic nature of the fiber particle [46]. Regarding the color parameters of PP, L^* and b^* values are high, which is a good indicator for enhancing the color when applied to food formulation.

3.2. Phenolic Compounds in Pomegranate Pomace Aqueous Extract (PP)

Phenolics are chemicals generated from the secondary metabolism of plants that are frequently used for their biological effects, notably their antioxidant properties. Phenolic compounds are antioxidants with biological and chemical actions, the most significant of which is free radical scavenging [47]. Figure 1 shows the significant ($p \leq 0.05$) phenolic content (total phenolic content and total flavonoids) in PP. The concentration of TPC and TF significantly ($p \leq 0.05$) increased in a concentration-dependent manner when PP (1 mg/mL) contained 205 mg of gallic acid equivalent/mL and 110 mg of quercetin equivalent/mL. There are no studies on the antioxidant content of PP extract; previous studies have focused on pomegranate peels and waste extracts with a high concentration of phenolics and flavonoids; hence, pomegranate peels can be considered a major source of antioxidants. The results may agree with Yassin et al. [48], who found that acetone extract of pomegranate peels contained 246 and 227.7 mg GAE/g. In comparison, the methanolic extract of pomegranate peels has a maximum total phenolic content of 18.89 g/100 g and a total flavonoid content of 13.95 mg QE/kg [42]. There is a preference for our PP in phenolic content, with a relative increase of 11%.

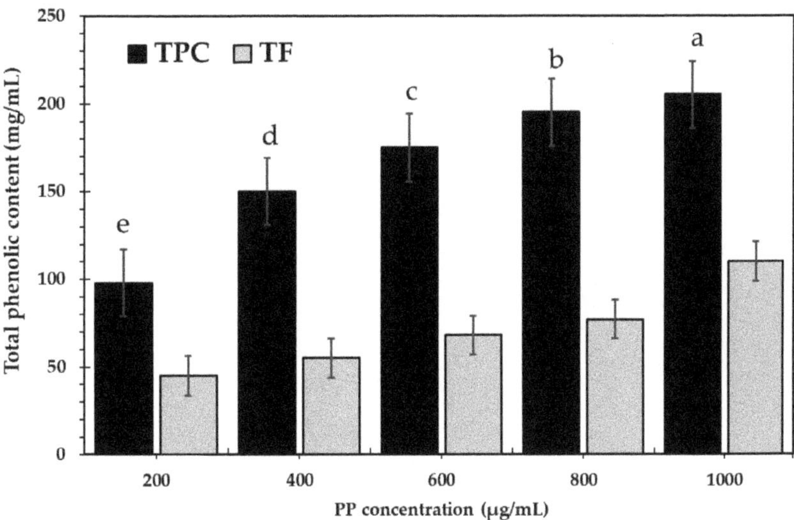

Figure 1. Total phenolic and flavonoid content in pomegranate pomace aqueous extract (PPE). TPC, total phenolic content; TF, total flavonoids; PP, pomegranate pomace. Data are presented as mean $\pm$ SE. Lowercase letters in the same columns indicate significant differences $p \leq 0.05$.

To determine the exact and active phenolic compounds in pomegranate pomace, high-resolution LC/MS was processed (Table 3). The detected compounds belong to three groups: ellagitannins (ETs), ellagic acid derivatives (EA), and gallic acid (GA); their values are 479.35, 24.29, and 1.20 g/L, respectively. The main compound in LC/MS profile is punicalagin (199 g/L), followed by granatin B (60 g/L) and pedunculagin A (52 g/L). In the LC/MS profile, causarinin, punicalin, galloyl-HHDP-hexoside, granatin B, pedunculagin A, and pedunculagin B were expressed as punicalagin (PC) equivalents. EA equivalents were EA hexoside, pentoside, and deoxyhexoside. Gallic acid was calculated with the corresponding standard. Total polyphenols accounted for 500 g/L; PC derivatives represent 95% of total phenolics, while EA derivatives and GA represent only 5%. Similar results were observed by Zivkovic et al. [49]; they found that pomegranate wastewater extract contains punicalin (197.13 mg/g), punicalagin (54.23 mg/g), and ellagic acid (25.42 mg/g) found to be dominant in peel extract. On the other hand, Coronado-Reyes et al. [50] found that pomegranate peel is a more important source of punicalagin and ellagic acid than aryls, where the punicalagin content of the crust ranged from 114.6 to 282 mg/g dry weight. The ellagic acid content varied from 1.07 to 2.49 mg/g dry weight. Additionally, a previous study on pomegranate aril showed that ellagic acid was detected in different extracts, and it was the highest detected phenolic compound with 34.5 (μg/mg) when a combination of ethanol: ether: water (8:1:1) was used [51]. These compounds can potentially be used in foods and medical applications [52].

Table 3. Phenolic compounds profile in pomegranate pomace extract (PP) achieved by LC/MS.

RT	Compound	Equivalent	MW	PP (g/L)
6.3	Gallic acid	-	169.014	1.20 ± 0.2 [j]
6.5	Punicalin	-	781.053	65.35 ± 0.9 [b]
7.8	Punicalagin	-	1083.059	199.85 ± 1.2 [a]
8.2	Pedunculagin A bis-Hexahydroxy diphenic acid (HHDP)	Punicalagin	783.069	52.98 ± 0.8 [c]
8.2	Causarinin	Punicalagin	935.080	22.96 ± 0.8 [f]
8.5	Galloyl-HHDP-hexoside	Punicalagin	633.073	48.15 ± 0.7 [d]
8.7	Ellagic acid hexoside	Ellagic acid	463.052	4.50 ± 0.6 [h]
8.8	Pedunculagin B (digalloyl- HHDP)	Punicalagin	785.084	31.66 ± 0.9 [e]
8.8	Ellagic acid dihexoside	Ellagic acid	625.105	1.21 ± 0.1 [ij]
9.0	Lagerstannin B	Punicalagin	949.059	1.93 ± 0.2 [i]
9.3	Granatin B	Punicalagin	951.075	62.42 ± 0.8 [b]
9.5	Ellagic acid pentoside	Ellagic acid	433.041	3.12 ± 0.1 [h]
9.6	Ellagic acid deoxyhexoside	Ellagic acid	447.057	3.89 ± 0.1 [h]
10.5	Ellagic acid	-	300.999	12.55 ± 0.3 [g]
	Total punicalagin derivatives			479.35 A
	Total ellagic acid derivatives			24.29 B
	Gallic acid			1.20 C
	Total phenolic compounds			504.84

RT, Retention time (min); molecular weight (MW), pomegranate pomace extract (PP). Lowercase letters in the same columns indicate significant differences between detected polyphenols $p \leq 0.05$. Uppercase letters indicate the differences ($p \leq 0.05$) between punicalagin derivatives, ellagic acid derivatives, and gallic acid.

3.3. Activity of Pomegranate Pomace Aqueous Extract

3.3.1. Antioxidant

The antioxidant potential of the pomegranate pomace aqueous extract was evaluated according to a 1,1-diphenyl-2-picrylhydrazyl (DPPH) assay; the SC_{50} (antioxidants needed to reduce 50% of the initial DPPH concentration) was identified. Effective free radical-scavenging activity was recorded in a concentration-dependent manner, with considerable SC_{50} = 200 μg/mL (Figure 2). Increase in the extract concentration resulted in increased inhibition percentage (decrease in the concentration of DPPH) and reached almost complete inhibition (95%) at 1 mg/mL of PP compared to ascorbic acid. Lower DPPH radical-scavenging activity values were found with 93% in PP ethanolic extract [42], 83.6% in PP ethyl acetate extract [53], and 78.23% in pomegranate waste methanolic extracts [54]. Bopitiya and Madhujith [55] concluded that significant amounts of phenolic and flavonoid compounds (such as ellagic and gallic acids) were attributed to the antioxidant properties of pomegranate peel extracts.

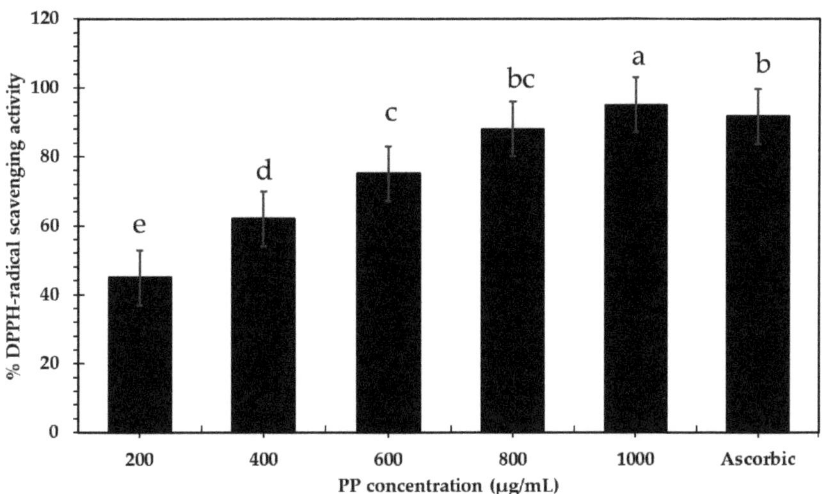

Figure 2. DPPH-radical-scavenging activity of pomegranate pomace aqueous extract (PP). Data are presented as mean ± SE. Lowercase letters in the same columns indicate significant differences $p \leq 0.05$.

Considerable antioxidant activity is essential to inhibit lipid oxidation in the food system. Free radicals are unstable and can react with biomolecules of living cells, causing various diseases such as cardiovascular diseases and cancer [56]; the antioxidants help the body to protect itself against this oxidative damage. The considerable amounts of hydrolyzable tannins in pomegranate peel are responsible for their antioxidant power [57]. Mekni et al. [17] found a significant correlation between the best antioxidant activity of various pomegranate extracts and the extract's highest phenolic and flavonoid contents; furthermore, this antioxidant activity can be related to the presence of different functional groups, such as hydroxyl and carbonyl groups.

3.3.2. Anticancer

In vivo anti-proliferative activity was assessed against lung carcinoma (A-549) using different concentrations of PP extract. The results indicated that the extract displayed potent anticancer activity and cytotoxic effects against A-549 lung carcinoma. The doxorubicin drug was also tested as a positive control for comparison. As illustrated in Figure 3A,B, increasing inhibition of cell viability with increasing extract concentrations was observed (dose-dependent manner). Maximum growth inhibition of 86% was detected for lung carcinoma cells at a dose of 1000 µg/mL. The 50% inhibition of cell viability (IC50) for A-549 was 250 µg/mL, representing a valuable result compared to the IC_{50} value of doxorubicin of 300 µg/mL. Compared to carcinoma cells, a less cytotoxic effect was recorded against normal lung cells. The anticancer activity of the pomegranate peel ethanolic extract was carried out against the oral cancer cell line (KB 3-1), and this peel extract exhibited promising anticancer properties. The MTT assay showed 94.53% inhibition on the oral cancer cell line, and the clonogenic assay showed a decrease in the colonies after treatment with the peel extract [58]. Moreover, against various kinds of cancer, for example, in prostate cancer and renal cell carcinoma tissues grown in vitro, pomegranate extracts inhibit the activity of NF-kB [59]. Punicalagin, generated from pomegranate extracts, has anti-proliferative activity by inducing apoptosis of cancer cells such as lung carcinoma and breast and cervical cancer cell lines [60]. Induction of the apoptosis process by punicalagin through caspase activation and poly ADP ribose polymerase (PARP) inhibition. Caspase-3, -8, and -9 protein expressions exhibited a dose-dependent increase after 50 and 75 µM punicalagin treatment. Furthermore, A549 cells treated with 50 and 75 µM concentrations of punicalagin showed cleavage of PARP protein [61].

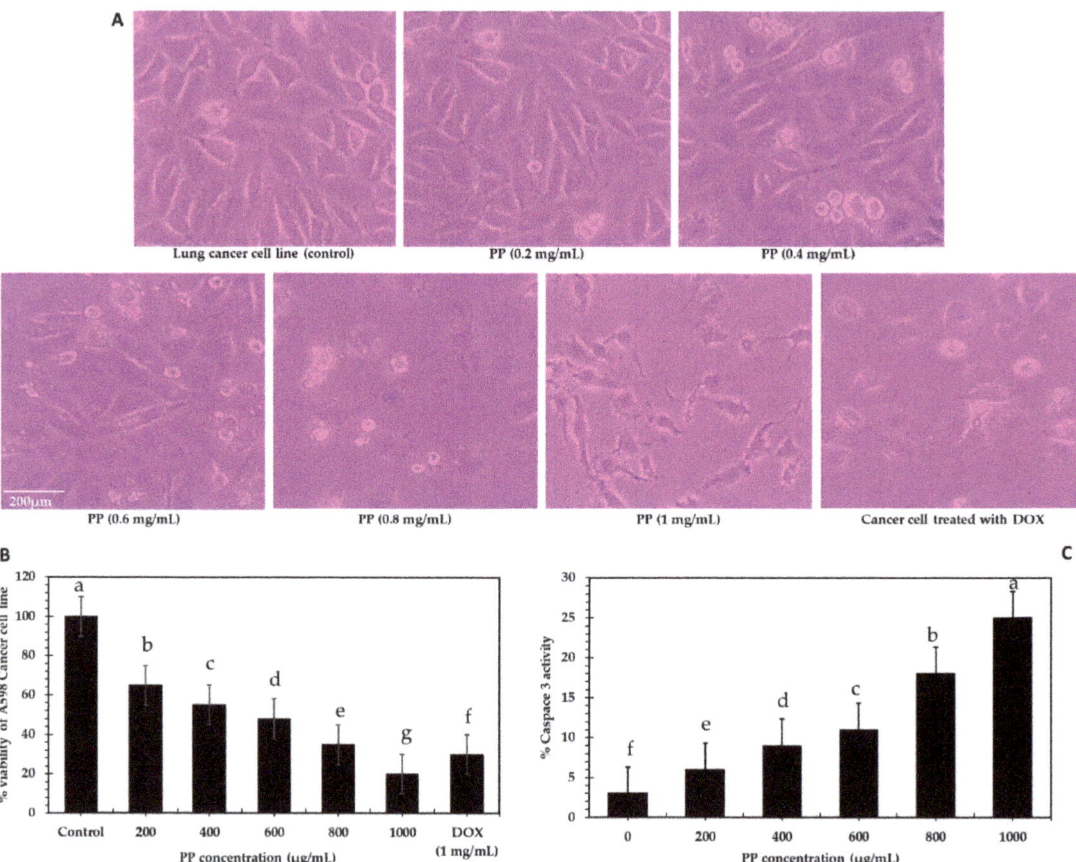

Figure 3. (**A**) Microscopic image of the effect of PP at different levels (0.2–1 mg/mL) on the lung cancer line viability compared to the negative and positive control, (**B**) histogram of PPE effect on cancer cell lines viability, (**C**) % caspase 3 activity in response to cancer cell death. Data are presented as mean ± SE. Lowercase letters in the same columns indicate significant differences $p \leq 0.05$.

Figure 3C shows that the activity of caspase-3 increased with cancer cell death, showing high activity when lung cancer cell lines were treated with PP extract (1 mg/mL). Caspase-3 is essential for normal brain development and other apoptotic conditions in a manner that is tissue-, cell-, or death-stimulus-specific. It is also necessary for apoptotic chromatin condensation and DNA fragmentation in every examined cell type. Caspase-3 is therefore required for cellular disintegration and the formation of apoptotic bodies. However, it may also function before or during cell death [62]. Pomegranate fruit and its extracts have been found to combat cancer in many in vitro and in vivo studies; modulating various signaling pathways is the mechanism of action against cancer growth. The introduction of targeted medicines has revolutionized the treatment of patients with A-549 carcinoma. When combined with natural extracts with stratagems, repurposing non-cancer drugs into new therapeutic niches presents a cost-effective and efficient technique with enhanced outcomes for discovering novel pharmacological activity to potentially increase cure rates for lung and liver cancers [63,64].

3.3.3. Antimicrobial

Table 4 shows the antibacterial and antifungal activities of PP extract against some foodborne and multidrug-resistant pathogens. The extract has a wide spectrum of antimicrobial activity; the inhibition zone diameters (IZDs) significantly increased $p \leq 0.05$ in a concentration-dependent manner and were 9–35 mm against the tested bacteria compared to penicillin. The most sensitive bacteria to PP extract (1 mg/mL) were *S. aureus* with IZD, 35 mm. However, *K. pneumonia* were the most resistant bacteria to PP extract (1 mg/mL), where IZD was 27 mm, with a relative decrease of about 30% of the *S. aureus* inhibition zone.

Table 4. Antimicrobial activity of PP extract expressed as IZDs (mm) against pathogenic microorganisms.

Pathogenic Bacteria	PP Concentration (µg/mL)						
	Inhibition Zone Diameters (mm)						
	100	200	400	600	800	1000	St * (1 mg/mL)
S. aureus	12 ± 0.1 [a]	18 ± 0.3 [a]	23 ± 0.2 [a]	27 ± 0.5 [a]	31 ± 0.4 [a]	35 ± 0.2 [a]	37 ± 0.1 [a]
L. monocytogenes	10 ± 0.2 [b]	14 ± 0.2 [cd]	19 ± 0.1 [c]	22 ± 0.3 [c]	26 ± 0.2 [c]	29 ± 0.1 [bc]	31 ± 0.5 [c]
B. cereus	11 ± 0.5 [ab]	16 ± 0.5 [b]	20 ± 0.2 [b]	25 ± 0.4 [b]	28 ± 0.5 [b]	30 ± 0.3 [b]	33 ± 0.3 [b]
P. aeruginosa	−	14 ± 0.1 [cd]	17 ± 0.6 [d]	22 ± 0.5 [c]	24 ± 0.3 [d]	27 ± 0.7 [cd]	29 ± 0.6 [d]
K. pneumonia	−	12 ± 0.3 [d]	16 ± 0.9 [d]	21 ± 0.2 [cd]	23 ± 0.3 [e]	25 ± 0.5 [d]	27 ± 0.6 [e]
E. coli	9 ± 0.2 [c]	15 ± 0.5 [c]	18 ± 0.6 [cd]	24 ± 0.1 [bc]	25 ± 0.5 [cd]	28 ± 0.1 [c]	30 ± 0.2 [cd]
Pathogenic fungi	Inhibition Zone Diameters (mm)						
A. niger	−	18 ± 0.6 [c]	21 ± 0.5 [bc]	25 ± 0.2 [bc]	26 ± 0.3 [c]	28 ± 0.3 [c]	30 ± 0.3 [bc]
A. flavus	16 ± 0.2 [ab]	21 ± 0.3 [ab]	25 ± 0.1 [a]	28 ± 0.3 [ab]	30 ± 0.1 [a]	31 ± 0.4 [ab]	34 ± 0.1 [ab]
P. expansum	17 ± 0.6 [a]	22 ± 0.2 [a]	25 ± 0.3 [a]	29 ± 0.1 [a]	30 ± 0.4 [a]	32 ± 0.3 [a]	35 ± 0.3 [a]
C. glabrata	−	17 ± 0.5 [c]	19 ± 0.2 [c]	21 ± 0.9 [c]	23 ± 0.6 [d]	25 ± 0.6 [d]	26 ± 0.2 [c]
C. albicans	13 ± 0.1 [c]	20 ± 0.7 [b]	22 ± 0.5 [b]	25 ± 0.5 [bc]	27 ± 0.4 [bc]	29 ± 0.9 [bc]	31 ± 0.0 [b]
C. davenportii	15 ± 0.3 [b]	21 ± 0.1 [ab]	24 ± 0.8 [ab]	27 ± 0.2 [b]	28 ± 0.3 [b]	30 ± 0.1 [b]	31 ± 0.1 [b]

Data are presented as mean ± SD, lowercase letters (a, b, c, d, …) in the same columns indicate significant differences (effect of each pp concentration on pathogenic bacteria and fungi) $p \leq 0.05$. * St, penicillin (1000 µg/mL) for bacteria and clotrimazole (1000 µg/mL) for fungi.

The resistant Gram-negative bacteria have lower IDZs than the Gram-positive ones because of the unique structure of the Gram-negative bacteria's membrane.

On the other hand, as shown in Table 4, PP extract exhibited considerable antifungal activity against *Candida*, *Aspergillus*, and *Penicillium* species in this study, with IZDs ranging from 13 to 35 mm compared to clotrimazole (1000 µg/mL). *A. niger* and *C. glabrata* were the most resistant fungi to PP extract, while *A. flavus* and *P. expansum* were the most vulnerable. Figure S1 shows the IZDs image of PP extract 0.6, 0.8, and 1.0 mg/mL against tested bacteria (*S. aureus*, *L. monocytogenes*, *B. cereus*, *P. aeruginosa*, *K. pneumonia*, *E. coli*), Candida (*C. glabrata*, *C. albicans*, *C. davenportii*), and fungi (*A. niger*, *A. flavus*, *P. expansum*). Kesur et al. [12] demonstrated that the high efficiency of the methanolic extract of pomegranate peel against the *Bacillus cereus*, *Bacillus megaterium*, *Bacillus subtilis*, *Escherichia coli*, and *Aspergillus niger* showed inhibition zones between 6 and 17 mm. At the same time, no activity was detected by chloroform extract against the same microorganisms. On the other hand, pomegranate waste extract has powerful antifungal activity against *Candida gelberta* and *Candida apis* [65].

The PP extract successfully inhibited the bacterial and fungal isolates in the 45–160 µg/mL concentration range while killing the tested isolates with 80–290 µg/mL concentrations

(Table 5). Kesur et al. [12] found that the minimum inhibitory concentration of the methanolic extracts of peels of pomegranate on *B. cereus*, *B. megaterium*, *P. vulgaris*, and *P. aeruginosa* was observed in the range of 50,0000 to 12,500 (µg/mL). In contrast, the MIC of peel extracts on *B. subtilis*, *S. typhi*, *S. typhi A*, and *S. typhi B* was found to be in the range of 50,000 to 25,000 (µg/mL), while MIC against fungi is higher than that. An efficient antimicrobial effect expressed as MIC of the crude extract was observed against the studied strain, for which we recorded 312.5 µg/mL. It was reported by Oliveira et al. [66] that strong inhibitors have MIC values below 500 µg/mL, moderate inhibitors have values between 600 and 1500 µg/mL, and weak inhibitors are above 1600 µg/mL. Hasan et al. [67] confirmed that pomegranate flavonoids and phenolic compounds are extracted more efficiently with methanol due to their polarity. Naziri et al. [68] showed that active antimicrobial substances have different solubility in various solvents and that using different solvents for extraction changes the extent of the antibacterial effect of the extract.

Table 5. The minimum concentration of PP extract that inhibits or kills bacteria or fungi.

Pathogenic Bacteria	MIC *	MBC *
S. aureus	50 [f]	90 [f]
L. monocytogenes	70 [d]	130 [d]
B. cereus	65 [e]	120 [e]
P. aeruginosa	110 [b]	190 [b]
K. pneumonia	150 [a]	250 [a]
E. coli	90 [c]	150 [c]
Pathogenic fungi	**MIC**	**MFC ***
A. niger	120 [b]	200 [b]
A. flavus	55 [d]	90 [e]
P. expansum	45 [e]	80 [f]
C. glabrata	160 [a]	290 [a]
C. albicans	75 [c]	130 [c]
C. davenportii	70 [cd]	120 [d]

Lowercase letters (a, b, c, d, …) in the same columns indicate significant differences (effect of each pp concentration on pathogenic bacteria and fungi) $p \leq 0.05$. * MIC, minimum inhibitory concentration; * MBC, minimum bactericidal concentration; * MFC, minimum fungicidal concentration of PP extract against tested pathogens.

Pomegranates are a good source of potentially bioactive phytochemicals such as phenolics, flavonoids, anthocyanins, alkaloids, and tannins [8]. The results of this study clearly showed a good relationship between the IDZs of the studied bacteria and pomegranate varieties. Rosas-Burgos et al. [69] consistently observed different inhibition levels for the studied pomegranate peel extracts, where the *Salmonella* strain was the most sensitive. They concluded that the highest punicalagin and ellagic acid concentrations were detected in sour and sweet pomegranate cultivars with the highest inhibitory activity. Vasconcelos et al. [70] numerous microbial enzymes in culture filtrates or pure forms are inhibited by tannic, and the toxicity of tannins may be connected to their effect on microbes' membranes. High antibacterial activity for the studied methanolic extracts was observed, while chloroform extracts were less effective, with highly significant differences. Suresh et al. [71] observed a gradual decline in the protein content of bacterial cells incubated with different concentrations of pomegranate peel extract (PPE) compared to the control.

Based on the polynomial model in (Figure 4), employing a higher concentration of PP, i.e., 2 mg/mL, will increase the IDZs by 58.6, 96.4, 20.3, and 42.5% against *Listeria monocytogenes*, *E. coli*, *C. albicans*, and *A. flavus*, respectively, compared to PP (1 mg/mL). We noticed that *C. albicans* were more sensitive to PP extract (2 mg/mL), followed by *E. coli*.

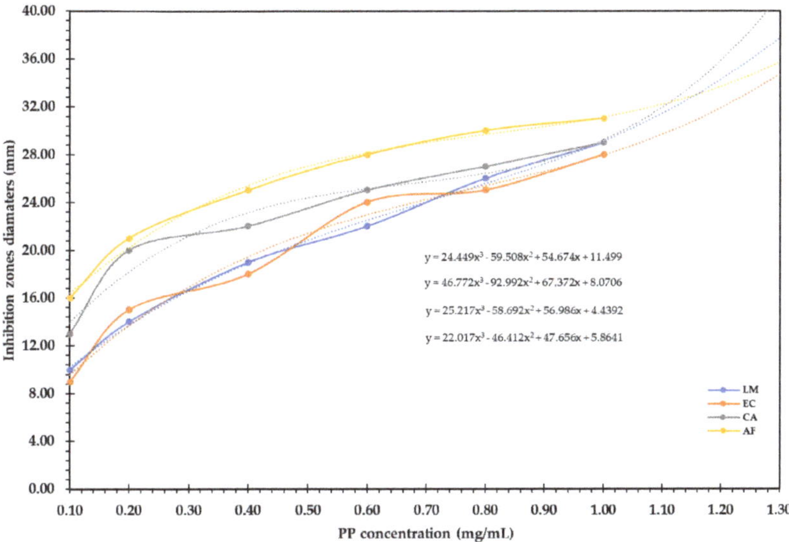

Figure 4. Polynomial model for forecasting the antimicrobial activity with increasing PP concentrations.

3.3.4. Antiviral

Figure 5 shows the inhibition activity of PP extract on the binding between Spike and ACE2 compared to a SARS-CoV-2 inhibitor test kit. At doses of 200, 400, 600, 800, and 1000 μg/mL, PP extract inhibited the interaction between Spike and ACE2 by as much as 75.6%. This impact was dose-dependent compared to the positive control, AC384, a monoclonal antibody that prevented the binding between Spike and ACE2 by identifying ACE2 itself and inhibiting 70% of the interaction. Surucic et al. [72] found that punicalagin in pomegranate peel extract inhibited 50% of S-glycoprotein and ACE2 contact only in the highest concentration sample (1 mg/mL).

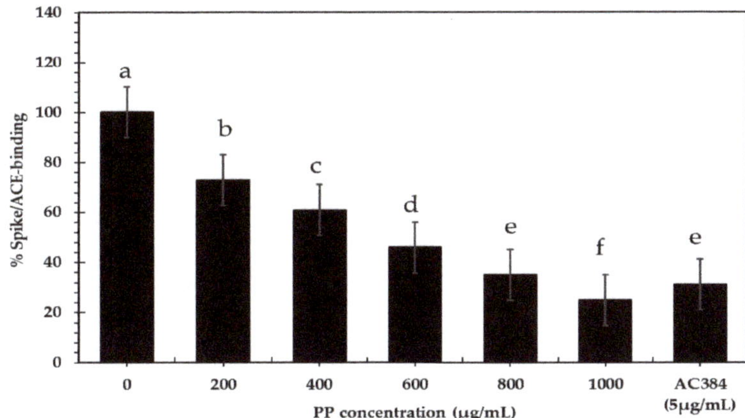

Figure 5. PP concentrations (200–1000 μg/mL) affect Spike/ACE2 binding compared with control and antibody inhibitor AC384 (5 μg/mL). ACE2, angiotensin-converting enzyme 2; PP, pomegranate pomace extract. Data are presented as mean ± SE. Lowercase letters in the same columns indicate significant differences $p \leq 0.05$.

There are no studies on the antiviral activity of PP extract; however, most earlier investigations focused on pomegranate peel extracts, which showed antibacterial and antiviral activity and suppressed influenza and herpes viruses, which can also be employed effectively as an antiviral agent against SARS-CoV-2. In vitro, the aqueous–alcoholic pomegranate peel extract stopped SARS-CoV-2 S-glycoprotein from binding to ACE2, showing that the extract can stop SARS-CoV-2 from entering the host cells [73].

3.4. Changes in Physiochemical Properties of PP-Strawberry-Yogurt Smoothie

3.4.1. Physiochemical Properties

The functional milk beverage fortified with pomegranate pomace extract during storage influences the pH and beverage acidity. During two months of storage at 4 °C, the pH values of PP-smoothie samples decreased by 8–13% compared to the control smoothie; therefore, the pp-smoothie acidity increased by 16–21% compared to the control. Thus, in agreement with Al-Hindi and Abd El Ghani [74], pH levels of the functional milk beverages with and without POPE decreased significantly from 4.5 to 4.1 with increasing storage time. The higher counts of viable bacteria at the end of the storage period resulted in lower pH levels in the functional milk beverages fortified with or without POPE. In addition, the beverage's acidity increased markedly from 0.2 to 0.35 with increasing storage time.

Massive decreases in vitamin C content in the control sample (133%) after 2 months of storage, but these reductions were considerably inhibited ($p \leq 0.05$) in PP-smoothies by 11–55% over control. The high phenolic content in PP extract inhibited the oxidation of vitamin C, and total sugars were decreased with storage period by up to 53% in the control sample, but these reductions decreased in PP-SYS1, PP-SYS2, PP-SYS3, and PP-SYS4 by 4–28% compared to the control. The high phenolic content in PP extract inhibited the fermentative action of microbes against sugars. The reductions in sugars were followed by an increase in TSS (28–38%) in PP smoothies compared to the control. In addition, the fat content was elevated based on pomegranate pomace concentrations (Table 6). During three weeks of cold storage, a substantial increment in fat content was detected between untreated and treated yogurt ($p < 0.05$) [3]. Saad et al. [75] observed the same trend in cucumber beverages supplemented with polyphenolic extracts, i.e., decrements in vitamin C and total sugars while increasing TSS percentage. They noticed that the decrements in vitamin C and total sugars were mitigated by polyphenolic extract addition.

Table 6. Changes in physiochemical parameters of smoothie supplemented with pomegranate pomace extract at different addition levels (0.4, 0.8, 1.2, 1.6 mg/mL) during two-month storage at 4 °C.

Smoothie Samples	Storage Period (Months)	pH	Acidity (mg/10 mL)	Vit. C (mg/mL)	Total Sugars (mg/mL)	Fat (%)	TSS (%)
SYSC	0	4.98 ± 0.4 [a]	55.32 ± 0.1 [d]	6.3 ± 0.1 [a]	1.20 ± 0.01 [a]	1.10 ± 0.2 [g]	9.30 ± 0.2 [e]
	1	4.75 ± 0.2 [b]	58.78 ± 0.2 [d]	4.0 ± 0.2 [cd]	0.91 ± 0.02 [cd]	1.30 ± 0.4 [f]	10.10 ± 0.1 [de]
	2	4.69 ± 0.3 [c]	61.22 ± 0.3 [cd]	2.7 ± 0.2 [e]	0.78 ± 0.01 [e]	1.50 ± 0.7 [e]	10.90 ± 0.3 [d]
PP-SYS1	1	4.68 ± 0.2 [c]	62.39 ± 0.7 [c]	4.2 ± 0.0 [c]	0.92 ± 0.09 [cd]	1.53 ± 0.5 [e]	10.90 ± 0.95 [d]
	2	4.60 ± 0.0 [c]	64.72 ± 0.4 [bc]	3.0 ± 0.1 [de]	0.81 ± 0.03 [d]	1.60 ± 0.2 [de]	11.60 ± 0.1 [c]
PP-SYS2	1	4.55 ± 0.1 [d]	62.59 ± 0.9 [c]	4.4 ± 0.3 [c]	0.93 ± 0.05 [c]	1.54 ± 0.1 [e]	11.02 ± 0.7 [cd]
	2	4.43 ± 0.3 [ef]	65.00 ± 0.1 [b]	3.5 ± 0.5 [d]	0.89 ± 0.06 [d]	1.72 ± 0.3 [c]	12.03 ± 0.0 [b]
PP-SYS3	1	4.51 ± 0.4 [d]	62.99 ± 0.3 [c]	4.8 ± 0.8 [c]	0.95 ± 0.04 [c]	1.62 ± 0.4 [de]	11.72 ± 0.0 [c]
	2	4.44 ± 0.6 [ef]	66.05 ± 0.7 [b]	3.8 ± 0.6 [d]	0.92 ± 0.01 [cd]	1.79 ± 0.7 [b]	12.55 ± 0.2 [ab]
PP-SYS4	1	4.49 ± 0.1 [e]	63.20 ± 0.5 [c]	5.2 ± 0.3 [b]	1.09 ± 0.00 [b]	1.69 ± 0.3 [d]	11.88 ± 0.1 [c]
	2	4.40 ± 0.5 [f]	67.02 ± 0.2 [a]	4.2 ± 0.1 [c]	1.00 ± 0.03 [bc]	1.82 ± 0.2 [a]	12.89 ± 0.2 [a]

Data are presented as mean ± SD, lowercase letters (a, b, c, d, …) in the same columns indicate significant differences (between the changes in physiochemical parameters of SYS samples during 2-month storage period) at $p \leq 0.05$.

Strawberry-yogurt smoothie (SYSC) supplemented with pomegranate pomace extracts 0.4 mg/mL (PP-SYS1), smoothie supplemented with pomegranate pomace extract 0.8 mg/mL (PP-SYS2), smoothie supplemented with pomegranate pomace extract 1.2 mg/mL (PP-SYS3), smoothie supplemented with pomegranate pomace extract 1.6 mg/mL (PP-SYS4).

3.4.2. Total Phenolic Content

Polyphenols have a crucial role in the color and flavor of foodstuffs while being very volatile and easily oxidized [76]. Adding PP extract with different levels (0.4, 0.8, 1.2, and 1.6 mg/mL) to the tested smoothie enhanced the antioxidant activity from 50% in the control smoothie to 88% in PP-SYS4 (data not shown). The antioxidant activity of smoothies was enhanced due to the presence of phenolics and flavonoids that were boosted by the PP extract. The total phenolic and flavonoid content in PP-SYS4 increased by 170% compared to the control at zero months of storage. However, after 2 months of storage, the activity further increased, estimated at 98%, despite decreasing the phenolic and flavonoid content by 14% (Figure 6). No studies have focused on using PP extract as an additive in foods. However, our findings may be correlated with Ahmed et al. [77], who added pomegranate peels to yogurt and observed their effects on total phenolic contents and antioxidant activity of the product; they found an average TPC in yogurt of (4.07 ± 0.37) mg GAE/g. However, the maximum content was detected in pomegranate yogurt (4.64 mg GAE/g compared to the control, which was 3.39 mg GAE/g). Antioxidant activity examination reveals that an average (70.58%) was calculated. Meanwhile, extreme activity was diagnosed in the pomegranate yogurt sample (83.87%), having 9% freeze-dried PPP.

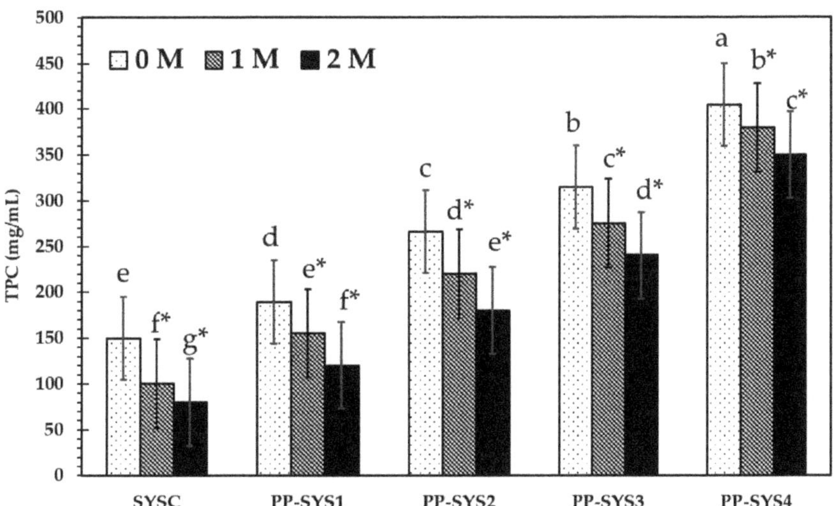

Figure 6. Fluctuations in total phenolic content (TPC) of smoothie supplemented with pomegranate pomace extract at different addition levels (0.4, 0.8, 1.2, 1.6 mg/mL) during two-month storage at 4 °C. Lowercase letters above columns indicate significant differences between SYS samples. Lowercase letters with a star indicate a significant change during the storage period for each SYS.

Due to the abundance of phenolic components in several plant extracts, developing natural alternatives to synthetic preservatives in food is of special interest because these compounds possess antioxidant and antimicrobial properties that inhibit the process of lipid peroxidation in fatty foods and scavenging free radicals and prevent the growth of microorganisms [75,78,79].

3.4.3. Color Difference and Sensorial Properties

Analysis of the nutritional values and the quality of ingredients of the beverages are very important to ensure the safety and quality of the beverages. Sensory investigation using color, taste, odor, and textural analysis efficiently assesses the final beverage quality [80]. Based on the color analysis of PP in Table 2, the extract has a high redness value that enhances the whiteness and red color of the strawberry-PP yogurt smoothie. Nonetheless, at the end of storage, the color parameters faded (data not shown), but the addition of PP extract mitigates the color reduction, and that is clear in Figure 7, where color change (ΔE) value was observed as limiting the color change in PP-supplemented smoothies; the ΔE value in the control was 6.14; however, in PP-SYS4, it decreased by 22%. Trigo et al. [81] found that with the addition of pomegranate peel extract to carrot beverage, the *whiteness* was not affected by PPE, but heat treatment led to higher L^* levels compared to the HPP treatment. The increase in L^* values was noticed in the first week of storage, but the L^* remained constant within the next few days. Concerning ΔE, this difference was evident throughout the entire storage period. For example, on day 7, the ΔE^* changed between 1.2 and 2.0 for HPP-treated juice and between 22.1 and 32.1 for TP-treated juice.

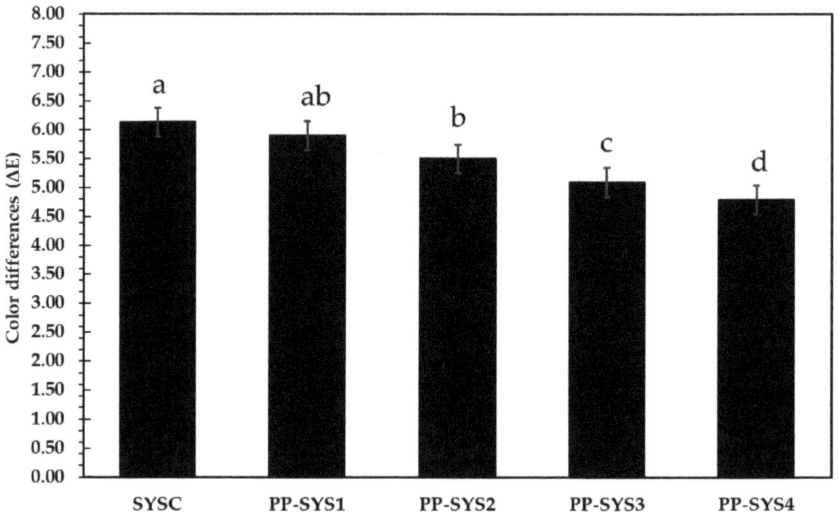

Figure 7. Changes in the color of the smoothie supplemented with pomegranate pomace extract at different addition levels (0.4, 0.8, 1.2, 1.6 mg/mL) during two months of storage at 4 °C. Different lowercase letters above columns indicate significant differences.

Table 7 shows the sensory properties of tested smoothies. The results indicated that PP-SYS4 gained the highest scores by panelists, i.e., flavor and color (9), texture (8.8), and over acceptability (8.9) after a month of storage at 4 °C, but a slight decrease was noticed and reflected on for acceptability values that declined by 3% compared to the control, which decreased by 9%. The results for the other smoothies are in between PP-SYS4 and the control. Ahmed et al. [77] noticed low sensorial scores in relation to the yogurt with pomegranate peels, which were reduced by 9%, but decreasing the addition level increases the sensorial scores. Sensory evaluation of food products is important in determining what consumers might like. The main way to judge a juice's quality is by color [82].

Table 7. Fluctuations in sensory traits of smoothie supplemented with pomegranate pomace extract at different addition levels (0.4, 0.8, 1.2, 1.6 mg/mL) during two months of storage at 4 °C.

Smoothie Samples	Storage Period (Months)	Flavor	Color	Texture	Over Acceptability
	0	8.4 ± 0.1 [bc]	8.3 ± 0.3 [d]	8.4 ± 0.5 [b]	8.3 ± 0.2 [c]
SYSC	1	8.1 ± 0.3 [c]	8.0 ± 0.2 [e]	8.0 ± 0.5 [c]	8.0 ± 0.1 [cd]
	2	7.5 ± 0.2 [d]	7.7 ± 0.1 [f]	7.5 ± 0.3 [d]	7.6 ± 0.3 [d]
PP-SYS1	1	8.6 ± 0.0 [b]	8.5 ± 0.2 [cd]	8.4 ± 0.4 [b]	8.5 ± 0.4 [bc]
	2	8.2 ± 0.3 [c]	8.3 ± 0.2 [d]	7.9 ± 0.6 [d]	8.1 ± 0.3 [cd]
PP-SYS2	1	8.8 ± 0.1 [b]	8.6 ± 0.0 [c]	8.4 ± 0.7 [b]	8.6 ± 0.2 [b]
	2	8.4 ± 0.2 [bc]	8.4 ± 0.0 [cd]	8.0 ± 0.3 [c]	8.2 ± 0.4 [c]
PP-SYS3	1	8.9 ± 0.3 [ab]	8.8 ± 0.2 [b]	8.5 ± 0.4 [b]	8.7 ± 0.3 [b]
	2	8.5 ± 0.5 [bc]	8.5 ± 0.1 [cd]	8.0 ± 0.6 [c]	8.3 ± 0.2 [c]
PP-SYS4	1	9.0 ± 0.0 [a]	9.0 ± 0.0 [a]	8.8 ± 0.1 [a]	8.9 ± 0.3 [a]
	2	8.6 ± 0.3 [b]	8.8 ± 0.3 [b]	8.5 ± 0.2 [b]	8.6 ± 0.4 [b]

Data are presented as mean $\pm$ SD. Lowercase letters in the same columns indicate significant differences $p \leq 0.05$. Smoothie supplemented with pomegranate pomace extract 0.4 mg/mL (PP-SYS1), smoothie supplemented with pomegranate pomace extract 0.8 mg/mL (PP-SYS2), smoothie supplemented with pomegranate pomace extract 1.2 mg/mL (PP-SYS3), smoothie supplemented with pomegranate pomace extract 1.6 mg/mL (PP-SYS4).

3.5. Microbial Load Changes during Storage of Smoothie

The microbial load in all smoothie samples was significantly increased with the storage period; however, it decreased with the gradual addition of levels of PP extract. The microbial load in control samples was 3.5 CFU/mL and increased by 56% after two months. However, in PP-SYS4, the microbial load significantly decreased ($p \leq 0.05$) by 37% compared to the control after 2 months of cold storage (Figure 8). Ebrahimain et al. [83] studied microbiological stability and sensory properties of traditional Iranian butter with pomegranate peel extract incorporated into it, and they found that PPE could reduce the bacterial count in butter from 9.25 log CFU to 6.30 log CFU after 90 days of storage, while with increasing storage time, the total number of bacteria was increased in the treatments; this increase is most evident when dairy products are made with poor safety quality.

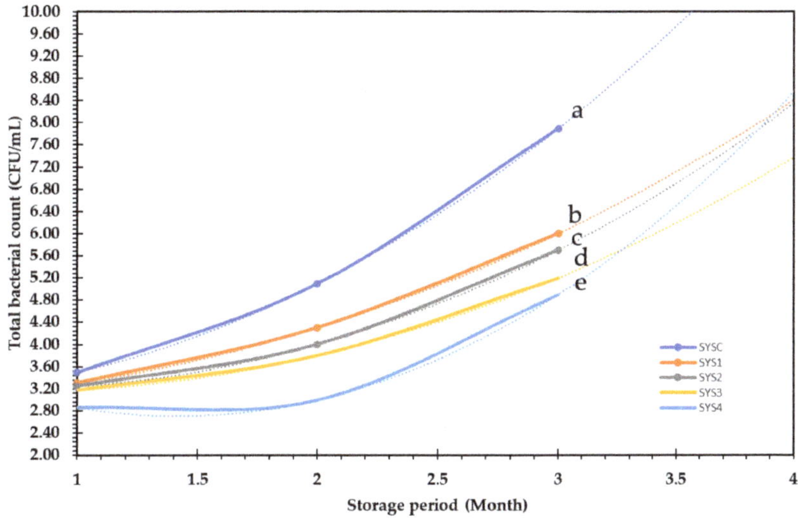

Figure 8. Total bacterial count (TBC) in smoothie supplemented with pomegranate pomace extract at different addition levels (0.4, 0.8, 1.2, 1.6 mg/mL) during two months of storage at 4 °C.

Many studies have indicated that phenolic compounds in PP extract contribute to the direct inhibition of bacterial pathogens and higher antioxidant activity [84]. The ultracellular damage of tested bacteria treated with sour pomegranate peel extract was examined using TEM, illustrating severe damage in the bacterial cells and confirming the extract's potential antibacterial and lethal activity. Karaman et al. [85] also showed that lipophilic parts in phenolic acids permeate through the cell membrane by passive diffusion, enhancing membrane permeability by reducing intracellular pH. The triggering of protein denaturation, explained by Li et al. [14], confirmed that damage to the bacterial cell membrane is one of the possible modes of action for the phytochemical elements, followed by releasing intracellular material and cell death.

4. Conclusions

There is a current need to renew attention in looking for natural resources of bioactive active compounds to be used as safe food additives. Isolation of such compounds from agro-industrial wastes will provide a cost-effective solution for controlling the spread of human pathogens and improve waste reduction. Given that, pomegranate pomace was found to exert a marked reduction in multidrug resistance and food pathogens and possesses great potential to scavenge DPPH free radicals with high antioxidant capacity, besides its potent antitumor activity in lung cell lines and potential cure of COVID-19. Adding PP extract (1.2 or 1.6 mg/mL) to SYS samples enhances color, texture, and quality and introduces a functional beverage that may benefit many patients. Considering all these details, we believe that pomace of *Punica granatum* may provide a safe and efficient alternative-perspective useful additive in functional food. We hope to continue further explorations of this attractive botanical species.

Supplementary Materials: The following supporting information can be downloaded at: https://www.mdpi.com/article/10.3390/bioengineering9120735/s1, Figure S1: Inhibition zones diameters of PP extracts (0.6, 0.8, and 1.0 mg/mL) against tested bacteria (*S. aureus, L. monocytogenes, B. cereus, P. aeruginosa, K. pneumonia, E. coli*); Candida (*C. glabrata, C. albicans, C. davenportii*), and fungi (*A. niger, A. flavus, P. expansum*).

Author Contributions: N.H.A., D.A.A.-Q. and A.S. designed the experiments. G.I.A., M.A. (Mona Alharbi), N.B., M.A. (Majidah Aljadani), and A.S.: Writing—original draft and editing. S.H.Q., F.A.J., J.A., A.N.A.E. and A.S.: Methodology. N.H.A., D.A.A.-Q. and A.S.: Conceptualization. G.I.A., M.A. (Mona Alharbi), N.B., M.A. (Majidah Aljadani), and A.S.: resources, formal analysis. S.H.Q., H.M.S., M.H., F.A.J., J.A., A.N.A.E. and A.S.: Methodology and Formal Analysis. N.H.A., D.A.A.-Q. and A.S.: Conceptualization and Revision. All authors have read and agreed to the published version of the manuscript the article.

Funding: This research received no external funding.

Institutional Review Board Statement: Not applicable.

Informed Consent Statement: Not applicable.

Data Availability Statement: Not applicable.

Conflicts of Interest: The authors declare no conflict of interest.

Sample Availability: Samples of the compounds are not available from the authors.

References

1. Abou-Kassem, D.E.; Mahrose, K.M.; El-Samahy, R.A.; Shafi, M.E.; El-Saadony, M.T.; Abd El-Hack, M.E.; Emam, M.; El-Sharnouby, M.; Taha, A.E.; Ashour, E.A. Influences of dietary herbal blend and feed restriction on growth, carcass characteristics and gut microbiota of growing rabbits. *Ital. J. Anim. Sci.* **2021**, *20*, 896–910. [CrossRef]
2. El-Saadony, M.T.; Abd El-Hack, M.E.; Swelum, A.A.; Al-Sultan, S.I.; El-Ghareeb, W.R.; Hussein, E.O.; Ba-Awadh, H.A.; Akl, B.A.; Nader, M.M. Enhancing quality and safety of raw buffalo meat using the bioactive peptides of pea and red kidney bean under refrigeration conditions. *Ital. J. Anim. Sci.* **2021**, *20*, 762–776. [CrossRef]
3. El-Saadony, M.T.; Sitohy, M.Z.; Ramadan, M.F.; Saad, A.M. Green nanotechnology for preserving and enriching yogurt with biologically available iron (II). *Innov. Food Sci. Emerg. Technol.* **2021**, *69*, 102645. [CrossRef]

4. Lansky, E.P.; Newman, R.A. *Punica granatum* (pomegranate) and its potential for prevention and treatment of inflammation and cancer. *J. Ethnopharmacol.* **2007**, *109*, 177–206. [CrossRef] [PubMed]
5. Saad, A.M.; El-Saadony, M.T.; Mohamed, A.S.; Ahmed, A.I.; Sitohy, M.Z. Impact of cucumber pomace fortification on the nutritional, sensorial and technological quality of soft wheat flour-based noodles. *Int. J. Food Sci. Technol.* **2021**, *56*, 3255–3268. [CrossRef]
6. FAO. Statistical Database. Food and Agriculture Organization of the United Nations, Codex Alimentarius Commission: Tunis, Tunisia. Available online: http://www.fao.org (accessed on 23 May 2012).
7. Akhtar, S.; Ismail, T.; Fraternale, D.; Sestili, P. Pomegranate peel and peel extracts: Chemistry and food features. *Food Chem.* **2015**, *174*, 417–425. [CrossRef] [PubMed]
8. Belkacem, N.; Djaziri, R.; Lahfa, F.; El-Haci, I.; Boucherit, Z. Phytochemical screening and *in vitro* antioxidant activity of various Punica granatum l. Peel extracts from Algeria: A comparative study. *Phytothérapie* **2014**, *12*, 372–379. [CrossRef]
9. Gözlekçi, Ş.; Saraçoğlu, O.; Onursal, E.; Özgen, M. Total phenolic distribution of juice, peel, and seed extracts of four pomegranate cultivars. *Pharmacogn. Mag.* **2011**, *7*, 161. [CrossRef] [PubMed]
10. Sorrenti, V.; Randazzo, C.L.; Caggia, C.; Ballistreri, G.; Romeo, F.V.; Fabroni, S.; Timpanaro, N.; Raffaele, M.; Vanella, L. Beneficial effects of pomegranate peel extract and probiotics on pre-adipocyte differentiation. *Front. Microbiol.* **2019**, *10*, 660. [CrossRef] [PubMed]
11. Khaleel, A.; Sijam, K.; Rashid, T. Determination of antibacterial compounds of punica Granatum peel extract by tlc direct bio-autography and GCMS analysis. *Biochem. Cell. Arch.* **2018**, *18*, 379–384.
12. Kesur, P.; Gahlout, M.; Chauhan, P.B.; Prajapati, H. Evaluation of antimicrobial properties of peels and juice extract of *Punica granatum* (Pomegranate). *Int. J. Res. Sci. Innov.* **2016**, *3*, 11–20.
13. Chaudhary, A.; Rahul, S.N. Antibacterial activity of *Punica granatum* (pomegranate) fruit peel extract against pathogenic and drug resistance bacterial strains. *Int. J. Curr. Microbiol. Appl. Sci.* **2017**, *6*, 3802–3807. [CrossRef]
14. Li, Y.; Yang, F.; Zheng, W.; Hu, M.; Wang, J.; Ma, S.; Deng, Y.; Luo, Y.; Ye, T.; Yin, W. *Punica granatum* (pomegranate) leaves extract induces apoptosis through mitochondrial intrinsic pathway and inhibits migration and invasion in non-small cell lung cancer in vitro. *Biomed. Pharmacother.* **2016**, *80*, 227–235. [CrossRef]
15. Zam, W.; Khaddour, A. Anti-virulence effects of aqueous pomegranate peel extract on E. coli urinary tract infection. *Progr. Nutr.* **2017**, *19*, 98–104.
16. Abou El-Nour, M.M. Functional properties and medical benefits of pomegranate (*Punica granatum* L.) peels as agro-industrial wastes. *Egypt. J. Exp. Biol.* **2019**, *15*, 377–392. [CrossRef]
17. Mekni, M.; Kharroubi, W.; Flamimi, G.; Garrab, M.; Mastouri, M.; Hammami, M. Comparative study between extracts of different pomegranate parts issued from five tunisian cultivars (*Punica granatum* L.): Phytochemical content, volatile composition and biological activity. *Int. J. Curr. Microbiol. Appl. Sci.* **2018**, *7*, 1663–1682. [CrossRef]
18. Ohshima, H.; Tazawa, H.; Sylla, B.S.; Sawa, T. Prevention of human cancer by modulation of chronic inflammatory processes. *Mutat. Res. Fundam. Mol. Mech. Mutagen.* **2005**, *591*, 110–122. [CrossRef]
19. Rocha, A.; Wang, L.; Penichet, M.; Martins-Green, M. Pomegranate juice and specific components inhibit cell and molecular processes critical for metastasis of breast cancer. *Breast Cancer Res. Treat.* **2012**, *136*, 647–658. [CrossRef]
20. Mahmoudi, R.; Servatkhah, M.; Fallahzadeh, A.R.; Abidi, H.; Shirazi, H.R.G.; Delaviz, H.; Nikseresht, M. Pomegranate seed oil shows inhibitory effect on invasion of human breast cancer cell lines. *J. Clin. Diagn. Res.* **2017**, *11*, 5–10. [CrossRef]
21. Mastrodi Salgado, J.; Baroni Ferreira, T.R.; de Oliveira Biazotto, F.; dos Santos Dias, C.T. Increased antioxidant content in juice enriched with dried extract of pomegranate (*Punica granatum*) peel. *Plant Foods Hum. Nutr.* **2012**, *67*, 39–43. [CrossRef] [PubMed]
22. Barros, Z.M.P.; Salgado, J.M.; Melo, P.S.; de Oliveira Biazotto, F. Enrichment of commercially-prepared juice with pomegranate (*Punica granatum* L.) peel extract as a source of antioxidants. *J. Food Res.* **2014**, *3*, 179. [CrossRef]
23. Altunkaya, A.; Hedegaard, R.V.; Harholt, J.; Brimer, L.; Gökmen, V.; Skibsted, L.H. Palatability and chemical safety of apple juice fortified with pomegranate peel extract. *Food Funct.* **2013**, *4*, 1468–1473. [CrossRef] [PubMed]
24. AOAC. *Official Methods of Analysis of AOAC International*, 19th ed.; AOAC, International: Gaithersburg, MA, USA, 2012.
25. Saad, A.M.; Sitohy, M.Z.; Ahmed, A.I.; Rabie, N.A.; Amin, S.A.; Aboelenin, S.M.; Soliman, M.M.; El-Saadony, M.T. Biochemical and functional characterization of kidney bean protein alcalase-hydrolysates and their preservative action on stored chicken meat. *Molecules* **2021**, *26*, 4690. [CrossRef] [PubMed]
26. Saad, A.M.; Mohamed, A.S.; El-Saadony, M.T.; Sitohy, M.Z. Palatable functional cucumber juices supplemented with polyphenols-rich herbal extracts. *LWT Food Sci. Technol.* **2021**, *148*, 111668. [CrossRef]
27. Namir, M.; Iskander, A.; Alyamani, A.; Sayed-Ahmed, E.T.A.; Saad, A.M.; Elsahy, K.; El-Tarabily, K.A.; Conte-Junior, C.A. Upgrading common wheat pasta by fiber-rich fraction of potato peel byproduct at different particle sizes: Effects on physicochemical, thermal, and sensory properties. *Molecules* **2022**, *27*, 2868. [CrossRef]
28. Colantuono, A.; Vitaglione, P.; Ferracane, R.; Campanella, O.H.; Hamaker, B.R. Development and functional characterization of new antioxidant dietary fibers from pomegranate, olive and artichoke byproducts. *Food Res. Int.* **2017**, *101*, 155–164. [CrossRef]
29. Abdel-Moneim, A.-M.E.; El-Saadony, M.T.; Shehata, A.M.; Saad, A.M.; Aldhumri, S.A.; Ouda, S.M.; Mesalam, N.M. Antioxidant and antimicrobial activities of *Spirulina platensis* extracts and biogenic selenium nanoparticles against selected pathogenic bacteria and fungi. *Saudi J. Biol. Sci.* **2022**, *29*, 1197–1209. [CrossRef]

30. El-Saadony, M.T.; Saad, A.M.; Elakkad, H.A.; El-Tahan, A.M.; Alshahrani, O.A.; Alshilawi, M.S.; El-Sayed, H.; Amin, S.A.; Ahmed, A.I. Flavoring and extending the shelf life of cucumber juice with aroma compounds-rich herbal extracts at 4°C through controlling chemical and microbial fluctuations. *Saudi J. Biol. Sci.* **2022**, *29*, 346–354. [CrossRef]

31. Mosmann, T. Rapid colorimetric assay for cellular growth and survival: Application to proliferation and cytotoxicity assays. *J. Immunol. Methods* **1983**, *65*, 55–63. [CrossRef]

32. Gomha, S.M.; Riyadh, S.M.; Mahmmoud, E.A. Synthesis and anticancer activities of thiazoles, 1, 3-thiazines, and thiazolidine using chitosan-grafted-poly (vinylpyridine) as basic catalyst. *Heterocycles: An Int. J. Rev. Comm. Heterocycl.Chem.* **2015**, *91*, 1227–1243.

33. Dai, Z.; Nair, V.; Khan, M.; Ciolino, H.P. Pomegranate extract inhibits the proliferation and viability of MMTV-Wnt-1 mouse mammary cancer stem cells in vitro. *Oncol. Rep.* **2010**, *24*, 1087–1091. [PubMed]

34. Dahham, S.S.; Ali, M.N.; Tabassum, H.; Khan, M. Studies on antibacterial and antifungal activity of pomegranate (*Punica granatum* L.). *Am. Eurasian J. Agric. Environ. Sci.* **2010**, *9*, 273–281.

35. El-Saadony, M.T.; Saad, A.M.; Najjar, A.A.; Alzahrani, S.O.; Alkhatib, F.M.; Shafi, M.E.; Selem, E.; Desoky, E.-S.M.; Fouda, S.E.; El-Tahan, A.M. The use of biological selenium nanoparticles to suppress *Triticum aestivum* L. crown and root rot diseases induced by *Fusarium* species and improve yield under drought and heat stress. *Saudi J. Biol. Sci.* **2021**, *28*, 4461–4471. [CrossRef] [PubMed]

36. El-Saadony, M.T.; El-Wafai, N.A.; El-Fattah, H.I.A.; Mahgoub, S.A. Biosynthesis, optimization and characterization of silver nanoparticles using a soil isolate of *Bacillus pseudomycoides* MT32 and their antifungal activity against some pathogenic fungi. *Adv. Anim. Vet. Sci* **2019**, *7*, 238–249. [CrossRef]

37. El-Saadony, M.T.; Alkhatib, F.M.; Alzahrani, S.O.; Shafi, M.E.; Abdel-Hamid, S.E.; Taha, T.F.; Aboelenin, S.M.; Soliman, M.M.; Ahmed, N.H. Impact of mycogenic zinc nanoparticles on performance, behavior, immune response, and microbial load in *Oreochromis niloticus*. *Saudi J. Biol. Sci.* **2021**, *28*, 4592–4604. [CrossRef] [PubMed]

38. El-Saadony, M.T.; Elsadek, M.F.; Mohamed, A.S.; Taha, A.E.; Ahmed, B.M.; Saad, A.M. Effects of chemical and natural additives on cucumber juice's quality, shelf life, and safety. *Foods* **2020**, *9*, 639. [CrossRef] [PubMed]

39. Namir, M.; Suleiman, A.R.; Hassanien, M.F.R. Characterization and functionality of alcohol insoluble solids from tomato pomace as fat substitute in low fat cake. *J. Food Meas. Charact.* **2015**, *9*, 557–563. [CrossRef]

40. El-Saadony, M.T.; Osama, S.F.; Osman, A.; Mashaeal, S.; Ayman, E.T.; Aboelenin, S.M.; Shukry, M.; Saad, A.M. Bioactive peptides supplemented raw buffalo milk: Biological activity, shelf life and quality properties during cold preservation. *Saudi J. Biol. Sci.* **2021**, *28*, 4581–4591. [CrossRef] [PubMed]

41. Ranjitha, J.; Bhuvaneshwari, G.; Terdal, D.; Kavya, K. Nutritional composition of fresh pomegranate peel powder. *Int. J. Chem. Stud.* **2018**, *6*, 692–696.

42. El-Hadary, A.E.; Taha, M. Pomegranate peel methanolic-extract improves the shelf-life of edible-oils under accelerated oxidation conditions. *Food Sci. Nutr.* **2020**, *8*, 1798–1811. [CrossRef]

43. Bchir, B.; Rabetafika, H.N.; Paquot, M.; Blecker, C. Effect of pear, apple and date fibres from cooked fruit byproducts on dough performance and bread quality. *J. Food Bioprocess Eng.* **2014**, *7*, 1114–1127. [CrossRef]

44. Dey, D.; Richter, J.K.; Ek, P.; Gu, B.-J.; Ganjyal, G.M. Utilization of food processing byproducts in extrusion processing: A review. *Front. Sustain. Food Syst.* **2021**, *4*, 603751. [CrossRef]

45. Viuda-Martos, M.; Ruiz-Navajas, Y.; Martin-Sánchez, A.; Sánchez-Zapata, E.; Fernández-López, J.; Sendra, E.; Sayas-Barberá, E.; Navarro, C.; Pérez-Álvarez, J. Chemical, physico-chemical and functional properties of pomegranate (*Punica granatum* L.) bagasses powder co-product. *J. Food Eng.* **2012**, *110*, 220–224. [CrossRef]

46. Jalal, H.; Pal, M.A.; Ahmad, S.R.; Rather, M.; Andrabi, M.; Hamdani, S. Physico-chemical and functional properties of pomegranate peel and seed powder. *J. Pharm. Innov.* **2018**, *7*, 1127–1131.

47. Ruan, J.-H.; Li, J.; Adili, G.; Sun, G.-Y.; Abuduaini, M.; Abdulla, R.; Maiwulanjiang, M.; Aisa, H.A. Phenolic Compounds and Bioactivities from Pomegranate (*Punica granatum* L.) Peels. *J.Agric. Food Chem.* **2022**, *70*, 3678–3686. [CrossRef]

48. Yassin, M.T.; Mostafa, A.A.-F.; Al Askar, A.A. *In Vitro* evaluation of biological activities and phytochemical analysis of different solvent extracts of *Punica granatum* L.(Pomegranate) Peels. *Plants* **2021**, *10*, 2742. [CrossRef] [PubMed]

49. Živković, I.; Šavikin, K.; Živković, J.; Zdunić, G.; Janković, T.; Lazić, D.; Radin, D. Antiviral effects of pomegranate peel extracts on human Norovirus in food models and simulated gastrointestinal fluids. *Plant Foods Hum. Nutr.* **2021**, *76*, 203–209. [CrossRef]

50. Coronado-Reyes, J.A.; Cortes-penagos, C.d.J.; Gonzalez-hernandez, J.C. Chemical composition and great applications to the fruit of the pomegranate (*Punica granatum*): A review. *Food Sci. Technol.* **2021**, *42*, 1–7. [CrossRef]

51. Singh, M.; Jha, A.; Kumar, A.; Hettiarachchy, N.; Rai, A.K.; Sharma, D. Influence of the solvents on the extraction of major phenolic compounds (punicalagin, ellagic acid and gallic acid) and their antioxidant activities in pomegranate aril. *J. Food Sci. Technol.* **2014**, *51*, 2070–2077. [CrossRef] [PubMed]

52. Ko, K.; Dadmohammadi, Y.; Abbaspourrad, A. Nutritional and bioactive components of pomegranate waste used in food and cosmetic applications: A review. *Foods* **2021**, *10*, 657. [CrossRef] [PubMed]

53. Gaber, N.B.; El-Dahy, S.I.; Shalaby, E.A. Comparison of ABTS, DPPH, permanganate, and methylene blue assays for determining antioxidant potential of successive extracts from pomegranate and guava residues. *Biomass Convers. Biorefinery* **2021**, 1–10. [CrossRef]

54. Khalil, A.A.; Khan, M.R.; Shabbir, M.A. In vitro antioxidant activity and punicalagin content quantification of pomegranate peel obtained as agro-waste after juice extraction. *Pak. J. Agric. Sci.* **2018**, *55*, 197–201.

55. Bopitiya, D.; Madhujith, T. Efficacy of pomegranate (*Punica granatum* L.) peel extracts in suppressing oxidation of white coconut oil used for deep frying. *Trop. Agric. Res.* **2014**, *25*, 298–306. [CrossRef]
56. Lobo, V.; Patil, A.; Phatak, A.; Chandra, N. Free radicals, antioxidants and functional foods: Impact on human health. *Pharmacogn Rev.* **2010**, *4*, 118. [CrossRef]
57. Tzulker, R.; Glazer, I.; Bar-Ilan, I.; Holland, D.; Aviram, M.; Amir, R. Antioxidant activity, polyphenol content, and related compounds in different fruit juices and homogenates prepared from 29 different pomegranate accessions. *J. Agric. Food Chem.* **2007**, *55*, 9559–9570. [CrossRef] [PubMed]
58. Jesse Joel, T.; Suluvoy, J.K.; Varghese, J. Evaluation of the secondary metabolites of the waste pomegranate rind and its cytotoxicity against oral cancer (KB 3-1). *J Pure Appl Microbiol.* **2019**, *13*, 1667–1672. [CrossRef]
59. An, J.; Guo, Y.; Wang, T.; Pantuck, A.J.; Rettig, M.B. Pomegranate extract inhibits EMT in clear cell renal cell carcinoma in a NF-κB and JNK dependent manner. *Asian J. Urol.* **2015**, *2*, 38–45. [CrossRef]
60. Aqil, F.; Munagala, R.; Vadhanam, M.V.; Kausar, H.; Jeyabalan, J.; Schultz, D.J.; Gupta, R.C. Anti-proliferative activity and protection against oxidative DNA damage by punicalagin isolated from pomegranate husk. *Food Res. Int.* **2012**, *49*, 345–353. [CrossRef] [PubMed]
61. Berköz, M.; Krosniak, M. Punicalagin induces apoptosis in A549 cell line through mitochondria-mediated pathway. *Gen. Physiol. Biophys.* **2020**, *39*, 557–567. [CrossRef] [PubMed]
62. Yadav, P.; Yadav, R.; Jain, S.; Vaidya, A. Caspase-3: A primary target for natural and synthetic compounds for cancer therapy. *Chem. Biol. Drug Des.* **2021**, *98*, 144–165. [CrossRef]
63. Jain, A.S.; Prasad, A.; Pradeep, S.; Dharmashekar, C.; Achar, R.R.; Silina, E.; Stupin, V.; Amachawadi, R.G.; Prasad, S.K.; Pruthvish, R. Everything old is new again: Drug repurposing approach for non-small cell lung cancer targeting MAPK signaling pathway. *Front.Oncol.* **2021**, *11*, 741326. [CrossRef] [PubMed]
64. Venkatachalapathy, D.; Shivamallu, C.; Prasad, S.K.; Thangaraj Saradha, G.; Rudrapathy, P.; Amachawadi, R.G.; Patil, S.S.; Syed, A.; Elgorban, A.M.; Bahkali, A.H. Assessment of chemopreventive potential of the plant extracts against liver cancer using HepG2 cell line. *Molecules* **2021**, *26*, 4593. [CrossRef]
65. Saad, A.M.; El-Saadony, M.T.; El-Tahan, A.M.; Sayed, S.; Moustafa, M.A.; Taha, A.E.; Taha, T.F.; Ramadan, M.M. Polyphenolic extracts from pomegranate and watermelon wastes as substrate to fabricate sustainable silver nanoparticles with larvicidal effect against *Spodoptera littoralis. Saudi J. Biol. Sci.* **2021**, *28*, 5674–5683. [CrossRef]
66. Oliveira, D.A.; Salvador, A.A.; Smânia Jr, A.; Smânia, E.F.; Maraschin, M.; Ferreira, S.R. Antimicrobial activity and composition profile of grape (*Vitis vinifera*) pomace extracts obtained by supercritical fluids. *J. Biotechnol.* **2013**, *164*, 423–432. [CrossRef] [PubMed]
67. Ali Redha, A.; Hasan, A.; Mandeel, Q. Phytochemical determinations of Pomegranate (*Punica granatum*) Rind and Aril extracts and their antioxidant, antidiabetic and antibacterial activity. *Nat. Prod. Chem. Res.* **2018**, *6*, 1–9.
68. Naziri, Z.; Rajaian, H.; Firouzi, R. Antibacterial effects of Iranian native sour and sweet pomegranate (Punica granatum) peel extracts against various pathogenic bacteria. 2012, 13, 282–288. *Iran Vet. Res. J.* **2012**, *13*, 282–288.
69. Rosas-Burgos, E.C.; Burgos-Hernández, A.; Noguera-Artiaga, L.; Kačániová, M.; Hernández-García, F.; Cárdenas-López, J.L.; Carbonell-Barrachina, Á.A. Antimicrobial activity of pomegranate peel extracts as affected by cultivar. *J. Sci. Food Agric.* **2017**, *97*, 802–810. [CrossRef] [PubMed]
70. Vasconcelos, L.C.d.S.; Sampaio, F.C.; Sampaio, M.C.C.; Pereira, M.d.S.V.; Higino, J.S.; Peixoto, M.H.P. Minimum inhibitory concentration of adherence of *Punica granatum* Linn (pomegranate) gel against *S. mutans, S. mitis* and *C. albicans. Braz. Dent. J.* **2006**, *17*, 223–227. [CrossRef] [PubMed]
71. Devatkal, S.K.; Jaiswal, P.; Jha, S.N.; Bharadwaj, R.; Viswas, K. Antibacterial activity of aqueous extract of pomegranate peel against *Pseudomonas stutzeri* isolated from poultry meat. *J. Food Sci. Technol.* **2013**, *50*, 555–560. [CrossRef] [PubMed]
72. Suručić, R.; Travar, M.; Petković, M.; Tubić, B.; Stojiljković, M.P.; Grabež, M.; Šavikin, K.; Zdunić, G.; Škrbić, R. Pomegranate peel extract polyphenols attenuate the SARS-CoV-2 S-glycoprotein binding ability to ACE2 Receptor: In silico and *in vitro* studies. *Bioorg. Chem.* **2021**, *114*, 105145. [CrossRef] [PubMed]
73. Hikal, W.M.; Said-Al Ahl, H.A.; Tkachenko, K.G.; Mahmoud, A.A.; Bratovcic, A.; Hodžić, S.; Atanassova, M. An Overview of Pomegranate Peel: A Waste Treasure for Antiviral Activity. *Trop. J. Nat. Prod. Res. (TJNPR)* **2022**, *6*, 15–19. [CrossRef]
74. Al-Hindi, R.R.; Abd El Ghani, S. Production of functional fermented milk beverages supplemented with pomegranate peel extract and probiotic lactic acid bacteria. *J. Food Qual.* **2020**, *2020*, 1–9. [CrossRef]
75. Kaderides, K.; Kyriakoudi, A.; Mourtzinos, I.; Goula, A.M. Potential of pomegranate peel extract as a natural additive in foods. *Trends Food Sci Technol.* **2021**, *115*, 380–390. [CrossRef]
76. Cao, H.; Saroglu, O.; Karadag, A.; Diaconeasa, Z.; Zoccatelli, G.; Conte-Junior, C.A.; Gonzalez-Aguilar, G.A.; Ou, J.; Bai, W.; Zamarioli, C.M. Available technologies on improving the stability of polyphenols in food processing. *Food Front.* **2021**, *2*, 109–139. [CrossRef]
77. Ahmed, M.; Ali, A.; Sarfraz, A.; Hong, Q.; Boran, H. Effect of freeze-drying on apple pomace and pomegranate peel powders used as a source of bioactive ingredients for the development of functional yogurt. *J. Food Qual.* **2022**, *2022*, 1–9. [CrossRef]
78. Reda, F.M.; El-Saadony, M.T.; El-Rayes, T.K.; Farahat, M.; Attia, G.; Alagawany, M. Dietary effect of licorice (*Glycyrrhiza glabra*) on quail performance, carcass, blood metabolites and intestinal microbiota. *Poult. Sci.* **2021**, *100*, 101266. [CrossRef] [PubMed]

79. Yaqoob, M.; Abd El-Hack, M.; Hassan, F.; El-Saadony, M.; Khafaga, A.; Batiha, G.; Yehia, N.; Elnesr, S.; Alagawany, M.; El-Tarabily, K. The potential mechanistic insights and future implications for the effect of prebiotics on poultry performance, gut microbiome, and intestinal morphology. *Poult. Sci.* **2021**, *100*, 101143. [CrossRef]
80. Aadil, R.M.; Madni, G.M.; Roobab, U.; ur Rahman, U.; Zeng, X.-A. Quality control in beverage production: An overview. *Q. Control Bever. Indust.* **2019**, 1–38. [CrossRef]
81. Trigo, J.P.; Alexandre, E.M.; Silva, S.; Costa, E.; Saraiva, J.A.; Pintado, M. Study of viability of high pressure extract from pomegranate peel to improve carrot juice characteristics. *Food Funct.* **2020**, *11*, 3410–3419. [CrossRef] [PubMed]
82. Pathare, P.B.; Opara, U.L.; Al-Said, F.A.-J. Colour measurement and analysis in fresh and processed foods: A review. *Food Bioproc. Tech.* **2013**, *6*, 36–60. [CrossRef]
83. Ebrahimian, M.; Mehdizadeh, T.; Aliakbarlu, J. Chemical and microbiological stability and sensorial properties of traditional Iranian butter incorporated with pomegranate peel extract. *Int. J. Dairy Technol.* **2022**, *21*, 1–9. [CrossRef]
84. Al-Huqail, A.A.; Elgaaly, G.A.; Ibrahim, M.M. Identification of bioactive phytochemical from two Punica species using GC–MS and estimation of antioxidant activity of seed extracts. *Saudi J. Biol. Sci.* **2018**, *25*, 1420–1428. [CrossRef] [PubMed]
85. Karaman, M.; Jovin, E.; Malbaša, R.; Matavuly, M.; Popović, M. Medicinal and edible lignicolous fungi as natural sources of antioxidative and antibacterial agents. *Phytother. Res.* **2010**, *24*, 1473–1481. [CrossRef] [PubMed]

 *bioengineering*

Review

Postharvest Operations of Cannabis and Their Effect on Cannabinoid Content: A Review

Pabitra Chandra Das , **Alec Roger Vista**, **Lope G. Tabil *** and **Oon-Doo Baik ***

Department of Chemical and Biological Engineering, University of Saskatchewan, 57 Campus Drive, Saskatoon, SK S7N 5A9, Canada; pcd476@usask.ca (P.C.D.); ajv715@usask.ca (A.R.V.)
* Correspondence: lope.tabil@usask.ca (L.G.T.); oon-doo.baik@usask.ca (O.-D.B.)

Abstract: In recent years, cannabis (*Cannabis sativa* L.) has been legalized by many countries for production, processing, and use considering its tremendous medical and industrial applications. Cannabis contains more than a hundred biomolecules (cannabinoids) which have the potentiality to cure different chronic diseases. After harvesting, cannabis undergoes different postharvest operations including drying, curing, storage, etc. Presently, the cannabis industry relies on different traditional postharvest operations, which may result in an inconsistent quality of products. In this review, we aimed to describe the biosynthesis process of major cannabinoids, postharvest operations used by the cannabis industry, and the consequences of postharvest operations on the cannabinoid profile. As drying is the most important post-harvest operation of cannabis, the attributes associated with drying (water activity, equilibrium moisture content, sorption isotherms, etc.) and the significance of novel pre-treatments (microwave heating, cold plasma, ultrasound, pulse electric, irradiation, etc.) for improvement of the process are thoroughly discussed. Additionally, other operations, such as trimming, curing, packaging and storage, are discussed, and the effect of the different postharvest operations on the cannabinoid yield is summarized. A critical investigation of the factors involved in each postharvest operation is indeed key for obtaining quality products and for the sustainable development of the cannabis industry.

Keywords: cannabis; cannabinoids; postharvest; drying; curing; storage; biosynthesis; pre-treatments; sorption isotherm

Citation: Das, P.C.; Vista, A.R.; Tabil, L.G.; Baik, O.-D. Postharvest Operations of Cannabis and Their Effect on Cannabinoid Content: A Review. *Bioengineering* **2022**, *9*, 364. https://doi.org/10.3390/bioengineering9080364

Academic Editors: Minaxi Sharma, Kandi Sridhar, Zeba Usmani and Xiaohu Xia

Received: 23 June 2022
Accepted: 28 July 2022
Published: 3 August 2022

Publisher's Note: MDPI stays neutral with regard to jurisdictional claims in published maps and institutional affiliations.

1. Introduction

Cannabis, scientifically known as *Cannabis sativa* L. from the family Cannabaceae, has been cultivated for ages and utilized as a potential source of fiber, oil, protein, and popularly, psychoactive and medicinal purposes [1]. Historically, cannabis products have been illicitly grown for recreational and social activities to appreciate their euphoric effects [2]. Cannabis contains more than 120 active compounds, which are classified to belong in a secondary metabolite group known as phytocannabinoids or cannabinoids [3] of which cannabidiol (CBD), Δ9-tetrahydrocannabinol (THC), cannabigerol (CBG), and cannabinol (CBN) are the most studied [4]. Terpenes, which are secondary metabolites usually associated with the flavor and smell profile of biological materials have recently been found to be relevant in terms of the chemical make-up of cannabis [5]. *Cannabis sativa* L. is often classified into three subspecies depending on physical or chemical characteristics, namely *Cannabis sativa* ssp. *sativa*, *Cannabis sativa* ssp. *indica*, *Cannabis sativa* ssp. *ruderalis* [6]. The cannabis plant (Figure 1) is holistically utilized to undergo different treatments after harvest for specific uses.

As shown in Figure 1, the cola or inflorescence of cannabis contains the most abundant amount of trichomes. Trichomes are glandular membranes containing secondary metabolites, phytocannabinoids, and terpenes, and are found to be most abundant on the surface of the cannabis inflorescence.

Figure 1. A cannabis plant: (**1**) mature bud/calyx/inflorescence, (**2**) stem, (**3**) fan leaves, (**4**) sugar leaves, and (**5**) trichomes (images were taken by M. Neufeldt and are used with permission).

Extracts or products of cannabis have been studied for the treatment of different illnesses including neurological and psychiatric disorders [2], prevention of vomiting and nausea of cancer patients, boosting of hunger in acquired immunodeficiency syndrome (AIDS) patients [7], and the treatment of muscle spasms, spasticity and neuropathic pain, and chronic pain [8]. On the other hand, hemp has been produced for textiles, biocomposites, papermaking, biofuel, functional foods, and cosmetics [6]. Hemp seeds have been used in the processing of different edible foods like cookies, chocolate, peanut-cannabis butter blends, coffee, etc. The benefits outweigh the detriments brought about by cannabis as it has been legalized by more than 50 countries and 12 states of the USA for production, processing, and use [9]. Today, the cannabis industry is developing around the world, resulting in economic gain. The total cannabis market was anticipated at $3.5 billion in 2019, and a consistent increase up to $20.2 billion is expected by 2025. In Canada, production and marketing of cannabis have been legalized since October 2018 for medicinal and recreational purposes, although special monitoring and licensing from Health Canada is required in many cases. Total cannabis sales in Canada were around $210 million from mid-October to December 2018 and this value can reach multi-billions by 2021 [10]. This growing business sector should be given special focus to ensure safety and quality for the consumers.

Aside from positive effects, cannabis has been shown to cause detrimental health impacts in humans. Continuous usage of cannabis has been found to cause a variety of immediate and long-term detrimental consequences, including cardiovascular and respiratory problems, behavioral impairment, psychosis, schizophrenia, and mental illnesses [2]. People with cannabis use disorder have problems such as extreme mood shifts and memory-loss-related and concentration issues [11].

After harvest, different treatments are involved in the processing of cannabis-based food or non-food products. Drying is the most important postharvest operation involved in cannabis along with storage and curing. However, the present industrial drying involves a hang-drying method, which is inefficient and can cause an inconsistent quality of products. Thus, it is urgent to identify suitable drying methods that can enhance consistency in terms of quality for large-scale operations. In this paper, the postharvest treatments involved in cannabis processing and preservation are critically discussed and interpreted. In a broader context, this paper includes the following sections: phytocannabinoids and their decarboxylation; the postharvest operations of cannabis, including trimming, drying of cannabis (and the technical factors associated with these operations, such as equilibrium moisture content and sorption isotherms), curing, and storage and packaging; and finally, the effects of different postharvest treatments on the cannabinoid profile of cannabis. The relevant published articles were collected using the search engines Google Scholar, Scopus, and the library website of the University of Saskatchewan by inputting the keywords "cannabis, postharvest, cannabinoids" and the published articles or books from 2011 to date were considered. However, for specific interests like biosynthesis, sorption, pre-treatments, etc., the keywords were changed, and a specific publication period was not considered for extracting basic information.

2. Phytocannabinoids and Decarboxylation

Because it has various natural ingredients, cannabis is a chemically complicated plant. It includes secondary metabolites or phytochemicals called cannabinoids, a special type of terpenophenolic molecule [12]. Apart from cannabis, few other plants, such as cinnamon, clove, oregano, cocoa, black truffles, and black pepper, among others, also have cannabinoids in minor amounts. There are currently more than 120 cannabinoids found in the genus Cannabis that have different properties, structures, and uses [13]. THC and CBD are the most popular.

Radwan et al. [12] grouped cannabinoid compounds into 11 subcategories: Δ9-THC type (23 compounds), Δ8-THC (5 compounds), CBG (16 compounds), cannabichromene (CBC, 9 compounds), CBD (7 compounds), cannabinodiol (CBND, 2 compounds), cannabielsoin (CBE, 5 compounds), cannabicyclol (CBL, 3 compounds), cannabinol (CBN, 11 compounds), and cannabitriol (CBT, 9 compounds), along with miscellaneous type cannabinoids (30 compounds). According to Addo et al. [4], Δ9-THC, CBD, CBG, and CBN are the major cannabinoid groups because of their distinct nature and application.

The biosynthesis process describes the formation of the neutral forms of major cannabinoid compounds from their acidic form(s). In the biosynthesis process (Figure 2), cannabigerolic acid (CBGA) is produced by combining the primary molecules olivetolic acid and geranyl-pyrophosphate (CBGA) with the action of geranyl-pyrophosphate-olivetolic acid geranyltransferase enzymes, which is then used to synthesize cannabinoids [14,15]. Oxidocyclase enzymes convert CBGA into acidic forms of THCA, CBDA, and CBCA [4].

All enzymatically generated cannabinoids (including CBG) start off at an acidic state, that is then decarboxylated through heating to yield the functional or active form. With the application of heat, THCA, CBDA, and CBGA change into their active forms of THC, CBD, and CBG, respectively. Active cannabinoids have several physiological or psychological functions that could be either beneficial or detrimental. THC is euphorigenic, hunger-triggering, pain-relieving, and works against vomiting and nausea. CBD is not euphorigenic but works as an anti-inflammatory and anticonvulsant agent. On the other hand, CBG has antibacterial properties as well as working against neurologic disorders and inflammatory bowel disease [15].

A process called decarboxylation, which is the removal of a carboxyl group in the presence of heat, is required to convert acidic cannabinoids to their pure form. Thus, decarboxylation can be defined as the chemical reaction that removes a carboxyl (COOH) group from THCA and CBDA, resulting in THC and CBD, respectively. This reaction is facilitated by heating and vaporization. Wang et al. [13] reported that the decarboxylation of THCA to THC is more pronounced and very abrupt in a period of 60 min for temperatures greater than 100 °C. The lower temperature limit used for this study was 80 °C and it was reported that the concentration of THCA is still decreasing, following an almost linear decreasing trend in a period of 1 h. CBDA also follows a similar pattern. Recalling the spontaneous deterioration of THC to CBN, which is a less psychoactive cannabinoid [16], a holistic approach is usually taken when drying cannabis; thus, the use of a low drying temperature is practiced.

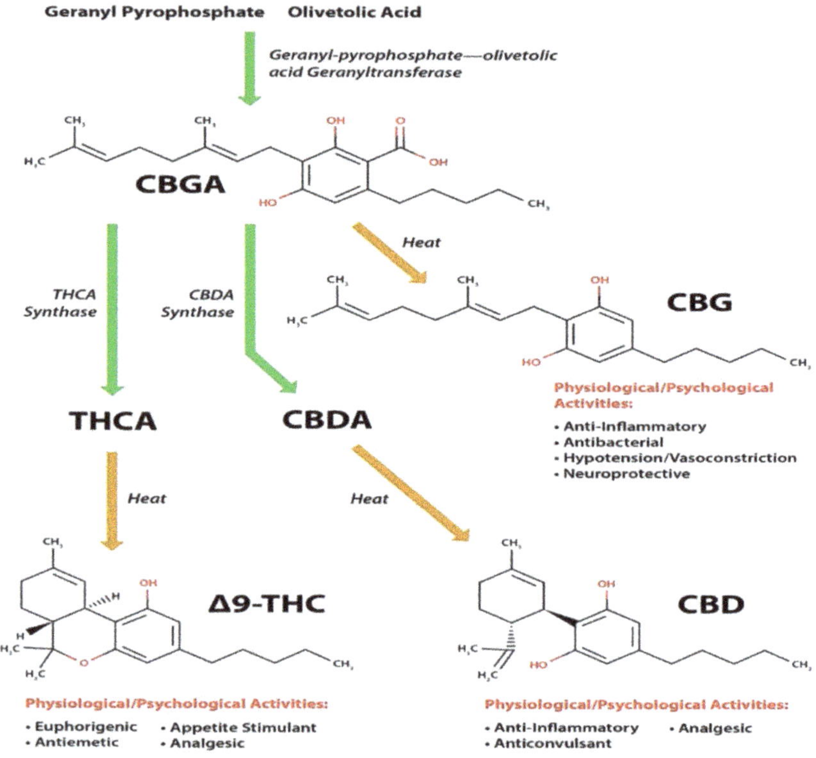

Figure 2. Biosynthesis of cannabinoids (adapted from Nachnani et al. [15] with permission).

Entourage effect refers to the joint effects of taking these cannabinoids together with other minor cannabinoids and phytochemicals called terpenes (volatile aromatic compounds) being more beneficial than just taking pure THC or pure CBD [17,18]. Trichomes (as shown in Figure 1), which are the microscopic glandular membranes found on the surface of the cannabis inflorescence, contain cannabinoids and terpenes that are vital for psychoactive and medicinal effects. The glands enclosing these cannabinoids and terpenes allow them to be protected from external environmental factors that could hasten degradation [8]. Trichomes, not the buds, contain the cannabinoids that are vital for psychoactive and medicinal use; the temperature sensitivity of the cannabinoid- and terpene-containing trichomes is of prime importance when considering the postharvest unit operation conditions [19–21]. Terpenes have also been closely investigated as a possible solution to the taxonomic classification challenges that arise with the rampant genetic modification of cannabis cultivars to produce more secondary metabolites, focusing on high THC or CBD content in the illicit market in the years that cannabis was illegal. The chemical classification, or more specifically, the metabolomics of cannabis terpenes, is in the forefront of this emerging solution [18,22].

The stability data of these cannabinoids and phytochemicals during harvesting, drying, and storage are important assets for the emerging industry. According to Kader et al. [23], there are biological and environmental factors that influence the deterioration of biological commodities. Biological factors include: (a) respiration, which is the breaking down of biological components, such as carbohydrates, proteins, and fats into simpler substances; (b) the production of ethylene, which encourages senescence or degradation; (c) changes in the composition (in the case of cannabis), which involves the spontaneous deterioration of THC to CBN; (d) further growth and development, which may do more harm than good to plant quality; (e) water loss or transpiration, although this could be beneficial for cannabis; (f) pathological breakdown due to the presence of moisture; (g) physiological breakdown; and h) physical damage. The physical factors that encourage deterioration are temperature, light, relative humidity (RH), the presence of ethylene, environmental conditions, and the addition of other chemicals, such as fertilizers and fungicides [16,23].

3. Postharvest Operations Involved in Cannabis

Maturity and harvesting of cannabis flowers depend on a number of factors similar to other plants, including day length, season, variety, soil fertility, temperature, etc. Generally, it takes six to sixteen weeks to harvest the necessary parts from cannabis plants [24]. The general operations involved in the processing of cannabis starting from harvesting and ending at marketing are depicted in Figure 3.

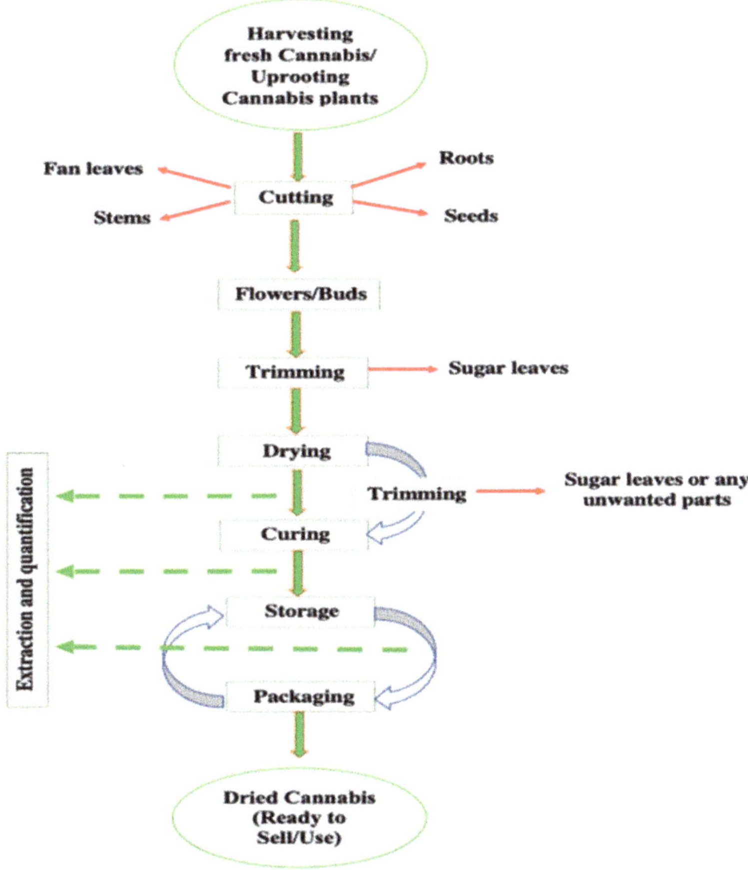

Figure 3. Postharvest operations involved in cannabis processing.

3.1. Trimming

During harvesting, different portions of cannabis are separated from the main plant and further processing is performed based on the target product types. The stems are used for the production of fiber, textiles, or animal bedding while the seeds are used for the extraction of oil, protein, and the processing of different food products. The cannabinoid-containing parts, the inflorescence along with stems, are removed from the main plant for pharmacological or recreational use. Stems and long fan or sugar leaves are cut to detach the inflorescence. This also helps in increasing the surface area for drying. Dried cannabis sometimes undergoes the trimming process to discard the less desirable parts, such as the sugar leaves, which have less cannabinoid content compared to the inflorescence. Size reduction processes in the food and agriculture industry often involve the use of mechanical force and the large material to separate physical parts or to divide larger particles into smaller more manageable ones. Size reduction processes are often classified based on the nature and direction of force applied to the process, such as compression, shear, or impact [25].

Trimming in the cannabis industry is a niche size reduction operation of removing the sugar leaves that cover the inflorescence, especially during drying. There are two iterations: trimming before the extensive drying process (wet trimming) or trimming after drying and/or curing process (dry trimming). An effective trimming process is delicate enough to preserve the structure of trichomes that house major cannabinoids, but effective enough to remove most of the sugar leaves [26]. Size reduction operations can be done before or after drying, depending on the producer's preference and discretion and ease of flow of operations.

Current trimmer designs commonly use shearing techniques to reduce the amount of sugar leaves on the inflorescence. The mechanism for trimming uses a rotating cylinder with perforated walls or a set of pressurized wires that are designed to shear off the sugar leaves but not cut or slice through the inflorescence. The rotating cylinder is then inclined downwards to allow continuous flow with options for manual feeding and output collection or extra accessories for conveying the product into and from the trimming operation [27]. The perforated cylinders are then surrounded by a closed shell attached to a vacuum to collect the discarded sugar leaves, where ambient air drying may occur [27]. Improvements in the equipment design and final output quality in trimming operations are critical for better industry standards. These improvements should focus on the effectiveness of the process of sugar leaf removal with minimal damage and abrasion to the trichomes on the surface of the inflorescence.

3.2. Drying of Cannabis

The main postharvest operation of cannabis processing is drying. As heat is the main agent for the decarboxylation of cannabinoids, the appropriate drying temperature and conditions are very important to obtain good quality product with a maintained amount of THC, CBD, or other cannabinoids. The inflorescence of raw cannabis contains around 78 to 80% moisture, which needs to be reduced for safe storage and production of ready-to-use dry products [28,29]. Researchers are trying to identify suitable drying technology, which can replace the existing hang-drying practice. In this section, insights on drying and the current industrial practices with their pros and cons are discussed. Additionally, equilibrium moisture content, sorption isotherm, sorption isotherm models, and possible pre-treatments for the improvement of drying are also discussed.

3.2.1. Mechanism of Drying

Drying is a popular unit operation usually characterized by the presence of a solid-liquid separation that is usually facilitated by the presence of heat, resulting in the evaporation of a liquid, which is water in the majority of cases. The definition of drying could also extend to solvent evaporation, dehydration of various biological materials such as in food and feed, the dehydration of salts, and the removal of hydroxyl groups from organic material [30–32].

Industrial drying involves the use of large amounts of energy usually in the form of heat, which can use different mechanisms, such as convection, conduction, radiation, or a combination of these. Certain amounts of sensible and latent heat need to be provided around the entirety of the material to achieve evaporation. Convection is characterized by a carrier gas or feed gas that is usually air, supplying the heat for evaporation of water or solvent. Conduction is the heat transfer mechanism during contact drying. The carrier gas in this context serves as the medium to which the solvent is evaporated. The radiation mechanism has different subtypes; penetrative radiation drying would consist of dielectric drying, such as radiofrequency or microwave drying, where heat is generated in the material rather than the heat diffusing into the material, which is non-penetrative [31,33].

Drying rate is the rate of moisture removal during a given drying process, obtained by the ratio of moisture removed to the time of drying. Total drying time is split into two major periods: (i) a constant rate period and (ii) one or more falling rate period(s). The constant rate period occurs when water is removed from the surface by evaporation and internal water movement is sufficient to keep the surface saturated. The surface temperature remains constant and is lesser than the circumambient drying air because the energy input rate is the same as the heat energy lost due to evaporation. In the falling rate period, the surface of a material is unsaturated, and the movement of interior moisture is lower than the rate of surface evaporation. During the first falling rate period, the drying rate is observed to fall as water content reduces due to increased resistance to evaporation, as well as a reduction in heat flowing into the material when the surface temperature rises to the heating medium temperature. However, the temperature of the material does not significantly increase or differ from the wet-bulb temperature. The second falling rate period commences when the partial pressure of moisture all throughout the material is lower than saturation. Heat flux from the hot air to the sample is very low as there is a small temperature gradient [32,33].

3.2.2. Current Industrial Drying Practices for Cannabis

Cannabis drying operations through the years have not evolved too far from the slow-drying method or low-temperature drying (Figure 4) that usually takes 5–6 d at near ambient conditions. The commonly used drying temperature is 18–21 °C and relative humidity is 50–55%. Currently, no model has yet been established for the prediction of the drying endpoint and total drying time. Postharvest processes are currently considered more of an art than a science as there are no standards for which the endpoint of drying is defined, and practices are based on word of mouth and are usually still subject to deviation. In the slow drying technique, cannabis stalks are hung on wire liners with the top buds or the cola upside down in isothermal room conditions until desired water activity is achieved. Screen drying is a variation of the slow drying process wherein the trimmed buds are dried in an enclosed and properly ventilated room by placing them on trays or screens. In this screen drying method, drying conditions are almost the same as in slow drying processes, but this process takes around 4 to 5 d to dry the cannabis completely [10].

Present industrial drying practices have several drawbacks such as a long processing time, chances of contamination by molds or fungi, and inconsistent quality in the final products. Researchers are trying to introduce modern and novel technologies in this sector. According to Chen et al. [34], drying conditions and methods dominate the drying time and quality characteristics of cannabis. Air drying, microwave-assisted convective drying, freeze drying, vacuum drying, microwave assisted freeze drying, intermittent (non-isothermal) drying, conveyor drying, radio frequency, and electrohydrodynamic drying, among others are the suggested modern drying techniques for cannabis [4,10,35]. However, critical identification of psychrometric behavior for devolvement of new drying technologies is equally important.

Psychrometric terms predominantly govern postharvest operations in the cannabis industry. Psychrometrics is the study of atmospheric air and its thermodynamic properties, typically focusing on psychrometric terms or parameters [36–38]. The interaction and control of these parameters are vital for success in an optimum quality-controlled process, storage of products, and packaging. In the cannabis industry, four humidity terms are used to describe the drying convective air: (a) vapor pressure, (b) relative humidity, (c) humidity ratio, and (d) water activity. These terms are usually common in psychrometric calculations together with dry and wet bulb temperature, enthalpy, dew point temperature, and specific volume [37]. However, relationships among the psychrometric parameters should be investigated as it would be helpful for cannabis drying and storage in different environmental conditions.

Figure 4. Hang-drying practice commonly used in large scale cannabis operations. Image taken by A. Vista (co-author).

3.2.3. Equilibrium Moisture and Sorption Isotherms

Equilibrium moisture content (EMC) is the amount of moisture in a biological material when equilibrium occurs. The corresponding relative humidity is the equilibrium relative humidity (ERH). The plot of product moisture versus air relative humidity at a constant temperature is called the equilibrium moisture content curve [32].

The EMC sorption curve appears in a sigmoidal shape due to how moisture is bound in a material. At low relative humidity values, shown in Figure 5 (region A), moisture is assumed to be bound tightly to the biological material's surface. In this region, water exhibits a monolayer behavior where the amount of tightly held water is approximately like when a single water molecule is being held on each hydrophilic site across the surface of a biological material. The Langmuir isotherm, which is obtained by the monomolecular adsorption of vapor by porous particles in a finite volume of voids, can be used to approximate these areas [39,40].

The intermediate region (region B in Figure 5) is where water is more accurately described as "multimolecular" because water molecules are bound to different sites at different distances from the center of the biological material, and thus is easily pictured as multilayers of moisture across the biological material. The point of transition between the monolayer and multilayer occurs at relative humidity values of around 5 to 20%. The third region (region C in Figure 5) is the free water region, usually associated with water activity values greater than 0.9, where properties exhibited by dilute solutions are observed and water is loosely attached to the binding sites and is readily available for reaction to other substances.

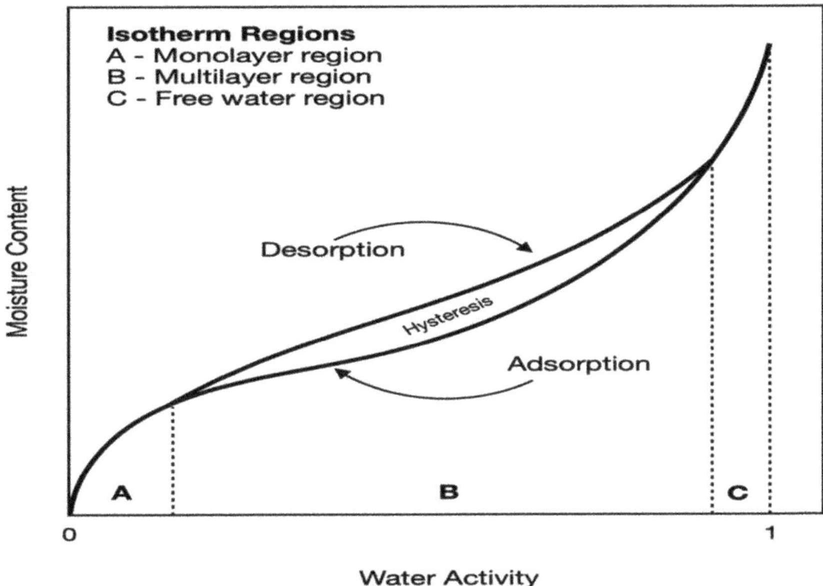

Figure 5. Sorption isotherms in biological materials (reproduced from [32,39,41]).

Hysteresis is a phenomenon related to the means on which water is distributed across the biological material, and it explains the nature of biological materials to undergo structural rearrangements that affect the availability of hydrophilic sites to bind to water at re-adsorption as a result of the presence of an offset, as seen in Figure 5 [32,39].

Moisture activities in biological materials for drying processes are usually predicted from standard water activity thresholds to prevent the proliferation of unwanted biological contaminants. The United States Food and Drug Administration (FDA) defined water activity as the ratio between the vapor pressure of the substance itself, when in a completely undisturbed balance or equilibrium with the circumambient air media, and the vapor pressure of distilled water under similar conditions. This means that at different temperatures, the water activity of a substance may vary [42]. The FDA also defined water activity as equivalent to the equilibrium relative humidity:

$$a_w = \frac{ERH}{100} \tag{1}$$

where ERH is the equilibrium relative humidity (%).

Sorption isotherms are equilibrium moisture content and equilibrium relative humidity values at a constant temperature and pressure. The knowledge of the stability of water in a biological material such as cannabis is important in designing and optimizing drying operations, designing packaging materials, predicting quality and stability, and estimating the loss of moisture during storage [39,43].

Sorption Isotherm Models

To predict EMC based on the measurements of ERH, isotherm equations are generated based on standardized equations. ASABE Standards ASAE D245.7 [44] summarized five isotherm equations.

- Modified Henderson equation:

$$\mathrm{ERH} = 1 - \exp\left[-A \times (T + C) \times (MC_D)^B\right] \qquad (2)$$

- Modified Chung–Pfost equation:

$$\mathrm{ERH} = 1 - \exp\left[-\frac{A}{T + C}\exp(-B \times MC_D)\right] \qquad (3)$$

- Modified Halsey equation:

$$\mathrm{ERH} = \exp\left[-\frac{\exp(A + B \times T)}{(MC_D)^C}\right] \qquad (4)$$

- Modified Oswin equation:

$$\mathrm{ERH} = \left[\left(\frac{A + B \times T}{MC_D}\right)^C + 1\right]^{-1} \qquad (5)$$

- Guggenheim–Anderson–de Boer (GAB) equation:

$$\mathrm{ERH} = \frac{A \times B \times C \times ERH}{(1 - B \times ERH) \times (1 - B \times ERH + B \times C \times ERH)} \qquad (6)$$

where ERH = equilibrium relative humidity (decimal value); T = temperature (°C); MC_D = dry-basis moisture content (decimal value); and A, B, and C = equation constants, and the related data specifications based on which these constants were obtained are summarized in Table 2 of ASABE Standards ASAE D245.7 [44].

3.2.4. Possible Pre-Treatments for Improvement of Drying of Cannabis

Pre-treatments are the physical or chemical treatments used prior to any process to enhance the quality of the product as well as to improve the process. Drying pre-treatments can reduce the drying time as well as the energy consumption during the drying process [45]. A number of physical and chemical pre-treatments are frequently used for the drying of foods, medical plants, herbs, spices, etc. Physical pre-treatments include steaming, blanching, microwave heating, ohmic heating, ultrasonic treatment, pulsed electric field, high pressure processing, freezing, etc., while chemical pre-treatments are conducted by treating the samples with different chemicals, such as sulfur dioxide, carbon dioxide, ozone, alkali solution, acid solution, etc. [45]. Do these pre-treatments improve the cannabinoid profile of cannabis? Up to now, there is no clear answer and very little research is conducted so far on the analysis of the potential of pre-treatments for maintaining the level of cannabinoids and terpenes. However, these pre-treatments can reduce the drying time. As decarboxylation requires heat treatment for the conversion of cannabinoids from their acidic form into their neutral form, thermal or physical pre-treatments could accelerate the process. This section discusses some possible pre-treatments that can be applied in the cannabis industry.

Microwave Heating

Electromagnetic waves with wavelengths ranging from 0.001 to 1 m and frequencies ranging from 300 to 300,000 MHz are called microwaves [46]. The frequency of electromagnetic energy equaling 915 MHz is used in the industry due to better penetration depth. During MW heating, heat is generated throughout the volume simultaneously due to the absorption of electromagnetic waves of certain frequencies in the form of energy [47]. Microwave heating helps in the improvement of operational speed, product quality, and energy use efficiency, and it lowers the operational costs of a process [48]. In cannabis processing, investigation into MW heating as a pre-treatment is scarce. Few approaches to MW-assisted extraction of cannabinoids are available. MW pre-treatments prior to distillation were performed by Fiorini et al. [49] for the enrichment of cannabidiol in the essential oil of industrial hemp inflorescence. They reported that MW pre-treatment had a significant effect (F13,27 = 28.229, p < 0.0001) in increasing the CBD level of cannabis inflorescences. According to their findings, a 3 min MW treatment with a power of 450 W and a 1 min MW treatment with a power of 900 W yielded 8.90% and 9.00% CBD, respectively, and the values were nearly two-fold above the untreated inflorescences that yielded 4.9% CBD. Overall, concerning different bioactive compounds, such as CBD, (*E*)-β-caryophyllene, and caryophyllene oxide of the cannabis inflorescence, MW treatment for 1 min with a power of 900 W was the most useful strategy [49]. MW-assisted extraction of phytochemicals from different plants, leaves and seeds is practiced all over the world. MW pre-treatments on rosemary leaves resulted in almost two times the phytochemical concentration during storage compared to the untreated leaves [50]. Phytocannabinoid extraction with the assistance of MW shows a significant improvement of the process output [51–54]. To prompt the drying process and to improve the polyphenol, cannabinoid, and terpene profile of cannabis, MW heating could be an ideal pre-treatment. However, optimum MW power and time should be comprehensively investigated.

Cold Plasma

Cold plasma (CP) is a new, non-thermal technique that may eradicate germs, inactivate enzymatic processes, enhance product quality, and assure product safety [55,56]. Plasmas are ionized gases that include a variety of electrons, ions, and neutral reactive species [57,58]. At atmospheric pressure and low temperature, a variety of gases (air, nitrogen, helium, and argon) and methods (corona discharge, dielectric barrier discharge, and plasma jet) are employed to create plasma [59]. Figure 6 represents different types of cold plasma devices. Plasma sources can be delivered to the object either directly (sample comes into direct connection with the plasma area) or indirectly (sample contacts the plasma or plasma-activated media) [60]. Cold plasma has been investigated for enhancement of the drying efficiency and quality of different foods, medicinal plants, and herbs, such as shiitake mushrooms [58], chili peppers [61], corn kernels [62], wolfberries [63], jujube slices [64], saffron [65], grapes [66], lemon verbena [67], coneflower leaves [68], and many others. All of these authors remarked that CP pre-treatment can significantly increase the drying efficiency and product quality concerning phytochemicals or microbial safety. In medicinal cannabis inflorescences, CP pre-treatment resulted in a decrease in fungal count. In medicinal cannabis inflorescences, both uninoculated and artificially infected by *B. cinerea*, cold plasma treatment for 10 min led to a 5-log-fold decrease in overall yeast and mold levels both in the control and the inflorescence of medicinal cannabis infected with the fungus [69]. Thus, the treatment of cannabis prior to drying, namely CP treatment in this case, may help to improve the shelf stability of dried cannabis with a shorter time required for drying. The time required for plasma treatment and the optimum distance between the plasma flume and inflorescence need to be comprehensively studied before industrial application of the technology for cannabis.

Pulsed Electric Field

Pulsed electric field (PEF) is a non-thermal approach that includes applying extremely short and high voltage pulses to a sample(s) positioned within electrodes, causing the development of new membrane pores, cellular membrane breakdown, and intracellular fluid release [70–72]. As this technique does not rely on heating, it may be used as a superior alternative to typical heat-transfer operations during the processing of biological materials. Although the mechanics of this technique is unknown, it is commonly referred to as electroporation or electropermeabilization. The term electroporation means either the formation of holes in the cellular membrane of the material or increase in the size of existing pores in the material, or a breakdown of the cell membrane continuity of the material [73].

Liu et al. [74] mentioned two important components of a typical batch PEF system: (a) a PEF generator and (b) a treatment chamber, as shown in Figure 7.

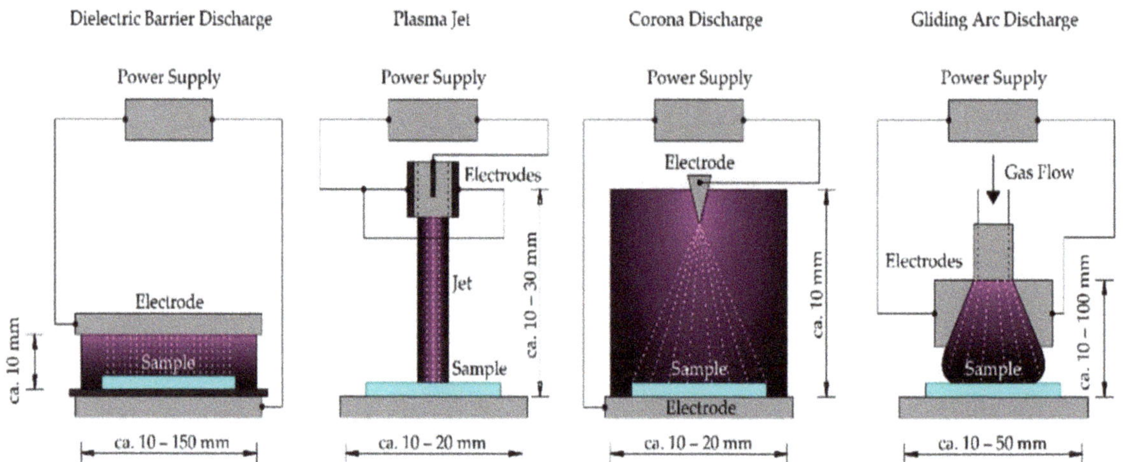

Figure 6. Schematic overview of cold plasma devices (adapted from Domonkos et al. [59] with permission).

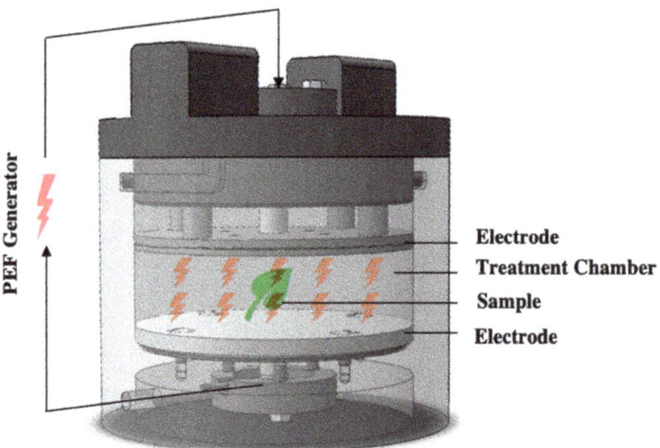

Figure 7. Schematic diagram of a typical PEF chamber (reproduced from Liu et al. [74] with permission).

The drying rate, one of the important targeting properties of a drying operation, is improved by PEF as it changes the cellular structure of the plant materials. The treatment of apple tissues with 10 kV/cm and 50 pulses reduced the drying time up to 12% [72], while the application of an electric field strength of 1000 V/cm and 30 pulses with a pulse width of 120 ms resulted in the improvement of productivity per unit area and drying rate by 28.50% and 27.02%, respectively, as well as in the reduction of specific energy consumption and drying time by 20.46% and 22.50%, respectively [75]. Similar desired outcomes were also observed for carrots [72,76,77], potatoes [78], onions [79], Thai basil leaves [80], and red bell peppers [81]. In addition to drying, PEF also helps in the recovery of more bioactives and phytochemicals. Several studies are available on the PEF-assisted extraction of bioactive components from different foods and plant materials. When a plant cell is exposed to an external electromagnetic field, pores develop in the membrane, permitting the contents of the cell, such as phenolic components, to be left out [82]. The PEF-assisted extraction of polyphenol from fresh tea leaves at an electric field intensity of 1.00 kV/cm with 100 pulses of 100 s pulse length and 5 s pulse repetition resulted in an approximately two times higher extraction rate without significant alteration of the polyphenol profile [74]. For extracting intracellular components from microalgae [83], red beetroot pigment [84], anthocyanins from purple-fleshed potato [85], phenolic compounds from Merlot grapes [86], polyphenols from grape skin [87], and many more, PEF has shown beneficial results. In the case of cannabis, application of PEF for drying or extraction is rarely noticed. A study on the PEF-assisted extraction of oil and phenolics from cannabis with a PEF intensity of 3 kV/cm and a press speed of 20 rpm resulted in an increased oil extraction rate and the highest phenolic concentration (2036 ppm) [88]. PEF has also been used as a pre-treatment for ultrasonic extraction to improve the extraction yield of polyphenols from a defatted hemp seed cake [54]. Thus, PEF can potentially be employed to improve drying efficiency due to decreased drying times as well as the retention of important phytoconstituents of cannabis. Further detailed studies on this matter are suggested.

Ultrasound

Sound is a mechanical wave that is created by the vibration of materials. Mechanical waves have a frequency range of less than 16 Hz to more than 1 GHz and are grouped into four different categories based on the frequency of waves. If the mechanical waves range between 1 to 16 Hz, they are infrasound; acoustic or audible sounds range between 16 to 20 kHz; ultrasound has a frequency range of 20 kHz to 1 GHz; and hypersound has a frequency of 1 GHz and above [89]. Ultrasounds are created using a variety of processes, including mechanical (aero- and hydrodynamics), thermal (electric discharge), optical (high-power laser impulse), and reversible electric and magnetic technologies (piezoelectric, electrostriction, and magnetostriction). Recently, the application of ultrasound for the processing and preservation of food and biomaterials has gained significant interest from scientists. Few applications of ultrasound in food processing include thermal processing (sterilization and pasteurization), freezing and thawing, emulsification, viscosity modification, polymerization, extraction of different compounds, tissue interruptions, aggregate dispersion, crystallization, etc. [89]. For many years, ultrasonically aided drying has piqued curiosity. This technology is being applied as a pre-treatment to improve drying kinetics as well as to minimize the energy cost associated with drying [90–92]. Ultrasonic waves in a solid cause a quick succession of alternate contractions and expansions of the substance through which they travel [93,94]. This alternating stress forms tiny channels, aiding in moisture transfer. Furthermore, highly intense acoustic waves cause the detonation of strongly bound water within the solid material, assisting in its evaporation [93]. There is a plethora of research on ultrasound-assisted drying to enhance drying parameters and product quality. The ultrasound-assisted vacuum drying (USVD) of red peppers resulted in a 25% reduction in drying time and an 89% enhancement of the effective moisture diffusivity without a substantial loss of bioactive components [95]. In the case of carrot slices, USVD reduced drying time by 41–53% and helped in retaining more beta-carotene, ascorbic acid, color, and texture [91]. For papaya, an enhanced drying rate with the lowest loss of ascorbic acid (41.3%) was observed when the samples were dried with the aid of ultrasound and vacuum [96]. In the case of apple tissue, ultrasonic treatment resulted in a lower drying time (31%) and density (6–20%) as well as higher shrinkage (9–11%) and porosity (9–14%) compared to the untreated apple tissue [92]. The addition of solvents like ethanol to ultrasound treatments constitutes another technique as it has resulted in the greatest reduction in both the drying time (59%) and energy consumption (44%) for pumpkin [97]. Ultrasound also improved the drying time for other foods, such as: (a) pineapple with a drying time reduction of 8% [98], (b) persimmon [93] and raspberries with a drying time reduction of 54–64% [99], and (c) purple-fleshed sweet potato with a reduction in the drying process energy consumption of up to 34.60% [74]. Several studies are also available on the extraction of bioactive compounds with the assistance of ultrasound from different sources, such as wild ginger [100], pomegranate peel [101], myrtle (*Myrtus communis* L.) [102], green tea [103], tobacco waste [104], etc. In the case of cannabis, this technology can play a very important role, especially for the reduction of drying time and energy consumption as well as increasing the cannabinoid profile. Ultrasound as a pre-treatment for the drying process has not yet been investigated for cannabis, although few studies were performed to improve the extraction of cannabinoids. Ultrasound-assisted extraction (UAE) was used to collect volatile chemicals from cannabis inflorescences. Ultrasonic treatment of *Cannabis sativa* L. cultivars for less than 5 min resulted in higher terpene concentrations than maceration, while ultrasonic treatment for more than 5 min resulted in higher THC concentrations [105]. Furthermore, Agarwal et al. [106] concluded that ultrasound considerably improved the extraction of cannabinoids present in cannabis. Ultrasound also can increase the recovery of oil from cannabis [107,108]. Considering the above outcomes from the literature, ultrasound can be an important treatment for the cannabis industry to enhance the drying process and extraction of cannabinoids. Thus, a feasibility study of this technique along with optimization in case of cannabis is highly recommended.

Irradiation

Exposing materials to ionizing radiation is known as irradiation. Ionizing radiation can remove electrons from atoms and molecules, causing them to become ions. Ionizing radiation is made up of high-energy charged particles, such as electrons, and photons, such as X-rays and γ-rays. However, due to limited penetrating power (alpha particles) or because they create radioactivity in the irradiated substance, all types of ionizing radiations are not used for irradiation; only three types of ionizing radiations (X-rays, γ-rays, and electron beams) are used for irradiation [109]. Irradiation treatment for the decontamination of cannabis is very common. More than 80% of licensed growers of cannabis in Canada are using irradiation treatment [110]. Medical cannabis treated with a $\geq$10 kGy irradiation dose from Cobalt-60 as radiation source resulted in a significant decontamination of the cannabis without largely affecting the phytocannabinoids [111]. However, irradiation has a significant use for enhancing the drying rate of different foods either using gamma irradiation or electron beam irradiation. Most of the research has been carried out using 60-Co γ-rays with a dose of up to 12.0 kGy. Irradiation pre-treatments resulted in a higher drying rate for potato and apple [112], carrot, potato and beetroot [113], and tofu protein [114]. A few researchers have performed irradiation as a pre-treatment for cannabis [110,111,115]. The level of irradiation, optimization, and its effect on phytocannabinoids are yet to be investigated.

3.3. Curing

Curing in agricultural industries often refers to the maturation of the chemical profile of a biological material over long periods of time to obtain optimum quality and flavor. This usually involves unique iterations of methods [104,116–120]. For the cannabis industry, curing involves essentially marking the end of a drying process and the beginning of the storage process, where the cannabis inflorescence is being maintained at an ideal moisture content and water activity. It is an important step to maintain the quality of cannabis. Interactions between the storage, method of curing, water activity, potency, chemical profile, and maturity of cannabis in this postharvest step are still not widely investigated. Color and texture changes of tobacco are determinants for curing [121] and a similar method to detect the more subtle changes in color, aroma, and texture in cannabis curing might be worth looking into.

For cannabis, this process includes keeping the dried inflorescence in a closed container for a certain period with a specific temperature and humidity. When a plant is cut, enzymes and aerobic bacteria start their activity to break down the extra sugars and starches created by chlorophyll breakdown. When smoking cannabis that has not been properly cured, the presence of these residual sugars and minerals causes a burning sensation in the throat. Before the breakdown of the remaining sugars, carbohydrates, and nutrients, the plant is forced to utilize these compounds in flavored items during the curing process. Thus, this resting period introduces less odor and sensation of throat burning when smoking or inhaling. It can also result in an increase in the level of THC and CBN by completing the decarboxylation process, as well as extending the storability by limiting fungal growth [29]. According to Jin et al. [28], maintaining a temperature of 18 °C and an RH of 60% for a period of 2 weeks, with the opening of the lid after 6 h, is the best curing condition. Although curing is one of the most significant postharvest operations, it has been overlooked and is not investigated properly. The time, temperature, humidity, and storage container during curing may be varied and should be investigated more to get the best product for storage and marketing.

3.4. Packaging and Storage

The cannabis inflorescence, with its high initial moisture content at harvest ranging from 75–78% (wet basis, wb), is considered to be a highly perishable material before drying. A common problem arising in the methods for the drying of cannabis is the inadequate control and prediction methods used to dry and store cannabis, resulting in the formation of mold [10,64,122].

Important factors to consider for quality, in terms of the packaging material, are water permeability to prevent moisture inconsistencies and rigidity in preventing mechanical injury. The ability to predict the movement of moisture throughout a biological material such as cannabis is critical to further improve quality problems in the industry [10]. Diffusion calculations through a membrane are similar to the calculations that use Fick's law as outlined by Geankoplis [25] and Srikiatden and Roberts [33]. Permeabilities for packaging materials are often widely available for consumers to check. A holistic approach considering the diffusion coefficient, sorption isotherms, and the permeability of packaging materials allows producers to make critical and efficient decisions to further improve their production process.

Common packaging materials used in the cannabis industry are Mylar bags, PET (polyethylene terephthalate) and/or metal cases and tubes, and glass jars, as found in online catalogs [123,124]. Water vapor permeability, or the measure of how fluid easily flows through a substance [125], plays a role in achieving the optimum quality measures needed for the industry. Mylar bags have a permeability ranging from 0.02 to 1.18 cm^3 mm/m^2 d atm [126]. This means that dried cannabis inflorescence being stored in bags made of Mylar gradually lose moisture as cannabis is stored in inventory for months and even years.

Storage conditions would also contribute to the degradation of cannabis and the cannabinoids and volatile compounds present in them. Trofin et al. [127–129] have extensively investigated the stability of various cannabis products, focusing on major cannabinoids via kinetic models and in-vitro studies, and have found that the storage of cannabis at 22 °C with the presence of light is highly detrimental to cannabis compared to samples stored in darkness at 4 °C. These results are also significantly similar to studies by Taschwer and Schmid [130] and Chen et al. [34], where the total amount of major cannabinoids present in cannabis, such as THC and CBD, decreased as the storage and drying temperatures increased. Temperature control during storage therefore plays an important role in best quality practices in maintaining both moisture content and cannabinoid concentration in dried cannabis inflorescence.

4. Effect of Postharvest Processing on Phytoconstituents of Cannabis

The main thermal treatment for cannabis is drying. In addition to that, few other treatments have been investigated by researchers to determine their effects on product quality. Though research in this area is hardly available in the literature, this section will summarize the recent output of different studies focused on temperature dependence of cannabinoids during drying. Table 1 represents the effect of postharvest treatments on the cannabinoid profile of cannabis. From the overall findings, it can be remarked that the cannabinoid content varies with the drying parameters, drying methods, storage temperature, extraction solvents, and other treatments.

Chen et al. [34] studied the effect of different drying techniques, e.g., freeze drying (FD); ambient air drying at 22.3 °C and 37.9%; hot air (HA) drying at 40, 50, 60, 70, and 90 °C (with air flow rate of 1.4 m/s); sequential infrared and hot air (SIRHA) drying (dried for either 1 or 2 min by IR and sequentially dried by HA drying at 60 °C or 40 °C for 1 or 2 min, respectively) on three different varieties (Pipeline, Maverick, and Queen Dream CBD) of *Cannabis sativa* inflorescences. The total THC content of the sample was below 0.3% while the CBD content varied due to the variation of treatments. FD resulted in 7.76 ± 0.03 to 13.93 ± 0.03 g total CBD/100 g of dry matter. HA drying resulted in a higher CBDA conversion rate or more decarboxylation than that of FD. Application of IR drying improved the drying rate but it was found that it can cause the loss of CBD [34]. Challa [131] worked on *Cannabis sativa* buds using FD, HA drying (25, 32, 40, 50, 60, and 75 °C with an air flow rate of 1 ± 0.1 m/s and 45–60% RH), and non-isothermal drying (initially dried at 40 °C and increased to 70 °C at 45% or 25% moisture content (MC); again, dried initially at 40°C and increased to 60 °C at 45% or 25% MC) to determine the effect of the different drying methods on the cannabinoid profile. They obtained 1.91% of total CBD in fresh cannabis buds, which gradually increased with the increasing drying temperature of HA drying; however, they did not observe any significant difference ($p > 0.05$) in terms of total CBD content due to the changes in the drying conditions. They observed that decarboxylation did not largely affect the CBD:CBDA ratio, although the total CBD content varied due to drying methods and conditions. The maximum total CBD was obtained from non-isothermal drying starting at 40 °C but increased to 70 °C when the MC decreased to 25% [131].

A study on *Cannabis sativa* inflorescences collected from three different portions of the stem (upper, middle, and lower) was conducted by Namdar et al. [132] to explore the effect of three different drying methods: stream of nitrogen, vacuum, and rotary evaporator, along with three different extraction solvents, such as ethanol, n-hexane, and a hexane + ethanol solution. They reported that the upper portion of cannabis plants contained more cannabinoids, followed by middle and lower portion. They also reported that the extraction capacity of ethanol is higher than n-hexane and the hexane + ethanol solution. Thus, the extraction techniques and solvents have a significant role on the quantity of the cannabinoids. Valizadehderakhshan [133] reviewed the modern extraction methods along with possible pre-treatment practices to enhance the extraction efficiency for large-scale processing. These techniques may include microwave-assisted extraction, supercritical fluid extraction, maceration, etc. Thermal treatments other than drying can also alter the cannabinoid profile of cannabis. The effect of sterilization at a temperature of 62.5–70 °C for 10–20 s was studied by Jerushalmi et al. [134] and revealed that sterilization can reduce the contained phytocannabinoids but not by more than 20%. In situ decarboxylation of *Cannabis sativa* was analyzed by Nuapia et al. [135] at a temperature of 80–150 °C for 5–60 min using pressurized hot water extraction techniques. The findings revealed that decarboxylation increased with the temperature and time, but after 150 °C the decarboxylation rate decreased with time. Microwave heating on size reduced cannabis resulted in almost 1.8 times higher CBD content than the untreated cannabis sample, which reveals that microwave pre-treatment can increase the cannabinoid content [49]. Irradiation can also help in increasing the active compounds of cannabis. Work has been conducted on four different cultivars of *Cannabis sativa* sp. *indica* to determine the effect of irradiation on THC levels and it was reported that THC significantly increased due to irradiation treatment [110]. Storage analysis revealed that the higher the storage temperature, the higher was the degradation of phytocannabinoids and it was suggested that a 4 °C temperature could be the optimum for storage of cannabis along with olive oil [136,137]. Considering these few pieces of literature, it can be summarized that decarboxylation of cannabinoids to their neutral forms depends on thermal treatments applied during post-harvest processing. However, more investigations have yet to be conducted to determine the optimum drying methods and conditions, pre-treatments, extraction methods and time, storage, and many others.

Table 1. The effect of postharvest treatments on the cannabinoid profile of cannabis.

Sample	Moisture Level (wb)	Experimental Details	Findings	Source
Inflorescence and leaves of hemp (three varieties: Pipeline, Maverick and Queen Dream CBD)	Initial: 75–78%, Final: 9–13%	Freeze drying, Ambient drying, Hot air drying, and Sequential infrared and hot air (SIRHA) drying	• Increased drying rate with the increasing of temperature • Enhanced decarboxylation of CBDA from 0.2% to 14.1% and reduced terpene retention from 82.1% to 29.9% when the drying temperature increased from ambient to 90 °C • SIRHA drying resulted in significant losses of CBD and terpene up to 16.2% and 72.3%, respectively	[34]
Hemp buds	Initial: 65%, Final: 10%	Freeze drying, Hot air drying, Non-isothermal (stepwise) drying	• Phytocannabinoids and drying time were significantly affected by drying techniques and conditions • Drying temperatures over 40 °C significantly reduced terpene concentration, but no effect on the CBD level • Decarboxylation (CBDA to CBD) increased with temperature and maximum at 70 °C • Highest amount of CBD in non-isothermal drying samples	[131]
Inflorescences of medicinal cannabis (*Cannabis sativa*)	Not mentioned	Steam sterilization for 10s at 62.5°C, 15 s at 65 °C and 20 s at 70 °C	• Steaming caused minor reduction of terpenes and CBD (<20%)	[134]
Powder of *Cannabis sativa* seeds	Initial: Not mentioned, Final: 3.5–5.1%	In situ decarboxylation using pressurized hot water extraction technique at temperature (80 to 150 °C) for 5 to 60 min	• Decarboxylation to CBD and THC increased with time and temperature but THC decreased with time at 150 °C • Optimal decarboxylation time and temperature were 42.2 min and 149.9 °C, respectively	[135]
Inflorescences of hemp cv Felina 32	Not mentioned	Steam distillation (SD) or hydro distillation (HD) of fresh sample; HD of ambiently dried inflorescences; HD of blended and powdered inflorescences; HD of powdered and heated (120 °C for 1, 3, or 6 min) inflorescences; HD of powdered and microwaved (900 and 450 W) inflorescences	• HD recovered higher cannabinoids over SD • Pretreatments and drying triggered the cannabinoid profile; • Microwave heating resulted almost double CBD • MW heating for 1 min at 900 W was the most effective approach for the best quality products	[49]
Cannabis sativa inflorescence collected from upper, middle, and lower portion of stem	Not mentioned	Solvent extraction (ethanol, n-Hexane, mixture of hexane and ethanol (7:3, *v:v*)) of undried and dried (using gentle stream of nitrogen, vacuum dryer and rotary evaporator)	• Extraction ability of solvents: ethanol > n-hexane > hexane and ethanol solution • Cannabinoid and terpene quantity was influenced by drying methods and declined as the sampled flower moved from upper to middle to lower	[132]

Table 1. *Cont.*

Sample	Moisture Level (wb)	Experimental Details	Findings	Source
Inflorescences of *Cannabis sativa* (cultivars: Pink, RBS, RMS, and GSC)	Not mentioned	Dried and irradiated with 5 kGy emitted from a 10 MeV accelerator	• Irradiation affected the THC and terpenes • Except for RMS, three of the four cultivars examined showed a significant rise in THC levels • Three of the four extracts studied had their anti-cancer capabilities modified after being irradiated	[110]
Two *Cannabis sativa* strains combined together, sieved through a 355-mm sieve, and homogenized (one strain contained primarily THCA/THC and the other contained CBDA/CBD).	Not mentioned	Stored in 66-L microbiological incubators with ±0.2 °C consistency for up to 52 weeks at different temperature (20 °C, +4 °C, +20 °C, +32 °C, +37 °C, and +40 °C).	• Cannabinoids followed 1st order reaction kinetics during storage and also affected by temperature • Lowering temperature by 5 °C doubled the shelf life of 85% cannabinoids.	[136]
Inflorescences of medicinal cannabis (one is Δ9-THC-rich and another is CBD rich)	Initial: Not mentioned Final: 9 ± 0.3%	Samples stored in the dark condition for 12 months at 4 distinct temperatures (−80, −30 °C, 4 °C, and 25 °C) and in 2 physical forms (whole or ground).	• Storage at 25 °C affected mostly on phytocannabinoid concentrations over time • Dissolving the whole inflorescences or extracts in olive oil and stored at 4 °C was the ideal postharvest storage conditions	[137]
Inflorescences and leaves of hemp	Initial: 65.7% Final: up to constant moisture ratio	Convective drying at constant (40, 50 and 60 °C) and time varying temperature rise (1.5, 2.5 and 4 °C/h) at temperature in the 40–60 °C range	• Time varying temperature drying resulted in significantly higher CBD mean value for inflorescences (+46.7%) and leaves (+65.3%), but not significant for THC level	[138]

CBD = Cannabidiol; THC = tetrahydrocannabinol; THCA = tetrahydrocannabinolic acid, CBDA = cannabidiolic acid; wb = wet weight basis; SIRHA = sequential infrared and hot air; SD = steam distillation; HD = hydro distillation; MW = microwave.

5. Summary and Future Perspectives

The ongoing demand for cannabis due to its medical and recreational uses, and its consequent legalization by the governments of several countries, enhances the development of the cannabis industry. It has come to the attention of scientists and engineers to understand its chemistry and uses. Harvested cannabis goes through different operations, such as trimming, drying, curing, storage, and packaging, in different forms of the final product. Previous work on postharvest treatments indicates that slow drying is the most practiced drying method in the cannabis industry, thought it has several drawbacks like long drying time, chances of microbial contamination, and undefined equilibrium moisture level. Researchers are trying to find suitable postharvest operations for cannabis. Still, a complete guideline for the postharvest processing of cannabis is lacking. This review suggests practical research on finding the best suitable drying method for cannabis along with optimum conditions, equilibrium moisture content of the final product, suitable pretreatments, curing conditions, and packaging materials. To improve the drying profile and product quality, different modern technologies such as microwave, cold plasma, pulse electric field, radio frequency, etc., can be applied. The determination of a suitable sorption isotherm model(s) and the optimum equilibrium moisture content considering cannabis' physical characteristics is also important and should be considered for proper drying and storage management.

Author Contributions: Conceptualization, P.C.D., A.R.V. and L.G.T.; methodology, P.C.D., A.R.V. and L.G.T.; validation, L.G.T. and O.-D.B.; investigation, P.C.D. and A.R.V.; resources, L.G.T. and O.-D.B.; data curation, L.G.T. and O.-D.B.; writing—original draft preparation, P.C.D. and A.R.V.; writing—review and editing, P.C.D., A.R.V., L.G.T. and O.-D.B.; project administration, L.G.T.; funding acquisition, L.G.T. and O.-D.B. All authors have read and agreed to the published version of the manuscript.

Funding: We acknowledge the funding support of the University of Saskatchewan (CGPS Dean's Scholarship) and the Natural Sciences and Engineering Research Council of Canada (CREAT/543319-2020) through the project titled Quality Assurance and Quality Control for Cannabis (QAQCC).

Institutional Review Board Statement: Not applicable.

Informed Consent Statement: Not applicable.

Data Availability Statement: The authors confirm that the data supporting this study are available within the article.

Conflicts of Interest: The authors declare no conflict of interest.

References

1. Mansouri, H.; Bagheri, M. Induction of Polyploidy and Its Effect on *Cannabis sativa* L. In *Cannabis sativa L.-Botany and Biotechnology*; Chandra, S., Lata, H., ElSohly, M.A., Eds.; Springer International Publishing: Cham, Switzerland, 2017; pp. 365–383.
2. Cohen, K.; Weizman, A.; Weinstein, A. Positive and Negative Effects of Cannabis and Cannabinoids on Health. *Clin. Pharmacol. Ther.* **2019**, *105*, 1139–1147. [CrossRef] [PubMed]
3. Morales, P.; Hurst, D.P.; Reggio, P.H. Molecular Targets of the Phytocannabinoids: A Complex Picture. In *Phytocannabinoids: Unraveling the Complex Chemistry and Pharmacology of Cannabis sativa*; Kinghorn, A.D., Falk, H., Gibbons, S., Kobayashi, J., Eds.; Springer International Publishing: Cham, Switzerland, 2017; pp. 103–131.
4. Addo, P.W.; Desaulniers Brousseau, V.; Morello, V.; MacPherson, S.; Paris, M.; Lefsrud, M. Cannabis Chemistry, Post-Harvest Processing Methods and Secondary Metabolite Profiling: A Review. *Ind. Crops Prod.* **2021**, *170*, 113743. [CrossRef]
5. Cox-Georgian, D.; Ramadoss, N.; Dona, C.; Basu, C. Therapeutic and Medicinal Uses of Terpenes. *Med. Plants* **2019**, 333–359. [CrossRef]
6. Salentijn, E.M.J.; Zhang, Q.; Amaducci, S.; Yang, M.; Trindade, L.M. New Developments in Fiber Hemp (*Cannabis sativa* L.) Breeding. *Ind. Crops Prod.* **2015**, *68*, 32–41. [CrossRef]
7. Tramèr, M.R.; Carroll, D.; Campbell, F.A.; Reynolds, D.J.M.; Moore, R.A.; McQuay, H.J. Cannabinoids for Control of Chemotherapy Induced Nausea and Vomiting: Quantitative Systematic Review. *BMJ* **2001**, *323*, 16. [CrossRef] [PubMed]
8. Pertwee, R.G. Targeting the Endocannabinoid System with Cannabinoid Receptor Agonists: Pharmacological Strategies and Therapeutic Possibilities. *Philos. Trans. R. Soc. B Biol. Sci.* **2012**, *367*, 3353–3363. [CrossRef]
9. Wartenberg, A.C.; Holden, P.A.; Bodwitch, H.; Parker-Shames, P.; Novotny, T.; Harmon, T.C.; Hart, S.C.; Beutel, M.; Gilmore, M.; Hoh, E.; et al. Cannabis and the Environment: What Science Tells Us and What We Still Need to Know. *Environ. Sci. Technol. Lett.* **2021**, *8*, 98–107. [CrossRef]
10. Challa, S.K.R.; Misra, N.N.; Martynenko, A. Drying of Cannabis—State of the Practices and Future Needs. *Dry. Technol.* **2021**, *39*, 2055–2064. [CrossRef]
11. Patel, B.; Wene, D.; Fan, Z.T. Qualitative and Quantitative Measurement of Cannabinoids in Cannabis Using Modified HPLC/DAD Method. *J. Pharm. Biomed. Anal.* **2017**, *146*, 15–23. [CrossRef] [PubMed]
12. Radwan, M.M.; Wanas, A.S.; Chandra, S.; ElSohly, M.A. Natural Cannabinoids of Cannabis and Methods of Analysis. In *Cannabis sativa L. -Botany and Biotechnology*; Chandra, S., Lata, H., ElSohly, M.A., Eds.; Springer International Publishing: Cham, Switzerland, 2017; pp. 161–182.
13. Wang, M.; Wang, Y.-H.; Avula, B.; Radwan, M.M.; Wanas, A.S.; van Antwerp, J.; Parcher, J.F.; ElSohly, M.A.; Khan, I.A. Decarboxylation Study of Acidic Cannabinoids: A Novel Approach Using Ultra-High-Performance Supercritical Fluid Chromatography/Photodiode Array-Mass Spectrometry. *Cannabis Cannabinoid Res.* **2016**, *1*, 262–271. [CrossRef] [PubMed]
14. Gülck, T.; Møller, B.L. Phytocannabinoids: Origins and Biosynthesis. *Trends Plant Sci.* **2020**, *25*, 985–1004. [CrossRef] [PubMed]
15. Nachnani, R.; Raup-Konsavage, W.M.; Vrana, K.E. The Pharmacological Case for Cannabigerol. *J. Pharmacol. Exp. Ther.* **2021**, *376*, 204–212. [CrossRef] [PubMed]
16. Turner, C.E.; Elsohly, M.A. Constituents of *Cannabis sativa* L. XVI. A Possible Decomposition Pathway of Δ9-Tetrahydrocannabinol to Cannabinol. *J. Heterocycl. Chem.* **1979**, *16*, 1667–1668. [CrossRef]
17. Hecksel, R.; LaVigne, J.; Streicher, J.M. In Defense of the Entourage Effect: Terpenes Found in *Cannabis sativa* Activate the Cannabinoid Receptor 1 In Vitro. *FASEB J.* **2020**, *34*, 1. [CrossRef]
18. Mudge, E.M.; Brown, P.N.; Murch, S.J. The Terroir of Cannabis: Terpene Metabolomics as a Tool to Understand *Cannabis sativa* Selections. *Planta Med.* **2019**, *85*, 781–796. [CrossRef] [PubMed]
19. Kim, E.S. Cannabinoid Synthesis and Accumulation in Glandular Trichomes of *Cannabis sativa*. *J. Acupunct. Meridian Stud.* **2020**, *13*, 77. [CrossRef]
20. Tanney, C.A.S.; Backer, R.; Geitmann, A.; Smith, D.L. Cannabis Glandular Trichomes: A Cellular Metabolite Factory. *Front. Plant Sci.* **2021**, *12*, 721986. [CrossRef] [PubMed]
21. Turner, J.C.; Hemphill, J.K.; Mahlberg, P.G. Trichomes and Cannabinoid Content of Developing Leaves and Bracts of *Cannabis sativa* L. (Cannabaceae). *Am. J. Bot.* **1980**, *67*, 1397–1406. [CrossRef]
22. Henry, P. Cannabis Chemovar Classification: Terpenes Hyper-Classes and Targeted Genetic Markers for Accurate Discrimination of Flavours and Effects. *PeerJ Prepr.* **2017**, *5*, e3307v1.

23. Kader, A.A.; Kasmire, R.F.; Mitchell, F.G.; Reid, M.S.; Sommer, N.F.; Thompson, J.F. *Postharvest Technology of Horticultural Crops*; University of California Agriculture & Natural Resources: Davis, CA, USA, 1985.

24. Available online: https://Weedmaps.Com/Learn/the-Plant/When-to-Harvest-Cannabis (accessed on 3 March 2022).

25. Geankoplis, C.J. *Transport Processes and Separation Process Principles: (Includes Unit Operations)*, 4th ed.; Prentice Hall Professional Technical Reference: Upper Saddle River, NJ, USA, 2003.

26. Sparks, B.D. The Latest Updates in Cannabis Trimming Innovation. Available online: https://www.greenhousegrower.com/production/the-latest-updates-in-cannabis-trimming-innovations/ (accessed on 5 March 2022).

27. Cresp, H. Best Bud Trimmers—Top Trimming Machines of 2020. Grow Light Central. Available online: https://growlightcentral.com/blogs/news/best-bud-trimmers (accessed on 16 March 2022).

28. Jin, D.; Jin, S.; Chen, J. Cannabis Indoor Growing Conditions, Management Practices, and Post-Harvest Treatment: A Review. *Am. J. Plant Sci.* **2019**, *10*, 925–946. [CrossRef]

29. Lazarjani, M.P.; Young, O.; Kebede, L.; Seyfoddin, A. Processing and Extraction Methods of Medicinal Cannabis: A Narrative Review. *J. Cannabis Res.* **2021**, *3*, 32. [CrossRef] [PubMed]

30. Brooker, D.B. *Drying and Storage of Grains and Oilseeds*; Van Nostrand Reinhold: New York, NY, USA, 1992.

31. van 't Land, C.M. *Drying in the Process Industry*; WILEY: Hoboken, NJ, USA, 2011.

32. Stroshine, R.L. *Physical Properties of Agricultural Materials and Food Products*; R. Stroshine: West Lafayette, IN, USA, 2004.

33. Srikiatden, J.; Roberts, J.S. Moisture Transfer in Solid Food Materials: A Review of Mechanisms, Models, and Measurements. *Int. J. Food Prop.* **2007**, *10*, 739–777. [CrossRef]

34. Chen, C.; Wongso, I.; Putnam, D.; Khir, R.; Pan, Z. Effect of Hot Air and Infrared Drying on the Retention of Cannabidiol and Terpenes in Industrial Hemp (*Cannabis sativa* L.). *Ind. Crops Prod.* **2021**, *172*, 114051. [CrossRef]

35. AL Ubeed, H.M.; Wills, R.B.H.; Chandrapala, J. Post-Harvest Operations to Generate High-Quality Medicinal Cannabis Products: A Systemic Review. *Molecular* **2022**, *27*, 1719. [CrossRef] [PubMed]

36. Balmer, R.T. Chapter 12-Mixtures of Gases and Vapors. In *Modern Engineering Thermodynamics*; Balmer, R.T., Ed.; Academic Press: Boston, MA, USA, 2011; pp. 405–446.

37. Dincer, I.; Rosen, M.A. Chapter 6-Exergy Analyses of Psychrometric Processes. In *Exergy*, 3rd ed.; Dincer, I., Rosen, M.A., Eds.; Elsevier: Amsterdam, The Netherlands, 2021; pp. 101–123.

38. Hall, M.R.; Allinson, D. 1—Heat and Mass Transport Processes in Building Materials. In *Materials for Energy Efficiency and Thermal Comfort in Buildings*; Hall, M.R., Ed.; Woodhead Publishing Series in Energy; Elsevier: Amsterdam, The Netherlands, 2010; pp. 3–53.

39. Andrade, R.D.; Lemus, R.; Perez, C.E. Models of Sorption Isotherms for Food: Uses and Limitations. *Vitae* **2011**, *18*, 325–334.

40. Mathlouthi, M.; Rogé, B. Water Vapour Sorption Isotherms and the Caking of Food Powders. *Food Chem.* **2003**, *82*, 61–71. [CrossRef]

41. Troller, J.A.; Christian, J.H.B. Water Activity-1. Basic Concepts. In *Water Activity and Food*; Troller, J.A., Christian, J.H.B., Eds.; Academic Press: Cambridge, MA, USA, 1978; pp. 1–12.

42. U.S. Food and Drug Administration. *Water Activity (Aw) in Foods*; U.S. Food and Drug Administration: Silver Spring, MD, USA, 2014.

43. Labuza, T.P.; Acott, K.; TATiNl, S.R.; Lee, R.Y.; Flink, J.; McCALL, W. Water Activity Determination: A Collaborative Study of Different Methods. *J. Food Sci.* **2008**, *41*, 910–917. [CrossRef]

44. ASABE (American Society of Agricultural and Biological Engineers). *D245.7 (R2021): Moisture Relationships of Plant Based Agricultural Products*; ASABE: St. Joseph, MI, USA, 2017.

45. Deng, L.-Z.; Mujumdar, A.S.; Zhang, Q.; Yang, X.-H.; Wang, J.; Zheng, Z.-A.; Gao, Z.-J.; Xiao, H.-W. Chemical and Physical Pretreatments of Fruits and Vegetables: Effects on Drying Characteristics and Quality Attributes–A Comprehensive Review. *Crit. Rev. Food Sci. Nutr.* **2019**, *59*, 1408–1432. [CrossRef] [PubMed]

46. Chandrasekaran, S.; Ramanathan, S.; Basak, T. Microwave Food Processing—A Review. *Food Res. Int.* **2013**, *52*, 243–261. [CrossRef]

47. Rahath Kubra, I.; Kumar, D.; Jagan Mohan Rao, L. Emerging Trends in Microwave Processing of Spices and Herbs. *Emerg. Trends Microw. Process. Spices Herbs* **2016**, *56*, 2160–2173. [CrossRef] [PubMed]

48. Gunasekaran, S. Pulsed Microwave-Vacuum Drying of Food Materials. *Dry. Technol.* **1999**, *17*, 395–412. [CrossRef]

49. Fiorini, D.; Molle, A.; Nabissi, M.; Santini, G.; Benelli, G.; Maggi, F. Valorizing Industrial Hemp (*Cannabis sativa* L.) by-Products: Cannabidiol Enrichment in the Inflorescence Essential Oil Optimizing Sample Pre-Treatment Prior to Distillation. *Ind. Crops Prod.* **2019**, *128*, 581–589. [CrossRef]

50. Sui, X.; Liu, T.; Ma, C.; Yang, L.; Zu, Y.; Zhang, L.; Wang, H. Microwave Irradiation to Pretreat Rosemary (*Rosmarinus officinalis* L.) for Maintaining Antioxidant Content during Storage and to Extract Essential Oil Simultaneously. *Food Chem.* **2012**, *131*, 1399–1405. [CrossRef]

51. Drinić, Z.; Vladić, J.; Koren, A.; Zeremski, T.; Stojanov, N.; Kiprovski, B.; Vidović, S. Microwave-Assisted Extraction of Cannabinoids and Antioxidants from *Cannabis sativa* Aerial Parts and Process Modeling. *J. Chem. Technol. Biotechnol.* **2020**, *95*, 831–839. [CrossRef]

52. Matešić, N.; Jurina, T.; Benković, M.; Panić, M.; Valinger, D.; Gajdoš Kljusurić, J.; Jurinjak Tušek, A. Microwave-Assisted Extraction of Phenolic Compounds from *Cannabis sativa* L.: Optimization and Kinetics Study. *Sep. Sci. Technol.* **2021**, *56*, 2047–2060. [CrossRef]

53. Radoiu, M.; Kaur, H.; Bakowska-Barczak, A.; Splinter, S. Microwave-Assisted Industrial Scale Cannabis Extraction. *Technol.* **2020**, *8*, 45. [CrossRef]
54. Teh, S.-S.; Niven, B.E.; Bekhit, A.E.-D.A.; Carne, A.; Birch, E.J. The Use of Microwave and Pulsed Electric Field as a Pretreatment Step in Ultrasonic Extraction of Polyphenols from Defatted Hemp Seed Cake (*Cannabis sativa*) Using Response Surface Methodology. *Food Bioprocess Technol.* **2014**, *7*, 3064–3076. [CrossRef]
55. Muhammad, A.I.; Li, Y.; Liao, X.; Liu, D.; Ye, X.; Chen, S.; Hu, Y.; Wang, J.; Ding, T. Effect of Dielectric Barrier Discharge Plasma on Background Microflora and Physicochemical Properties of Tiger Nut Milk. *Food Control.* **2019**, *96*, 119–127. [CrossRef]
56. Muhammad, A.I.; Liao, X.; Cullen, P.J.; Liu, D.; Xiang, Q.; Wang, J.; Chen, S.; Ye, X.; Ding, T. Effects of Nonthermal Plasma Technology on Functional Food Components. *Compr. Rev. Food Sci. Food Saf.* **2018**, *17*, 1379–1394. [CrossRef] [PubMed]
57. Amini, M.; Ghoranneviss, M. Effects of Cold Plasma Treatment on Antioxidants Activity, Phenolic Contents and Shelf Life of Fresh and Dried Walnut (*Juglans regia* L.) Cultivars during Storage. *LWT* **2016**, *73*, 178–184. [CrossRef]
58. Shishir, M.R.I.; Karim, N.; Bao, T.; Gowd, V.; Ding, T.; Sun, C.; Chen, W. Cold Plasma Pretreatment–A Novel Approach to Improve the Hot Air Drying Characteristics, Kinetic Parameters, and Nutritional Attributes of Shiitake Mushroom. *Dry. Technol.* **2020**, *38*, 2134–2150. [CrossRef]
59. Domonkos, M.; Tichá, P.; Trejbal, J.; Demo, P. Applications of Cold Atmospheric Pressure Plasma Technology in Medicine, Agriculture and Food Industry. *Appl. Sci.* **2021**, *11*, 4809. [CrossRef]
60. Brány, D.; Dvorská, D.; Halašová, E.; Škovierová, H. Cold Atmospheric Plasma: A Powerful Tool for Modern Medicine. *Int. J. Mol. Sci.* **2020**, *21*, 2932. [CrossRef]
61. Zhang, X.-L.; Zhong, C.-S.; Mujumdar, A.S.; Yang, X.-H.; Deng, L.-Z.; Wang, J.; Xiao, H.-W. Cold Plasma Pretreatment Enhances Drying Kinetics and Quality Attributes of Chili Pepper (*Capsicum annuum* L.). *J. Food Eng.* **2019**, *241*, 51–57. [CrossRef]
62. Li, S.; Chen, S.; Han, F.; Xv, Y.; Sun, H.; Ma, Z.; Chen, J.; Wu, W. Development and Optimization of Cold Plasma Pretreatment for Drying on Corn Kernels. *J. Food Sci.* **2019**, *84*, 2181–2189. [CrossRef]
63. Zhou, Y.-H.; Vidyarthi, S.K.; Zhong, C.-S.; Zheng, Z.-A.; An, Y.; Wang, J.; Wei, Q.; Xiao, H.-W. Cold Plasma Enhances Drying and Color, Rehydration Ratio and Polyphenols of Wolfberry via Microstructure and Ultrastructure Alteration. *LWT* **2020**, *134*, 110173. [CrossRef]
64. Bao, T.; Hao, X.; Shishir, M.R.I.; Karim, N.; Chen, W. Cold Plasma: An Emerging Pretreatment Technology for the Drying of Jujube Slices. *Food Chem.* **2021**, *337*, 127783. [CrossRef]
65. Tabibian, S.A.; Labbafi, M.; Askari, G.H.; Rezaeinezhad, A.R.; Ghomi, H. Effect of Gliding Arc Discharge Plasma Pretreatment on Drying Kinetic, Energy Consumption and Physico-Chemical Properties of Saffron (*Crocus sativus* L.). *J. Food Eng.* **2020**, *270*, 109766. [CrossRef]
66. Miraei Ashtiani, S.-H.; Rafiee, M.; Mohebi Morad, M.; Khojastehpour, M.; Khani, M.R.; Rohani, A.; Shokri, B.; Martynenko, A. Impact of Gliding Arc Plasma Pretreatment on Drying Efficiency and Physicochemical Properties of Grape. *Innov. Food Sci. Emerg. Technol.* **2020**, *63*, 102381. [CrossRef]
67. Ebadi, M.-T.; Abbasi, S.; Harouni, A.; Sefidkon, F. Effect of Cold Plasma on Essential Oil Content and Composition of Lemon Verbena. *Food Sci. Nutr.* **2019**, *7*, 1166–1171. [CrossRef] [PubMed]
68. Mildaziene, V.; Pauzaite, G.; Naucienė, Z.; Malakauskiene, A.; Zukiene, R.; Januskaitiene, I.; Jakstas, V.; Ivanauskas, L.; Filatova, I.; Lyushkevich, V. Pre-Sowing Seed Treatment with Cold Plasma and Electromagnetic Field Increases Secondary Metabolite Content in Purple Coneflower (*Echinacea purpurea*) Leaves. *Plasma Process. Polym.* **2018**, *15*, 1700059. [CrossRef]
69. Jerushalmi, S.; Maymon, M.; Dombrovsky, A.; Freeman, S. Effects of Cold Plasma, Gamma and e-Beam Irradiations on Reduction of Fungal Colony Forming Unit Levels in Medical Cannabis Inflorescences. *J. Cannabis Res.* **2020**, *2*, 12. [CrossRef] [PubMed]
70. Knorr, D.; Angersbach, A.; Eshtiaghi, M.N.; Heinz, V.; Lee, D.-U. Processing Concepts Based on High Intensity Electric Field Pulses. *Trends Food Sci. Technol.* **2001**, *12*, 129–135. [CrossRef]
71. Rastogi, N.K.; Eshtiaghi, M.N.; Knorr, D. Accelerated Mass Transfer During Osmotic Dehydration of High Intensity Electrical Field Pulse Pretreated Carrots. *J. Food Sci.* **1999**, *64*, 1020–1023. [CrossRef]
72. Wiktor, A.; Nowacka, M.; Dadan, M.; Rybak, K.; Lojkowski, W.; Chudoba, T.; Witrowa-Rajchert, D. The Effect of Pulsed Electric Field on Drying Kinetics, Color, and Microstructure of Carrot. *Dry. Technol.* **2016**, *34*, 1286–1296. [CrossRef]
73. Saulis, G. Electroporation of Cell Membranes: The Fundamental Effects of Pulsed Electric Fields in Food Processing. *Food Eng. Rev.* **2010**, *2*, 52–73. [CrossRef]
74. Liu, Z.; Esveld, E.; Vincken, J.-P.; Bruins, M.E. Pulsed Electric Field as an Alternative Pre-Treatment for Drying to Enhance Polyphenol Extraction from Fresh Tea Leaves. *Food Bioprocess Technol.* **2019**, *12*, 183–192. [CrossRef]
75. Wu, Y.; Guo, Y.; Zhang, D. Study of the Effect of High-Pulsed Electric Field Treatment on Vacuum Freeze-Drying of Apples. *Dry. Technol.* **2011**, *29*, 1714–1720. [CrossRef]
76. Alam, M.R.; Lyng, J.G.; Frontuto, D.; Marra, F.; Cinquanta, L. Effect of Pulsed Electric Field Pretreatment on Drying Kinetics, Color, and Texture of Parsnip and Carrot. *J. Food Sci.* **2018**, *83*, 2159–2166. [CrossRef] [PubMed]
77. Gachovska, T.K.; Simpson, M.V.; Ngadi, M.O.; Raghavan, G. Pulsed Electric Field Treatment of Carrots before Drying and Rehydration. *J. Sci. Food Agric.* **2009**, *89*, 2372–2376. [CrossRef]
78. Lebovka, N.I.; Shynkaryk, N.V.; Vorobiev, E. Pulsed Electric Field Enhanced Drying of Potato Tissue. *J. Food Eng.* **2007**, *78*, 606–613. [CrossRef]

79. Ostermeier, R.; Giersemehl, P.; Siemer, C.; Töpfl, S.; Jäger, H. Influence of Pulsed Electric Field (PEF) Pre-Treatment on the Convective Drying Kinetics of Onions. *J. Food Eng.* **2018**, *237*, 110–117. [CrossRef]

80. Thamkaew, G.; Gómez Galindo, F. Influence of Pulsed and Moderate Electric Field Protocols on the Reversible Permeabilization and Drying of Thai Basil Leaves. *Innov. Food Sci. Emerg. Technol.* **2020**, *64*, 102430. [CrossRef]

81. Ade-Omowaye, B.I.O.; Rastogi, N.K.; Angersbach, A.; Knorr, D. Combined Effects of Pulsed Electric Field Pre-Treatment and Partial Osmotic Dehydration on Air Drying Behaviour of Red Bell Pepper. *J. Food Eng.* **2003**, *60*, 89–98. [CrossRef]

82. Donsì, F.; Ferrari, G.; Pataro, G. Applications of Pulsed Electric Field Treatments for the Enhancement of Mass Transfer from Vegetable Tissue. *Food Eng. Rev.* **2010**, *2*, 109–130. [CrossRef]

83. Goettel, M.; Eing, C.; Gusbeth, C.; Straessner, R.; Frey, W. Pulsed Electric Field Assisted Extraction of Intracellular Valuables from Microalgae. *Algal Res.* **2013**, *2*, 401–408. [CrossRef]

84. Fincan, M.; DeVito, F.; Dejmek, P. Pulsed Electric Field Treatment for Solid–Liquid Extraction of Red Beetroot Pigment. *J. Food Eng.* **2004**, *64*, 381–388. [CrossRef]

85. Puértolas, E.; Cregenzán, O.; Luengo, E.; Álvarez, I.; Raso, J. Pulsed-Electric-Field-Assisted Extraction of Anthocyanins from Purple-Fleshed Potato. *Food Chem.* **2013**, *136*, 1330–1336. [CrossRef]

86. Delsart, C.; Ghidossi, R.; Poupot, C.; Cholet, C.; Grimi, N.; Vorobiev, E.; Milisic, V.; Mietton Peuchot, M. Enhanced Extraction of Phenolic Compounds from Merlot Grapes by Pulsed Electric Field Treatment. *Am. J. Enol. Vitic.* **2012**, *63*, 205–211. [CrossRef]

87. Takaki, K.; Hatayama, H.; Koide, S.; Kawamura, Y. Improvement of Polyphenol Extraction from Grape Skin by Pulse Electric Field. In Proceedings of the 2011 IEEE Pulsed Power Conference, Chicago, IL, USA, 19–23 June 2011; pp. 1262–1265.

88. Haji-Moradkhani, A.; Rezaei, R.; Moghimi, M. Optimization of Pulsed Electric Field-Assisted Oil Extraction from Cannabis Seeds. *J. Food Process Eng.* **2019**, *42*, e13028. [CrossRef]

89. Musielak, G.; Mierzwa, D.; Kroehnke, J. Food Drying Enhancement by Ultrasound—A Review. *Trends Food Sci. Technol.* **2016**, *56*, 126–141. [CrossRef]

90. Chemat, F.; Zill-e-Huma; Khan, M. K. Applications of Ultrasound in Food Technology: Processing, Preservation and Extraction. *Ultrason. Sonochemis* **2011**, *18*, 813–835. [CrossRef]

91. Chen, Z.-G.; Guo, X.-Y.; Wu, T. A Novel Dehydration Technique for Carrot Slices Implementing Ultrasound and Vacuum Drying Methods. *Ultrason. Sonochemis* **2016**, *30*, 28–34. [CrossRef]

92. Nowacka, M.; Wiktor, A.; Śledź, M.; Jurek, N.; Witrowa-Rajchert, D. Drying of Ultrasound Pretreated Apple and Its Selected Physical Properties. *J. Food Eng.* **2012**, *113*, 427–433. [CrossRef]

93. Cárcel, J.A.; García-Pérez, J.V.; Riera, E.; Mulet, A. Influence of High-Intensity Ultrasound on Drying Kinetics of Persimmon. *Dry. Technol.* **2007**, *25*, 185–193. [CrossRef]

94. Gallego-Juarez, J.A.; Rodriguez-Corral, G.; Gálvez Moraleda, J.C.; Yang, T.S. A New High-Intensity Ultrasonic Technology for Food Dehydration. *Dry. Technol.* **1999**, *17*, 597–608. [CrossRef]

95. Tekin, Z.H.; Baslar, M. The Effect of Ultrasound-Assisted Vacuum Drying on the Drying Rate and Quality of Red Peppers. *J. Therm. Anal. Calorim.* **2018**, *132*, 1131–1143. [CrossRef]

96. Vieira da Silva Júnior, E.; Lins de Melo, L.; Batista de Medeiros, R.A.; Pimenta Barros, Z.M.; Azoubel, P.M. Influence of Ultrasound and Vacuum Assisted Drying on Papaya Quality Parameters. *LWT* **2018**, *97*, 317–322. [CrossRef]

97. Rojas, M.L.; Silveira, I.; Augusto, P.E.D. Ultrasound and Ethanol Pre-Treatments to Improve Convective Drying: Drying, Rehydration and Carotenoid Content of Pumpkin. *Food Bioprod. Process.* **2020**, *119*, 20–30. [CrossRef]

98. Fernandes, F.A.N.; Linhares, F.E.; Rodrigues, S. Ultrasound as Pre-Treatment for Drying of Pineapple. *Ultrasonics Sonochemis* **2008**, *15*, 1049–1054. [CrossRef]

99. Szadzińska, J.; Łechtańska, J.; Pashminehazar, R.; Kharaghani, A.; Tsotsas, E. Microwave- and Ultrasound-Assisted Convective Drying of Raspberries: Drying Kinetics and Microstructural Changes. *Dry. Technol.* **2019**, *37*, 1–12. [CrossRef]

100. Tomšik, A.; Pavlić, B.; Vladić, J.; Ramić, M.; Brindza, J.; Vidović, S. Optimization of Ultrasound-Assisted Extraction of Bioactive Compounds from Wild Garlic (*Allium ursinum* L.). *Ultrason. Sonochem.* **2016**, *29*, 502–511. [CrossRef]

101. Sharayei, P.; Azarpazhooh, E.; Zomorodi, S.; Ramaswamy, H.S. Ultrasound Assisted Extraction of Bioactive Compounds from Pomegranate (*Punica granatum* L.) Peel. *LWT* **2019**, *101*, 342–350. [CrossRef]

102. González de Peredo, A.V.; Vázquez-Espinosa, M.; Espada-Bellido, E.; Ferreiro-González, M.; Amores-Arrocha, A.; Palma, M.; Barbero, G.F.; Jiménez-Cantizano, A. Alternative Ultrasound-Assisted Method for the Extraction of the Bioactive Compounds Present in Myrtle (*Myrtus communis* L.). *Molecules* **2019**, *24*, 882. [CrossRef] [PubMed]

103. Xu, X.-Y.; Meng, J.-M.; Mao, Q.-Q.; Shang, A.; Li, B.-Y.; Zhao, C.-N.; Tang, G.-Y.; Cao, S.-Y.; Wei, X.-L.; Gan, R.-Y.; et al. Effects of Tannase and Ultrasound Treatment on the Bioactive Compounds and Antioxidant Activity of Green Tea Extract. *Antioxidants* **2019**, *8*, 362. [CrossRef]

104. Banožić, M.; Banjari, I.; Jakovljević, M.; Šubarić, D.; Tomas, S.; Babić, J.; Jokić, S. Optimization of Ultrasound-Assisted Extraction of Some Bioactive Compounds from Tobacco Waste. *Molecules* **2019**, *24*, 1611. [CrossRef] [PubMed]

105. Porto, C.; Decorti, D.; Natolino, A. Ultrasound-Assisted Extraction of Volatile Compounds from Industrial *Cannabis sativa* L. Inflorescences. *Int. J. Appl. Res. Nat. Prod.* **2014**, *7*, 8–14.

106. Agarwal, C.; Máthé, K.; Hofmann, T.; Csóka, L. Ultrasound-Assisted Extraction of Cannabinoids from *Cannabis sativa* L. Optimized by Response Surface Methodology. *J. Food Sci.* **2018**, *83*, 700–710. [CrossRef] [PubMed]

107. Esmaeilzadeh Kenari, R.; Dehghan, B. Optimization of Ultrasound-Assisted Solvent Extraction of Hemp (*Cannabis sativa* L.) Seed Oil Using RSM: Evaluation of Oxidative Stability and Physicochemical Properties of Oil. *Food Sci. Nutr.* **2020**, *8*, 4976–4986. [CrossRef] [PubMed]

108. Esposito, M.; Piazza, L. Ultrasound-Assisted Extraction of Oil from Hempseed (*Cannabis sativa* L.): Part 1. *J. Sci. Food Agric.* **2022**, *102*, 732–739. [CrossRef]

109. Thakur, B.R.; Singh, R.K. Food Irradiation-chemistry and Applications. *Food Rev. Int.* **1994**, *10*, 437–473. [CrossRef]

110. Kovalchuk, O.; Li, D.; Rodriguez-Juarez, R.; Golubov, A.; Hudson, D.; Kovalchuk, I. The Effect of Cannabis Dry Flower Irradiation on the Level of Cannabinoids, Terpenes and Anti-Cancer Properties of the Extracts. *Biocatal. Agric. Biotechnol.* **2020**, *29*, 101736. [CrossRef]

111. Hazekamp, A. Evaluating the Effects of Gamma-Irradiation for Decontamination of Medicinal Cannabis. *Front. Pharmacol.* **2016**, *7*, 108. [CrossRef] [PubMed]

112. Wang, J.; Chao, Y. Effect of Gamma Irradiation on Quality of Dried Potato. *Radiat. Phys. Chem.* **2003**, *66*, 293–297. [CrossRef]

113. Nayak, C.A.; Suguna, K.; Narasimhamurthy, K.; Rastogi, N.K. Effect of Gamma Irradiation on Histological and Textural Properties of Carrot, Potato and Beetroot. *J. Food Eng.* **2007**, *79*, 765–770. [CrossRef]

114. Ling, Y.; Wang, H.; Fei, X.; Yue, Q.; Huang, T.; Shan, Q.; Hei, D.; Chen, T.; Zhang, X.; Jia, W. Improved Dehydration Performance of Tofu Protein by Ionizing Irradiation Pretreatment. *J. Radioanal. Nucl. Chem.* **2021**, *327*, 575–583. [CrossRef]

115. Sung, Y.J.; Shin, S.-J. Compositional Changes in Industrial Hemp Biomass (*Cannabis sativa* L.) Induced by Electron Beam Irradiation Pretreatment. *Biomass Bioenergy* **2011**, *35*, 3267–3270. [CrossRef]

116. Anuradha, K.; Shyamala, B.N.; Naidu, M.M. Vanilla- Its Science of Cultivation, Curing, Chemistry, and Nutraceutical Properties. *Crit. Rev. Food Sci. Nutr.* **2013**, *53*, 1250–1276. [CrossRef] [PubMed]

117. Dignum, M.J.W.; Kerler, J.; Verpoorte, R. Vanilla Production: Technological, Chemical, and Biosynthetic Aspects. *Food Rev. Int.* **2001**, *17*, 119–120. [CrossRef]

118. Ko, E.Y.; Sharma, K.; Nile, S.H. Effect of Harvesting Practices, Lifting Time, Curing Methods, and Irrigation on Quercetin Content in Onion (*Allium cepa* L.) Cultivars. *Emir. J. Food Agric.* **2016**, *28*, 594–600.

119. Mogren, L.M.; Olsson, M.E.; Gertsson, U.E. Quercetin Content in Field-Cured Onions (*Allium cepa* L.): Effects of Cultivar, Lifting Time, and Nitrogen Fertilizer Level. *J. Agric. Food Chem.* **2006**, *54*, 6185–6191. [CrossRef]

120. Shi, L.; Kim, E.; Yang, L.; Huang, Y.; Ren, N.; Li, B.; He, P.; Tu, Y.; Wu, Y. Effect of a Combined Microwave-Assisted Drying and Air Drying on Improving Active Nutraceutical Compounds, Flavor Quality, and Antioxidant Properties of *Camellia sinensis* L. (Cv. Longjing 43) Flowers. *Food Qual. Saf.* **2021**, *5*, fyaa040. [CrossRef]

121. Wang, L.; Cheng, B.; Li, Z.; Liu, T.; Li, J. Intelligent tobacco flue-curing method based on leaf texture feature analysis. *Optik* **2017**, *150*, 117–130. [CrossRef]

122. Tabil, L.; Sokhansanj, S. Mechanical and Temperature Effects on Shelf Life Stability of Fruits and Vegetables. In *Food Shelf Life Stability*; CRC Press Inc.: Boca Raton, FL, USA, 2000; pp. 37–86.

123. Uline.ca. Uline Products in Stock-Uline.ca. Available online: https://www.uline.ca/cls_uline/Uline-Products (accessed on 14 March 2022).

124. NDSupplies.ca. All Child Resistant Packaging. ND Supplies—Manufacturing of Packaging Supplies. Available online: https://ndsupplies.ca/product-category/all-child-resistant-packaging/ (accessed on 14 March 2022).

125. Popham, N. 8-Resin Infusion for the Manufacture of Large Composite Structures. In *Marine Composites*; Pemberton, R., Summerscales, J., Graham-Jones, J., Eds.; Woodhead Publishing Series in Composites Science and Engineering; Woodhead Publishing: Sawston, UK, 2019; pp. 227–268.

126. McKeen, L.W. 6-Polyesters. In *Film Properties of Plastics and Elastomers 3rd ed*; McKeen, L.W., Ed.; Plastics Design Library: Norwich, NY, USA; William Andrew Publishing, 2012; pp. 91–123.

127. Trofin, I.G.; Dabija, G.; Vaireanu, I.; Filipescu, L. Long-Term Storage and Cannabis Oil Stability. *Rev. Chim.* **2012**, *63*, 293–297.

128. Trofin, I.G.; Dabija, G.; Văireanu, I.; Filipescu, L. The Influence of Long-Term Storage Conditions on the Stability of Cannabinoids Derived from Cannabis Resin. *Rev. Chim. Bucharest.* **2012**, *64*, 422–427.

129. Trofin, I.G.; Vlad, C.C.; Dabija, G.; Filipescu, L. Influence of Storage Conditions on the Chemical Potency of Herbal Cannabis. *Rev. Chim.* **2011**, *62*, 639–645.

130. Taschwer, M.; Schmid, M.G. Determination of the Relative Percentage Distribution of THCA and Δ9-THC in Herbal Cannabis Seized in Austria—Impact of Different Storage Temperatures on Stability. *Forensic Sci. Int.* **2015**, *254*, 167–171. [CrossRef] [PubMed]

131. Challa, S.K.R. Drying Kinetics and the Effects of Drying Methods on Quality (CBD, Terpenes and Color) of Hemp (*Cannabis sativa* L.) Buds. Master's Thesis, Dalhousie University, Halifax, NS, Canada, 2020.

132. Namdar, D.; Mazuz, M.; Ion, A.; Koltai, H. Variation in the Compositions of Cannabinoid and Terpenoids in *Cannabis sativa* Derived from Inflorescence Position along the Stem and Extraction Methods. *Ind. Crops Prod.* **2018**, *113*, 376–382. [CrossRef]

133. Valizadehderakhshan, M.; Shahbazi, A.; Kazem-Rostami, M.; Todd, M.S.; Bhowmik, A.; Wang, L. Extraction of Cannabinoids from *Cannabis sativa* L. (Hemp)—Review. *Agriculture* **2021**, *11*, 384. [CrossRef]

134. Jerushalmi, S.; Maymon, M.; Dombrovsky, A.; Regev, R.; Schmilovitch, Z.; Namdar, D.; Shalev, N.; Koltai, H.; Freeman, S. Effects of Steam Sterilization on Reduction of Fungal Colony Forming Units, Cannabinoids and Terpene Levels in Medical Cannabis Inflorescences. *Sci. Rep.* **2021**, *11*, 13973. [CrossRef] [PubMed]

135. Nuapia, Y.; Maraba, K.; Tutu, H.; Chimuka, L.; Cukrowska, E. In Situ Decarboxylation-Pressurized Hot Water Extraction for Selective Extraction of Cannabinoids from *Cannabis sativa*. Chemometric Approach. *Molecules* **2021**, *26*, 3343. [CrossRef] [PubMed]
136. Meija, J.; McRae, G.; Miles, C.O.; Melanson, J.E. Thermal Stability of Cannabinoids in Dried Cannabis: A Kinetic Study. *Anal. Bioanal. Chem.* **2022**, *414*, 377–384. [CrossRef] [PubMed]
137. Milay, L.; Berman, P.; Shapira, A.; Guberman, O.; Meiri, D. Metabolic Profiling of Cannabis Secondary Metabolites for Evaluation of Optimal Postharvest Storage Conditions. *Front. Plant Sci.* **2020**, *11*, 583605. [CrossRef] [PubMed]
138. Chasiotis, V.; Tsakirakis, A.; Termentzi, A.; Machera, K.; Filios, A. Drying and quality characteristics of *Cannabis sativa* L. inflorescences under constant and time-varying convective drying temperature schemes. *Therm. Sci. Eng. Prog.* **2022**, *28*, 101076. [CrossRef]

bioengineering

MDPI

Review

Attenuation of Hyperlipidemia by Medicinal Formulations of *Emblica officinalis* Synergized with Nanotechnological Approaches

Puttasiddaiah Rachitha [1,†], Krupashree Krishnaswamy [2,†], Renal Antoinette Lazar [1], Vijai Kumar Gupta [3], Baskaran Stephen Inbaraj [4], Vinay Basavegowda Raghavendra [1,*], Minaxi Sharma [5,*] and Kandi Sridhar [6,*]

1   P.G. Department of Biotechnology, Teresian College, Siddarthanagar, Mysuru 570011, India
2   Biochemistry Department, Central Food Technological Research Institute, Mysore 570006, India
3   Biorefining and Advanced Materials Research Center, Scotland's Rural College (SRUC), Edinburgh EH9 3JG, UK
4   Department of Food Science, Fu Jen Catholic University, New Taipei City 242 05, Taiwan
5   Haute Ecole Provinciale de Hainaut-Condorcet, 7800 Ath, Belgium
6   Department of Food Technology, Koneru Lakshmaiah Education Foundation Deemed to be University, Vaddeswaram 522502, India
*   Correspondence: viragh79@gmail.com (V.B.R.); minaxi86sharma@gmail.com (M.S.); sridhar4647@gmail.com (K.S.)
†   These authors contributed equally to this work.

Abstract: The ayurvedic herb *Emblica officinalis* (*E. officinalis*) is a gift to mankind to acquire a healthy lifestyle. It has great therapeutic and nutritional importance. *Emblica officinalis*, also known as Indian gooseberry or Amla, is a member of the Euphorbiaceae family. Amla is beneficial for treating illnesses in all its forms. The most crucial component is a fruit, which is also the most common. It is used frequently in Indian medicine as a restorative, diuretic, liver tonic, refrigerant, stomachic, laxative, antipyretic, hair tonic, ulcer preventive, and for the common cold and fever. Hyperlipidemia is also known as high cholesterol or an increase in one or more lipid-containing blood proteins. Various phytocompounds, including polyphenols, vitamins, amino acids, fixed oils, and flavonoids, are present in the various parts of *E. officinalis*. *E. officinalis* has been linked to a variety of pharmacological effects in earlier studies, including hepatoprotective, immunomodulatory, antimicrobial, radioprotective, and hyperlipidemic effects. The amla-derived active ingredients and food products nevertheless encounter challenges such as instability and interactions with other food matrices. Considering the issue from this perspective, food component nanoencapsulation is a young and cutting-edge field for controlled and targeted delivery with a range of preventative activities. The nanoformulation of *E. officinalis* facilitates the release of active components or food ingredients, increased bioaccessibility, enhanced therapeutic activities, and digestion in the human body. Accordingly, the current review provides a summary of the phytoconstituents of *E. officinalis*, pharmacological actions detailing the plant *E. officinalis*'s traditional uses, and especially hyperlipidemic activity. Correspondingly, the article describes the uses of nanotechnology in amla therapeutics and functional ingredients.

Keywords: *Emblica officinalis*; phytochemicals; hyperlipidemia; pharmacology; nanoformulation

Citation: Rachitha, P.; Krishnaswamy, K.; Lazar, R.A.; Gupta, V.K.; Inbaraj, B.S.; Raghavendra, V.B.; Sharma, M.; Sridhar, K. Attenuation of Hyperlipidemia by Medicinal Formulations of *Emblica officinalis* Synergized with Nanotechnological Approaches. *Bioengineering* **2023**, *10*, 64. https://doi.org/10.3390/bioengineering10010064

Academic Editors: Chengli Hou and Rossana Madrid

Received: 15 November 2022
Revised: 30 December 2022
Accepted: 3 January 2023
Published: 4 January 2023

1. Introduction

The most prevalent type of dyslipidemia, hyperlipidemia, is caused by increased blood lipid levels, including triglycerides and cholesterol. The early onset of atherosclerosis and its cardiovascular complications are predicted to continue to be significantly influenced by hyperlipidemia. Globally available cholesterol-lowering drugs may be expensive and have side effects, making herbal medicines an alternative in cases where cardiovascular diseases are the primary cause of death [1,2].

In the ancient Indian medical system known as Ayurveda, *Emblica officinalis* (*E. officinalis*) holds a revered place. It is the first tree ever made in the history of the universe. It belongs to the Euphorbiaceae family, according to ancient Indian mythology. commonly known as *Phyllanthus emblica*, Amla, or Indian gooseberry [3]. In Ayurveda, the fruit of the *E. officinalis* plant is frequently used to boost immunity. Cancer, diabetes, liver disease, ulcers, heart problems, anemia, and a number of other diseases can all be helped by the chemical constituents of *P. emblica*. Additionally, it helps with lowering cholesterol levels, ophthalmic disorders, and improving memory [4,5].

The hypoglycemic medications on the market all have one or more side effects. According to WHO recommendations, it can be difficult to find new cholesterol-lowering drugs made from herbal plants that have negligible or no side effects [6]. Interestingly, there is a paucity of information on the effectiveness of EO in lowering cholesterol. Nanocoatings made from Amla essential oil appear to extend the shelf life of fruits [7]. In addition, nanoencapsulated amla provides nutraceuticals and functional foods to improve human health. The therapeutic and traditional potential of amla can be fully utilized by nanoencapsulation of the active components of amla for target delivery, enhanced bioavailability, and increased bioactivity [8].

The present review covers hyperlipidemia, different types of hyperlipidemia, biochemical profile of *E. officinalis*, and attenuation of hyperlipidemic activity of *Emblica officinalis* phytochemicals. With these most significant advances, theranostic treatments for hyperlipidemia could become very efficient in the near future.

2. Hyperlipidemia

A prevalent condition of lipid metabolism known as hyperlipidemia is brought on by elevated triglycerides and total cholesterol levels. According to Panahi, et al. [9], hyperlipidemia is characterized by elevated triglycerides, free fatty acid (FFA), low-density lipoprotein (LDL), cholesterol, and apolipoprotein B (apoB) levels as well as a reduced plasma concentration of high-density lipoprotein (HDL) cholesterol. This condition is the primary risk factor for cardiovascular diseases (CVDs), such as coronary heart disease (CHDs) and the normal blood cholesterol level [10].

2.1. Causes of Hyperlipidemia

The plasma lipids result from either a primary genetic defect, a secondary diet, medication, or disease [11]. The vast majority of cholesterol is produced internally. 3-hydroxy-3-methyl-glutaryl coenzyme A (HMG CoA) reductase is the rate-limiting enzyme for the synthesis of endogenous cholesterol, and it offers a crucial approach to pharmacologic therapy [12]. The liver packages endogenously produced cholesterol and triglycerides into soluble particles. The core of soluble particles, which are rich in triglycerides and cholesterol ester, is encased in a phospholipid membrane made up of different apolipoproteins. The characteristics of apolipoproteins include receptor recognition sites, which facilitate the specific metabolism of these particles by lipoprotein lipase in conjunction with the metabolism of triglycerides. In the beginning, the liver creates triglyceride-rich, very-low-density lipoproteins (VLDL). Low-density lipoproteins (LDL) are made from these particles, which shrink in size, or from oxidized LDL, broken down by the lipoprotein lipase enzyme [13]. This LDL can be absorbed by particular receptors, used by macrophages or the liver, and removed from circulation. High-density lipoprotein (HDL) particles, which are cholesterol-rich and have antioxidant properties, start reverse cholesterol transport [14].

2.2. Types of Hyperlipidemia

The American Heart Association (AHA) divides hyperlipidemia into two categories: primary hyperlipidemia, which is based on genetics, and secondary hyperlipidemia, which is brought on by conditions such as diabetes, liver, kidney, thyroid, and Cushing's syndrome, as well as estrogen therapy, alcohol consumption, and drug use that alters lipid metabolism [15,16]. Additionally, in accordance with the "Fredrickson" classification, there have been five different types of hyperlipidemia: Type I (raised cholesterol with high triglyceride levels), Type II (high cholesterol with normal triglyceride levels), Type III (raised cholesterol and triglycerides), Type IV (raised triglycerides, atheroma, and raised uric acid), and Type V—raised triglycerides [17,18], as shown in Table 1.

Table 1. Classification of hyperlipidemia [1].

Type	Lipoprotein Abnormality	Total Cholesterol	LDL Cholesterol	Plasma TGs	Clinical Manifestation	Population Prevalence
Familia chylomicronemia (HLP type 1)	Excess chylomicrons	Elevated	Low or normal	Elevated	Lipemia retinalis, focal neurologic symptoms, failure to thrive, recurrent epigastric pain, hepatosplenomegaly	1 in 1 million
Combined hyperlipidemia (HLP type 2)	Excess LDL and VLDL	Elevated or normal	Elevated	Normal	Physical stigmata such as xanthomas or xanthelasmas are rare	1 in 40
Dysbetalipidemia (HLP type 3)	Excess IDL elevated chylomicron remnants	Elevated	Low or normal	Elevated	Tuberous and palmar xanthomata elevations in atherogenic IDL leads to increased risk for CVD	1 in 10,000
Primary simple hyperlipidemia (HLP type 4)	Excess VLDL	Elevated or normal	Normal	Elevated	Associated with increased risk of obesity, CVD, DM2, hypertension, insulin resistance, and hyperuricemia	1 in 20
Primary mixed hyperlipidemia (HLP type 5)	Excess chylomicron and VLDL	Elevated	Normal	Elevated	Similar clinical manifestation as type 1 but develops in adulthood	1 in 600

[1] Source: Table 1 is adapted with permission (Copyright © 2016 Taylor & Francis, Oxfordshire, England, UK) from Sharma et al. [10]. LDL, low-density lipoprotein; TGs, triglycerides; HLP, hyperlipoproteinemia; VLDL, Very-low-density lipoprotein; IDL, intermediate-density lipoprotein; CVD, cardiovascular disease; DM2, diabetes mellitus, type 2.

The specific classes of triglyceride-rich lipoprotein particles that accumulate in plasma, such as VLDL, chylomicrons, or intermediate density lipoprotein, help to distinguish these types of hyperlipidemia (IDL). Elevated VLDL density and chylomicron are characteristics of simple HTG, or HLP type 4. Elevated VLDL and LDL concentrations identify HLP type 2. Furthermore, patients with all types of HTG typically exhibit decreased HDL cholesterol [1]. Starc [19] represented epidemiological studies indicated the risk for coronary heart disease (CHD) starts to rise at cholesterol levels above 200 mg/dL; as a result, scientists think the relationship between atherogenesis and plasma cholesterol is linear over the entire range of cholesterol concentrations. Given the existence of a CHD risk threshold, concentrations above this point are referred to as hypercholesterolemia [20].

The connection between plasma fatty substances, such as triglyceride levels, and the onset of atherosclerosis has proven challenging to prove. According to Nelson, 2013 [14], triglyceride levels and the risk of CHD are closely related. Additionally, plasma triglyceride levels are positively correlated with risk for CHD. According to research from Harchaoui et al. [21], triglyceride concentrations lose their predictive power when data are subjected to multifactorial analysis, so we considered other risk factors. Patients with hypertriglyceridemia may experience accelerated atherosclerosis for unknown reasons. Furthermore, elevated triglyceride levels may be more highly predictive of CHD when present or absent than the degree of elevation. The fact that hypertriglyceridemia is present in the majority of patients with disproportionate CHD thus summarizes the situation [21]. The reduction of hepatic cholesterol level causes a decrease in VLDL cholesterol level in the blood [22].

2.3. Pathophysiology

Studies on the pathophysiology of hyperlipidemia fall into two categories at their core. Hypertriglyceridemia results from a defect in lipid metabolism, and hyperchylomicronemia is brought on by a defect in lipoprotein lipase activity or the absence of the surface apoprotein CII31 [23]. According to Karr 2017 [24], postprandial absorption of chylomicrons from the gastrointestinal tract occurs 30 to 60 min after the consumption of meals high in fat and raises serum triglycerides. Low LPL activity ultimately causes hyperlipidemia by increasing the production of VLDL cholesterol. Patients with diabetes mellitus reported seeing this.

Additionally, hypercholesterolemia manifests nephrotic syndrome, a common pathway for the synthesis of cholesterol, and albumin lowers oncotic pressure, resulting in increased cholesterol synthesis [25]. Several medications, including progestins, thiazide diuretics, glucocorticoids, Î2 blockers, isotretinoin, protease inhibitors, cyclosporine, mirtazapine, and sirolimus, have been reported by Hill and Bordoni [26] to increase lipid levels as secondary forms of hyperlipidemia. The lack of LDL degradation by cells as a result of hypercholesterolemia is due to the LDL-R complex's inability to bind LDL and the uncontrolled production of cholesterol.

3. Biochemical Profiling of Amla

Medicinal remedies have served as the foundation for sophisticated traditional medicine and are extremely important to humankind in terms of basic therapeutics. Natural products are a gift for improving health and are great starting points for developing new drugs. International public health is a specialty of the United Nations and is the focus of the World Health Organization [27]. As stated by the WHO, conventional medicine is the primary source of care for 80% of the world's population, which is significant for health care. The World Health Organization (WHO) is enjoining researchers to encourage the effective use of herbal medicine in developing nations' health programs. Currently, herbal plants are the "local heritage of global significance", contributing more to the healthcare system than chemical drugs [28].

The Ayurvedic valuable tree *E. officinalis* or amla has long been recognized for its therapeutic and pharmacological significance. *E. officinalis* is a member of the Euphorbiaceae family; it is also known by the common names Amla and *Phyllanthus emblica* in the botanical world [29,30]. The summary of the taxonomy of common names of *E. officinalis* is presented in Table 2. Specifically, central and southern India, Sri Lanka, southern China, Pakistan, Bangladesh, the Mascarene Islands, Malaysia, and tropical Southeastern Asia are home to the *E. officinalis* species. The *E. officinalis* appears to be a large tree with a height of 8 to 18 m, but in India, *E. officinalis* trees can be found throughout tropical forests that rise up to 4500 feet on hills [31,32]. The botanical description of *E. officinalis* is examined in Table 3 [29,33], which lists various parts of the plant, including fruits, leaves, seeds, bark, and flowers, that are used for various pharmacological effects. The most studied plant is *E. officinalis*, and reports indicate that it contains a variety of chemical components, such as gallic acid, amino acids, flavone glycosides, phenolic glycosides, flavonol glycosides, sesquiterpenoids, nor sesquiterpenoids, and rich fiber, carbohydrate, iron, tannins, alkaloids, and phenolic compounds. According to Singh et al. [34], the nutritional value of *E. officinalis* fruit contains significantly more minerals, proteins, and amino acids such as glutamic acid, proline, aspartic acid, alanine, cystine, and lysine than other fruits such as apple, lime, pomegranate, and grape. It is also said to be the richest source of vitamin C compared to other fruits, as represented in Table 4 [35–38].

Table 2. Taxonomical classification of *E. officinalis* [1].

Kingdom	Plantae (Plants)
Subkingdom	*Tracheobionta* (vascular plants)
Super division	*Spermatophyta* (seed plants)
Division	Angiospermae (flowering plants)
Class	Magnoliopsida
Subclass	Rosidae
Order	Euphorbiales
Family	Euphorbiaceae
Genus	*Emblica*
Species	*officinalis* Geartn

[1] Source: Table 2 is adapted from Hasan et al. [29], which is an open access article (Copyright © 2016 by authors) distributed under the terms and conditions of the Creative Commons Attribution (CC BY) license.

Table 3. Botanical description of *E. officinalis* [1].

Feature	Description
Habitat	Central and southern India, Pakistan, Bangladesh, Sri Lanka, Malaysia, southern China, the Mascarene Islands, Southeast Asia, and Uzbekistan.
Appearance	Medium sized deciduous tree, 8–18 m height, with thin light gray bark exfoliating in small, thin, irregular flakes.
Used parts	Dried fruits, fresh fruit, seed, leaves, root bark, and flowers.
Leaves	Simple, subsessile, closely set along the branchlets, light green, having the appearance of pinnate leaves.
Fruits	15–20 mm long and 18–25 mm wide, nearly spherical or globular, wider than long and with a small and slight conic depression on both apexes. Mesocarp is yellow, and endocarp is yellowish brown in ripened condition. Globose, fleshy, pale yellow with six obscure vertical furrows enclosing six trigonous seeds in three 2-seeded crustaceous cocci. Seedlings start bearing fruits in 7–8 years after planting, while the budded clones will start bearing fruits from the 5th year onward. Fresh fruits are light green, and ripe fruits turn light brown in color. The average weight of the fruit is 60–70 g.
Flowers	Greenish yellow, in axillary fascicles, unisexual, males numerous on short slender pedicels, females few, subsessile, ovary 3-celled.
Seeds	Four to six, smooth, dark brown.
Barks	Thick to 12 mm, shining grayish brown or grayish green
Flowering and fruiting	February–May and December–January
Edible part	Mesocarp and endocarp that forms the hard stone which encages the seed

[1] Source: Table 3 is adapted from Hasan et al. [29], which is an open access article (Copyright © 2016 by authors) distributed under the terms and conditions of the Creative Commons Attribution (CC BY) license.

Table 4. Nutritional value of *E. officinalis* [1].

Chemical Components	Amount
Fruits: moisture (%)	81.20
Protein (%)	0.50
Fat (%)	0.10
Mineral matter (%)	0.10
Fiber (%)	3.40
Carbohydrate (%)	14.10
Bulk elements (mg/100 g)	Net weight
Calcium (%)	0.05
Phosphorus (%)	0.02
Iron (mg/100 g)	1.20
Vitamin C (mg/100 g)	600
Nicotinic acid (mg/100 g)	0.20

[1] Source: Table 4 is adapted from Singh et al. [34], which is an open access article (Copyright © 2011 by the Open Science Publishers LLP, London, UK) distributed under the terms and conditions of the Creative Commons Attribution (CC BY) license.

3.1. Phytochemistry

The *E. officinalis* fruit's most prevalent ingredient is ascorbic acid. Phosphatides, fixed oils, phosphatides, essential oils, tannins, minerals, vitamins, and amino acids are some additional phytochemicals associated with this plant. The *E. officinalis* is good source of phytochemicals (Table 5). Methyl gallate, luteolin, corilagin, isostrictiniin, gallic acid, ellagic acid, chebulinic acid, and chebulagic acid are the major polyphenolic compounds in *E. officinalis* [5,39]. Fruit pulp has been found to contain tannins such as phyllaemblicin B, emblicanin A, B, punigluconoin, and pedunculagin [40]. There have been reports of linolenic, oleic, stearic, palmitic, and myristic acids in *E. officinalis*, as well as by-products of organic acids such as citric acids and sugars such as glucose, fructose, myo-inositol, galacturonic acid, arabinosyl, rhamnosyl, xylosyl, and glucosyl. Many compounds' isolates obtained from these plants include amlaic acid, arginine, aspartic acid, astragallin, -carotene,

-sitosterol, chebulagic acid, chebulic acid, chebulinic acid, corilagic acid, corilagin, ellagic acid, emblicol, gibberellins, glutamic acid, glycine, histidine, and isoleucine [41–46] and gallic acid, 1,6-di-O-galloyl-d-glucose, 3,6-di-O-galloyl-d-glucose, corilagin, 3-ethylgallic acid (3-ethoxy-4,5-dihydroxybenzoic acid), isostrictiniin, kaempferol-3-O-L-(6'-methyl)-rhamnopyranoside [47–49]. In addition to isolating -rhamnopyranoside from *E. officinalis*, researchers have also isolated a number of phenolic compounds from the fruit juice of the plant, including mucic acid 2-O-gallate, L-malic acid 2-O-gallate, mucic acid 1,4-lactone 2-O-gallate, mucic acid 1,4-lactone 5-O-gallate, mucic acid 1,4-lactone 3-O-gallate, and mucic acid [50]. The phenolic glycosides 2-carboxylmethylphenol1-O-d-glucopyranoside and 2,6-dimethoxy-4-(2-hydroxyethyl) phenol 1-O-d-glucopyranoside were combined with phyllaemblicin-A, B, and C, phyllaemblic acid (methyl ester of highly oxygenated nor bisabolane), and phyllaemblic acid C and phyllaemblicin D [51,52]. Gallic acid, methyl gallate, 1,2,3,4,6-penta-O-galloylglucose, two newly discovered acrylated flavanones glycosides, (S)-eriodictyol 7-O-(6'-O-trans-p coumaroyl)—d-glucopyranoside, and luteolin-4'-O-neohesperidoside were also isolated from the leaves of *E. officinalis*. Furthermore, *E. officinalis* contains six ellagitannins and phyllanemblinins A–F [53,54]. Several new sterols, including trihydroxysitosterol and 5,6,7-acetoxysitosterol, were discovered in the branches and leaves of *E. officinalis* [55,56]. According to analytical reports, the pulp of *E. officinalis* seeds has a high phenolic content, and its seeds are primarily made up of tannins. In the pulp and seeds of *E. officinalis*, coumaric acid, myricetin, caffeic acid, gallic acid, and quercetin have all been identified [57]. Among them, gallic acid, myricetin, kaempferol, Emblicanin A and B, chebulagic acid, ellagic acid, pedunculagin, and corilagin are the major compounds that protect from hyperlipidemia, as shown in Figure 1.

Table 5. Major phyto-constituents present in *E. officinalis* and their pharmaceutical effects beyond cardioprotective and anti-hyperlipidemia [1].

Source	Main Active Compounds	Biological Activity
Fruit	Gallic acid	Cardioprotective, anti-hyperlipidemia
Fruit	Ellagic acid	Cardioprotective, anti-hyperlipidemia
Fruit	Emblicanin A and B	Cardioprotective, anti-hyperlipidemia
Fruit	Myricetin and Kaempferol	Cardioprotective, anti-hyperlipidemia
Fruit	Punigluconin pedunculagin	Cardioprotective, anti-hyperlipidemia
Fruit	Chebulagic acid	Cardioprotective, anti-hyperlipidemia
Fruit	Geraniin and corilagin	Cardioprotective, anti-hyperlipidemia
Fruit	Quercetin and rutin	Anti-inflammatory
Fruit	Tannins and gallic acid	Gastrointestinal activity
Fruit	Flavonoids	Antidiabetic activity
Fruit	Polyphenols	Neuroprotective
Fruit	Gallic acid	Anticancer activity

[1] Source: Table 5 is adapted with permission (Copyright © 2016 Elsevier Ltd., Amsterdam, the Netherlands) from Variya et al. [5].

3.2. Pharmacological Activity of Amla

In light of the medicinal and pharmaceutical qualities of *E. officinalis*, every part is beneficial. According to research by Krishnaveni and Mirunalini [3], *E. officinalis* has antioxidant, antimicrobial, anti-inflammatory, anticancer, antiulcer, antidiabetic, memory enhancer, cardioprotective, neuroprotective, neuroprotective, antidiarrheal, renoprotective, and immunomodulatory potential, as shown in Figure 2, and major phytoconsituents present in *E. officinalis* and its pharmaceutical effects point toward anti-hyperlipidemia properties, as shown in Table 5 [5,58]. It also has positive effects on hyperlipidemia, osteoporosis, and a number of reactive oxygen species that can lead to oxidative stress and fundamental cell damage in the body.

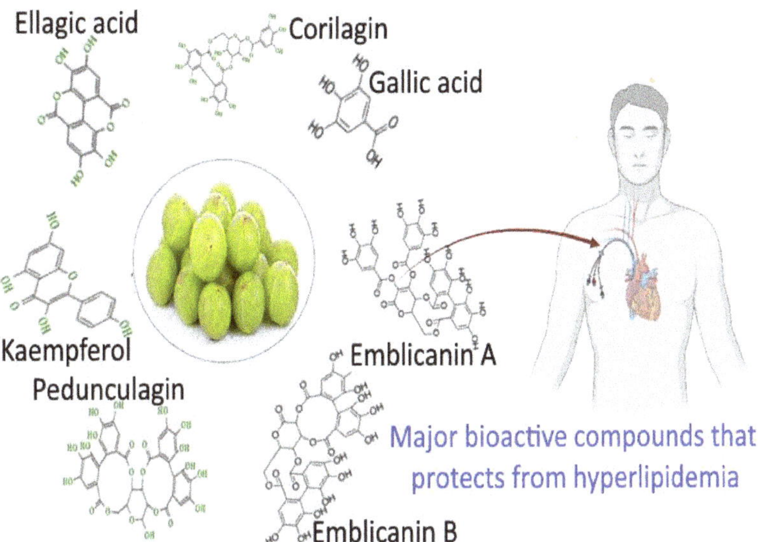

Figure 1. Major phytochemical constituents of *E. officinalis* that protect from hyperlipidemia.

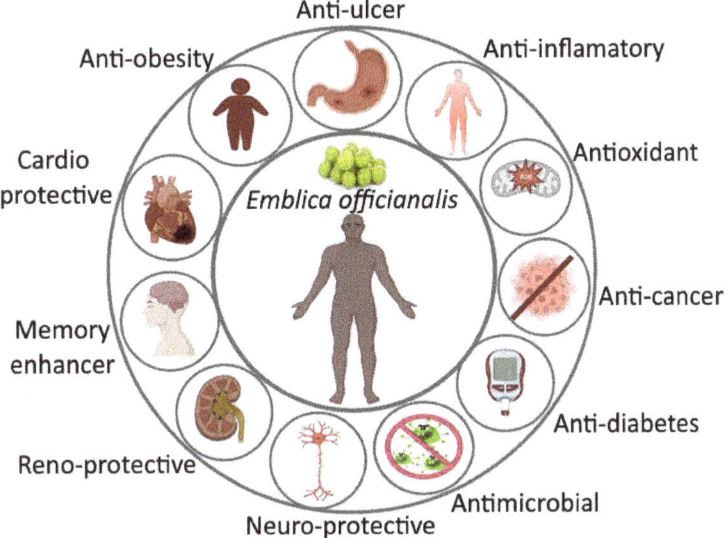

Figure 2. Different pharmaceutical activities of *E. officinalis*.

Large quantities of polyphenols, tannins, and other phytochemicals found in *E. officinalis* can lessen oxidative damage to cells. The natural antioxidants of *E. officinalis* play a significant role in the activity of free radical scavengers, and methanolic seed extract and pulp extract of *E. officinalis* show promising 1,1, diphenyl-2-picryl-hydrazil (DPPH) free radical scavenging activity in a concentration-dependent manner [36,59]. There is strong considerable potential for ferric reduction, free radical scavenging, and ROS (reactive oxygen species) inhibition in the water extract of *E. officinalis* fruit [60].

Different solvent systems were used to test *E. officinalis'* antimicrobial activity, and it was found to have antifungal properties against *Aspergillus* sps. [61]. Fruit ethanolic and acetone extract demonstrated activity against *Candida albicans* and *Fusarium equiseti*. The antibacterial activity against *Staphylococcus* was demonstrated using the zone inhibition method, and the tube dilution method significantly reduced the colony counts of *Escherichia coli, Staphylococcus aureus, Klebsiella pneumoniae,* and *Pasteurella multocida* [62–64]. The phytochemical in *E. officinalis* called pentagalloyl glucose has anti-influenza properties. WST-1 assay, plaque-forming unit assay, time of-addition assay, and hemagglutination inhibition (HI) assay were used to evaluate a virus replication with a dual mode of action [65].

In Sprague-Dawley rats exposed to both acute and chronic inflammation caused by carrageenan and cotton pellets, the water extract of *E. officinalis* was discovered to exhibit anti-inflammatory effects by minimizing paw volume in the case of acute inflammation and myeloperoxidase activity, granulomatous tissue mass, and plasma extravasation in the case of chronic inflammation [66]. Histopathological studies were used to investigate the hepatoprotective activity of *E. officinalis*. Liver-protective behavior HepG2 cells were used to test the efficacy of *E. officinalis* against tert-butyl hydroperoxide (t-BH)-induced toxicity, and rats were used to test the efficacy of a 50% hydroalcoholic extract of fresh *E. officinalis* fruit against chronic toxicity brought on by carbon tetrachloride and thioacetamide. Nephroprotective properties reduced the elevated levels of thiobarbituric acid-reactive substance in the serum, renal homogenate, and creatinine and urea nitrogen in aged rats [67,68]. The forced swim test (FST) and tail suspension test (TST) with Swiss albino mice were used to test the aqueous extract of fruits from *E. officinalis* for its antidepressant activity. The results revealed a significant decrease in depression. The result was that aged mice with improved memory (elevated plus maze and passive avoidance apparatus) had lower total serum cholesterol levels and higher brain cholinesterase activity [69]. In albino rats, *E. officinalis* was found to have immunomodulatory activity as evidenced by increases in hemagglutination antibody titer, macrophage migration index, hypersensitivity reaction, respiratory burst activity of the peritoneal macrophages, total leukocyte count, percentage lymphocyte distribution, serum globulin, and relative lymphoid organ weight. It is also able to stimulate humoral and cell mediated immunity as well as macrophage phagocyte [70]. On type II diabetes, triglycerides (TG), and the liver-specific enzyme alanine transaminase, the aqueous fruit extract of *E. officinalis* was assessed (ALT). According to this study, alloxan-induced diabetic rats could significantly lower their blood glucose levels when given an aqueous fruit extract dose of 200 mg/kg body weight [71]. Compared to control and extract-treated diabetic rats, oral administration of the aqueous extract (350 mg/kg body weight) significantly decreased serum glucose levels, glycosylated hemoglobin, insulin, cholesterol, triglycerides, HDL-cholesterol, protein, urea, and creatinine [72].

E. officinalis has the potential to lower cholesterol because it naturally contains flavonoids and other phytochemicals. Several clinical studies showed significant drops in C-reaction protein (CRP), low-density lipoprotein, and total cholesterol [73]. Variya et al. [74] assessed the hypolipidemic effect of *E. officinalis* and compared it to the standard simvastatin in patients with type-II hyperlipidemia. Treatment with *E. officinalis* resulted in significantly lower levels of total cholesterol, LDL cholesterol, and triglycerides as well as noticeably higher levels of the common medication simvastatin [75].

Polyphenols from *E. officinalis* have also been shown to protect gastrointestinal organs. Because *Helicobacter pylori* is a pathogen, one of the potential effects of amla's bioactive compounds is the competitive inhibitor of clarithromycin-resistant strains in vitro [76]. Studies using animals reported relevant results as well. In order to induce gastrointestinal ulcers in mice, Al-Rehaily et al. [77] used a variety of techniques, including ligating the pylorus, administering indomethacin and necrotizing agents (25% NaCl, 0.2 M NaOH, and 80% ethanol), and inducing hypothermia. These techniques included studying the antisecretory and antiulcer activities of *E. officinalis* extract. Using the pylorus-ligated and necrotizing agent-intoxicated ulcer methods, both doses (250 and 500 mg/kg) decreased gastric secretion, intraluminal bleeding, ulcer index, and gastric lesions. Only the animals receiving treatment with 500 mg/kg for the indomethacin-induced ulcer method had a significantly lower ulcer index than animals in the control group (treated only with indomethacin).

3.3. Antihyperlipidemic Activity of Emblica officinalis

According to reports, *E. officinalis* fruit has significant anti-hyperlipidemic, hypolipidemic, and antiatherogenic effects [78]. In patients with type II hyperlipidemia, treatment with *E. officinalis* resulted in a significantly lower level of total cholesterol (TC), low-density lipoprotein (LDL), triglyceride (TG), and very-low-density lipoprotein (VLDL), as well as a significantly higher level of high-density lipoprotein (HDL) (Figure 3). Studies conducted in vitro and in vivo using cholesterol-fed rats and Cu^{2+}-induced LDL-oxidation established the anti-hyperlipidemic activity of extract from *E. officinalis* and demonstrated a significant reduction in total and free cholesterol levels in a dose-dependent manner [79]. Given that oxidized LDL is a key enzyme in atherosclerosis, administration of *E. officinalis'* potential antioxidant property resulted in a decrease in oxidized LDL levels in cholesterol-fed subjects, as shown in Figure 3 [80].

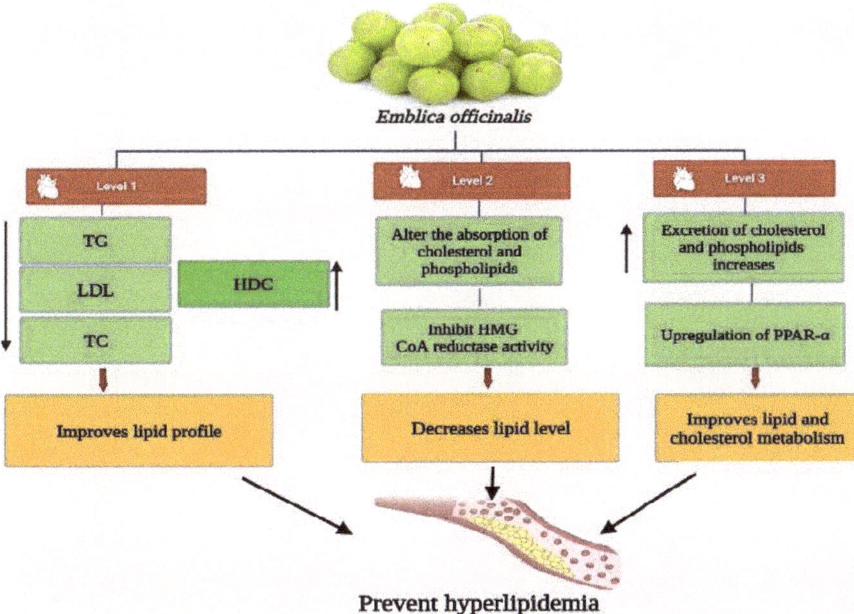

Figure 3. Anti-hyperlipidemia activity of amla bioactive compounds. Figure 3 is adapted from Gul et al. [80] (Copyright © 2022 by authors), which is an open access article distributed under the terms and conditions of the Creative Commons Attribution (CC BY) license. TG, triglyceride; HDL, high-density lipoprotein; LDL, low-density lipoprotein; TC, total cholesterol; PPAR-α, peroxisome proliferator-activated receptors-α.

Another study revealed elevated lecithin-cholesterol acyltransferase and hepatic 3-hydroxy 3-methylglutaryl coenzyme A (HMG-CoA) reductase inhibition as anti-hyperlipidemic effects (LCAT). Flavonoids, which prevent the synthesis and deterioration of lipids, are responsible for this effect [81]. Oral administration of *E. officinalis* extracts in a dose of 500 mg twice daily for 12 weeks significantly lowered total cholesterol, LDL cholesterol, and high-sensitive creatine kinase (hr-CRP) levels, according to a study on animals that used class one obese subjects with body weights between 250 and 350 g. *E. officinalis* showed significant antiatherosclerotic action through a decrease in serum and hepatic cholesterol content, serum triglyceride, phospholipids, and LDL cholesterol in a high-cholesterol-fed rabbit model. Platelet aggregation induced by ADP and collagen was also decreased, and diabeto-cardiac malaise was eventually reduced [73]. Growing fructose consumption in the diet has been linked to a higher risk of obesity and related metabolic syndromes in Western countries. Increased intake of fructose alters several signaling cascades, including NF-, TNF-, JNK-1, PTP-1B, PTEN, LXR, FXR, and SREBP-1c [82]. The SREBP-1 expression, total cholesterol, TG level, and metabolic issues associated with high fructose levels are all improved by the supplement's high polyphenol content from *E. officinalis*. According to reports, *E. officinalis* inhibits the level of MDA in the liver and controls the expression of the COX-2, bax, NF-, and bcl-2 markers [83]. Previous findings have shown that *E. officinalis* therapies increase lipid metabolism and regulate the expression of proteins such as PPAR-α, which is involved in fatty acid-oxidation, FXR, and LXR involved in lipid metabolism, as well as insulin-induced gene-2 to reduce fructose-induced metabolic syndrome and prevent the maturation of steroyl CoA desaturase-1 and SREBP-1, which are involved in the synthesis of TG. Additionally, RAW 264.7 cell lines' expression of CD36 scavenger receptor was markedly reduced to prevent foam cell formation by these mechanisms [57]. The major contributors to *E. officinali's* antihyperlipidemic activity are its polyphenols and functional products. Gallic acid, Vitamin C, Emblicanin A and B, apigenin, ellagic acid, and 3-hydroxy-3-methylglutaryl-CoA myricitine, among other antioxidants, polyphenols, and phenolic acids, play a significant role in lowering hyperlipidemia [84–87]. As these molecules are less stable with different environmental conditions or pH and less soluble in water, they are reported to have less bioavailability or poor assimilation of major constituents. This is the main drawback of these polyphenols and functional products [87,88]. Consequently, nanoformulation strategies can increase the bioavailability of phenolic acids and functional products.

3.4. Nanoparticulate Carrier System for the Treatment of Hyperlipidemia

Nanotechnology is one of the emerging technologies that influence human life in different approaches that assist in overcoming the multiple limitations of various diseases, especially hyperlipidemia. Nanoformulations have become a novel profitable approach for increasing the bioavailability of poor soluble drugs [84,89]. These nanoformulations have several unique qualities that make them more valuable for the drug delivery system. Diverse nanostructures include solid lipid nanoparticles (SLNs), nanoliposomes, phytosomes, noisome polymer nanoparticles, nanomicelles, and carbon nanotubes, which are used in drug delivery systems, significantly increase the effectiveness, and improve the pharmacokinetics of drugs with reduced side effects [90]. Many reports revealed that several nanoformulations from a number of natural products, such as emblicanin-A and emblicanin-B, quercetin, curcumin, piperine, nigella, etc., have become a promising technology for the use of nanoformulation from natural products, as shown in Figure 4 [88,91].

3.5. Nanoformulation of Emblica officinalis and Its Applications

Nanomedicine has developed into a successful platform that incorporates various modalities, including therapeutics and diagnostics. It might provide individualized medical treatment to manage fatal illnesses such as cancer and diabetes. Indeed, nanocarriers act as precise delivery mechanisms for desired phytochemicals to the target site and are biocompatible, biodegradable, and less toxic. The site-specific, controlled, slow, and sustained delivery of phyto-based drugs by nanoparticles (NPs) with exceptional entrapment efficiency may enhance both their pharmacokinetics and bioavailability, while also increasing membrane permeability and preventing drug efflux through gastrointestinal mucosa [91]. Recently, there has been an increase in interest in using nanotechnology to boost phytochemical effectiveness. In the healthcare industry, the advent of nanotechnology in medical therapeutic strategies has raised hopes for the delivery of better treatments with greater efficacy and precision [92–94]. Nanoformulation of *E. offficinalis* is the major research area of interest due to its synergistic and improved bioavailability efficacy. The study by Omran et al. [95] focused on the synthesis of nanoemulgel by adding Carbopol 940 along with *E. officinalis* and other extracts to improve the synergistic efficacy of the extract for their antimicrobial property. Silver nanoformulation amla were studied for their antiproliferative and cytotoxic activity by Rosario et al. [96] and Abitha et al. [97]. Biosynthesis of nanocomposites using silver and graphene oxide and *E. officinalis* were characterized and studied for their antibacterial and cytotoxicity activities [98]. A recent study by Ranjani et al. [99] shows the significant cytotoxicity and antibacterial activity of amla-mediated graphene oxide and silver nanocomposites against oral pathogens. Another survey by Naik et al. [100] exhibits the anticancer and antidiabetic activity of phytofabricated silver and zinc-oxide conjugated *E. officinalis*. Considering its eco-friendly and safe aspects, the study recommends its use for pharmaceutical applications. In addition, green synthesized amla with magnesium oxide exhibited photocatalysis activity (Evans blue degradation) and antibacterial activity, thereby confirming amla's efficacy in the removal of water contaminants.

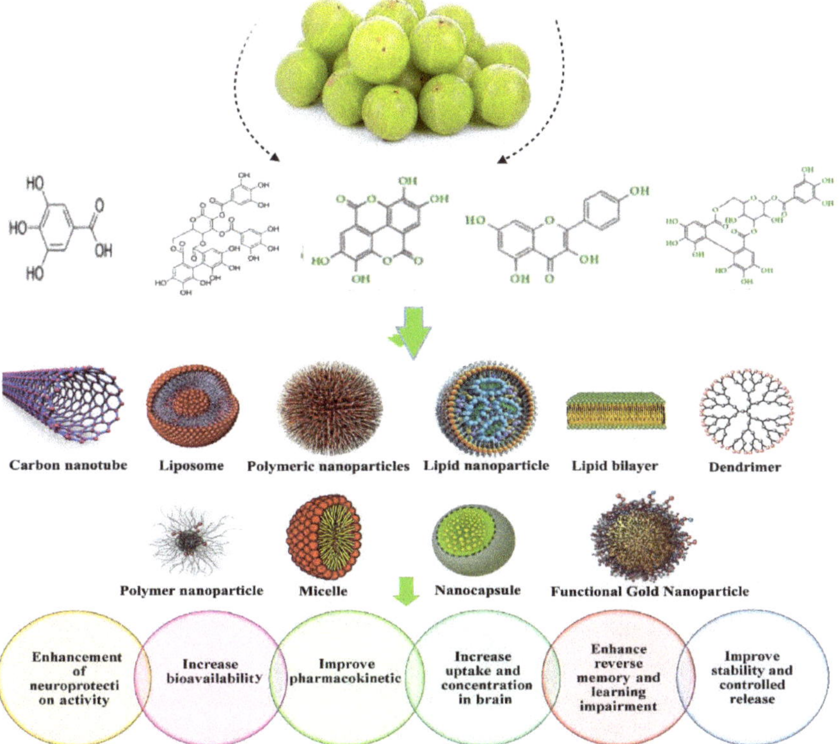

Figure 4. Nanoformulations from natural products and their advantages at a target delivery site. Figure 4 is adapted from Moradi et al. [91] (Copyright © 2020 by authors), which is an open access article distributed under the terms and conditions of the Creative Commons Attribution (CC BY) license.

In contrast, phytoconstituents of *E. officinalis* such as ellagic acid, gallic acid, quercetine, and chebulagic acid were nanoformulated for the amelioration of oral bioavailability and biocompatibility properties. Harakeh et al. [101] studied the antidiabetic property of novel nanoformulated ellagic acid. Another study conducted by Hosny et al. [102] developed the sustained release of ellagic acid nanotransferosomes for its antiproliferative activity. The amla fruit's active ingredient, gallic acid, is abundant naturally and has a variety of health benefits that makes it appealing for use in clinical settings. To increase amla's aqueous solubility and subsequently bioactivity, gallic acid was extracted and separated from it. Using a probe sonicator and a high-pressure homogenization method, glyceryl monooleate (GMO), chitosan, and poloxamer 407 were combined to create gallic acid nanoparticles. According to the study's findings, nanoparticles can be designed and manufactured to facilitate the extraction, manufacture, and sustained release of gallic acid, particularly in the colonic region [103]. Dendrimer nanodevices coated with gallic acid were developed to fight against chemoresistance in neuroblastoma cells [104]. Gallic acid and quercetine nanopolymers were synthesized to improve its bioavailability. Ongoing studies have concentrated on the pharmacological properties of gallic acid and its derivatives, as well as their biological effects on skin, with a particular focus on their use in (nano-)cosmetic formulations. Because the field of study is still developing, emphasis has been given to its advantages of various nanoformulations [105]. *E. officinalis* (leaves, stem, root, fruit, seeds) and its active compounds nanoformulation and delivery system are shown in Figure 5.

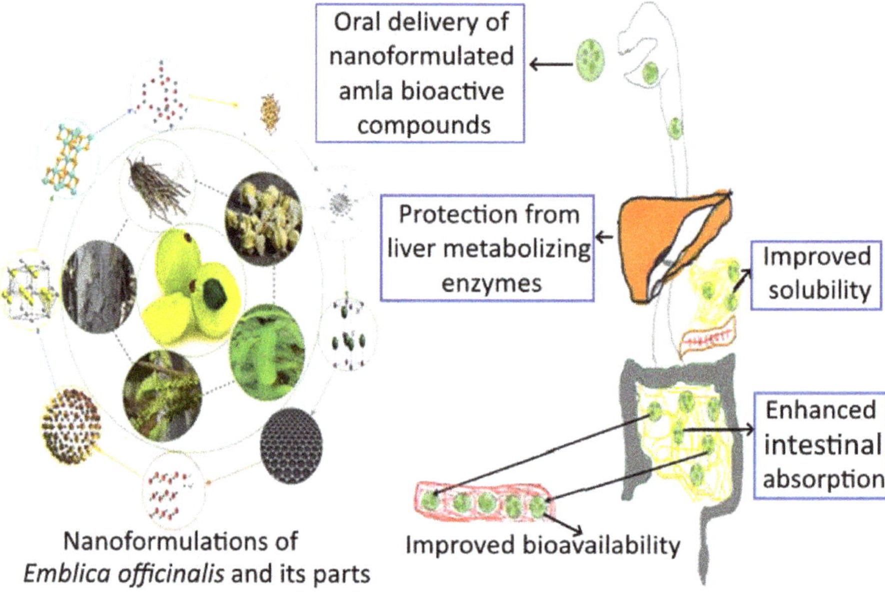

Figure 5. Nanoformulations of amla and its health benefits.

3.6. Health Care Application for Emblicanin-A and Emblicanin-B Nanoformulation

The fruit stands out from the competition due to its natural composition, which is thought to help the body fight off various illnesses and strengthen the immune system. With adequate amounts of fiber, carbohydrates, and iron, this is the best source of vitamin C. The herb *E. officinalis* is a powerful antioxidant. The fruit contains Emblicanin-A and Emblicanin-B, two hydrolyzable tannins. Ellagic acid (EA), gallic acid (GA), glucose, and EA glucose are produced during the hydrolysis of emblicanin-A and emblicanin-B [106].

The creation of a nanosized formulation has as its goal the achievement of high therapeutic efficacy with minimal toxicity. As it has better efficacy and fewer side effects, herbal medicine has long been accepted as a form of treatment by doctors for their patients. The scientific method for sustained drug delivery to the target site helps avoid repeated dosing and causes less harm to the other healthy cells or tissues. For herbal constituents, novel drug delivery systems reduce the need for repeated administration. The development of formulations with the aid of nanotechnology is one potential application. The development of herbal constituent nanoformulations relies heavily on nanocarriers [107]. The involved EA microdispersion was prepared to improve the EA's poor water solubility and low bioavailability. The content improved nearly 30 times the water solubility and 22% (w/w) drug loading by using only water and low methoxylated pectin as a food-compatible excipient (DL). Later, non-PAMAM was used. We were successful in creating two EA nanodispersions using hydrophilic and amphiphilic (polyamidoamine) dendrimers as nanocontainers, obtaining water-solubility 300–1000 times at (60–70 nm) with 46% and 53% (*w/w*) DL higher than the free EA's. Suitable for food and biomedical applications, this bioactive compound is a very effective antioxidant that is also nontoxic [108].

GA's nanoformulation was fabricated and measured. The GA units for peripheral esterification and a delivery system that is GA-enriched (GAD) with exceptional antioxidant capacity and significant potential were successful in preventing diseases from oxidative stress (OS). GA is highly efficient against the illness that OS causes [94]. It has very few clinical applications due to inadequate gastrointestinal absorption and pharmacokinetic drawbacks, fast metabolism, and strong backs. The ready dendrimer made of polyester GA has been manufactured with an absorbable carrier to protect and deliver it. The stability in solution with a tendency to form was indicated by a ZP of 25 mV low polydispersity index and megamers. It has been on display to demonstrate GAD has four times more intrinsic antioxidant power than the GA [109].

Ellagic acid-nanosponges (EA-NS) utilized cyclodextrin and cross-linked by dimethyl carbonate is a nanoformulation that improved the solubilization efficiency of EA and controlled its release to achieve better oral absorption bioavailability. The polyphenolic compound EA, which is naturally present in many fruits, has demonstrated antioxidant, anticancer, and antimutagenic properties; however, its disadvantage is that it has a low oral bioavailability by creating a nanoformulation [109]. The use of natural product-based nanoformulations in treating various metabolic syndromes has grown in popularity among researchers. The compounds' solubility, bioavailability, and efficacy were all improved through nanosizing. The effectiveness of a number of natural constituent nanoformulations in the treatment of numerous diseases has been observed. The molecular targets were pertinent to the way these compounds affect metabolic disorders. The natural bioactive substances emblicanin-A and emblicanin-B have high therapeutic potential and can be incorporated into systems for treating various diseases using nanotechnology [110].

The primary use of lipid-lowering medications such as statins and/or derivatives of fibric acid has been to treat elevated lipid levels and the negative effects that go along with them. It is likely that the modern medical system works to treat disease on the one hand while having negative side effects on the other [111]. The creation of lipid-lowering medications or formulations derived from natural sources has become more significant in recent years. As a result, there has been a lot of interest in using natural products with minimal side effects; *E. officinalis* is one such ingredient thought to have medicinal benefits. Recent research on nanoformulated gallic acid in in vitro models shows great lipid lowering activities [111]. Ellagic acid and 3-hydroxy-3-methylglutaryl-CoA nanoemulsion attenuates fat in in vivo models investigated by Harakeh et al. [112], and Dayar and Pechanova [113]. Because of its multimode cardio protective properties, *E. officinalis* has recently attracted new attention. It is also a powerful antioxidant that has been shown to affect how lipid metabolism is regulated.

3.7. Adversity and Toxicity of Nanoformulations

In recent years, the use of nanotechnology in medicine has significantly increased. When long-term or ongoing treatment is necessary for the management of metabolic diseases compliance has been regarded as a crucial factor. By providing a variety of administration methods, controlling release, enhancing biological stability, achieving target specificity, and reducing toxicity, nanoformulations have been found to increase patient compliance [114]. Accordingly, interest in creating nanoformulations to treat metabolic diseases has been dramatically increasing. Nevertheless, the majority of these studies have been limited by a lack of long-term exploratory statistics and insufficient data, particularly when it comes to the sustained resilience profiling, long-term therapeutic efficacy, and toxicological properties of the developed nanoformulations of plant-derived molecules to treat metabolic disorders. As a result, the majority of the findings are limited to the laboratory scale. Therefore, finding a solution to this problem requires considerable attention [115,116]. The toxicity evaluation of nanoscale materials and with their multiple delivery methods

55. Qi, W.-Y.; Li, Y.; Hua, L.; Wang, K.; Gao, K. Cytotoxicity and structure activity relationships of phytosterol from *Phyllanthus emblica*. *Fitoterapia* **2013**, *84*, 252–256. [CrossRef]
56. Chugh, C.A.; Bharti, D. Chemical characterization of antifungal constituents of *Emblica officinalis*. *Allelopath. J.* **2014**, *34*, 155–178.
57. Nambiar, S.S.; Paramesha, M.; Shetty, N.P. Comparative analysis of phytochemical profile, antioxidant activities and foam prevention abilities of whole fruit, pulp and seeds of *Emblica officinalis*. *J. Food Sci. Technol.* **2015**, *52*, 7254–7262. [CrossRef]
58. Goyal, R.; Patel, S. *Emblica officinalis* Geart.: A Comprehensive Review on Phytochemistry, Pharmacology and Ethnomedicinal Uses. *Res. J. Med. Plant* **2012**, *6*, 6–16. [CrossRef]
59. Priya, G.; Parminder, N.; Jaspreet, S. Antimicrobial and antioxidant activity on *Emblica officinalis* seed extract. *Int. J. Res. Ayur. Pharma.* **2012**, *3*, 591–596.
60. Usha, T.; Middha, S.K.; Goyal, A.K.; Lokesh, P.; Yardi, V.; Mojamdar, L.; Keni, D.S.; Babu, D. Toxicological Evaluation of *Emblica officinalis* Fruit Extract and its Anti-inflammatory and Free Radical Scavenging Properties. *Pharmacogn. Mag.* **2015**, *11*, S427–S433. [CrossRef]
61. Satish, S.; Mohana, D.C.; Raghavendra, M.P.; Raveesha, K.A. Antifungal activity of some plant extracts against important seed borne pathogens of *Aspergillus* sp. *J. Agric. Technol.* **2007**, *3*, 109–119.
62. Saini, R.; Sharma, N.; Oladeji, O.S.; Sourirajan, A.; Dev, K.; Zengin, G.; El-Shazly, M.; Kumar, V. Traditional uses, bioactive composition, pharmacology, and toxicology of *Phyllanthus emblica* fruits: A comprehensive review. *J. Ethnopharmacol.* **2022**, *282*, 114570. [CrossRef]
63. Patil, S.G.; Deshmukh, A.A.; Padol, A.R.; Kale, D.B. In vitro antibacterial activity of *Emblica officinalis* fruit extract by tube Dilution Method. *Int. J. Toxicol. Appl. Pharmacol.* **2012**, *2*, 49–51.
64. Kamal, R.; Yadav, S.; Mathur, M.; Katariya, P. Antiradical efficiency of 20 selected medicinal plants. *Nat. Prod. Res.* **2012**, *26*, 1054–1062. [CrossRef]
65. Liu, G.; Xiong, S.; Xiang, S.; Guo, C.W.; Ge, F.; Yang, C.R.; Zhang, Y.; Wang, Y.; Kitazato, K. Antiviral activity and possible mechanisms of action of pentagalloylglucose (PGG) against influenza A virus. *Arch. Virol.* **2011**, *156*, 1359–1369. [CrossRef]
66. Usharani, P.; Merugu, P.L.; Nutalapati, C. Evaluation of the effects of a standardized aqueous extract of *Phyllanthus emblica* fruits on endothelial dysfunction, oxidative stress, systemic inflammation and lipid profile in subjects with metabolic syndrome: A randomised, double blind, placebo controlled clinical study. *BMC Complement. Altern. Med.* **2019**, *19*, 97.
67. Malik, S.; Suchal, K.; Bhatia, J.; Khan, S.I.; Vasisth, S.; Tomar, A.; Goyal, S.; Kumar, R.; Arya, D.S.; Ojha, S.K. Therapeutic potential and molecular mechanisms of *Emblica officinalis* gaertn in countering nephrotoxicity in rats induced by the chemotherapeutic agent cisplatin. *Front. Pharmacol.* **2016**, *7*, 350. [CrossRef]
68. Purena, R.; Seth, R.; Bhatt, R. Protective role of *Emblica officinalis* hydro-ethanolic leaf extract in cisplatin induced nephrotoxicity in Rats. *Toxicol. Rep.* **2018**, *5*, 270–277. [CrossRef]
69. Dhingra, D.; Joshi, P.; Gupta, A.; Chhillar, R. Possible Involvement of Monoaminergic Neurotransmission in Antidepressant-like activity of *Emblica officinalis* Fruits in Mice. *CNS Neurosci. Ther.* **2012**, *18*, 419–425. [CrossRef] [PubMed]
70. Suja, R.S.; Nair, A.M.C.; Sujith, S.; Preethy, J.; Deepa, A.K. Evaluation of immunomodulatory potential of *Emblica officinalis* fruit pulp extract in mice. *Indian J. Anim. Res.* **2009**, *43*, 103–106.
71. Ansari, A.; Shahriar, S.Z.; Hassan, M.; Das, S.R.; Rokeya, B.; Haque, A.; Haque, E.; Biswas, N.; Sarkar, T. *Emblica officinalis* improves glycemic status and oxidative stress in STZ induced type 2 diabetic model rats. *Asian Pac. J. Trop. Med.* **2014**, *7*, 21–25. [CrossRef] [PubMed]
72. Akhtar, M.S.; Ramzan, A.; Ali, A.; Ahmad, M. Effect of Amla fruit (*Emblica officinalis* Gaertn.) on blood glucose and lipid profile of normal subjects and type 2 diabetic patients. *Int. J. Food Sci. Nutr.* **2011**, *62*, 609–616. [CrossRef] [PubMed]
73. Gopa, B.; Bhatt, J.; Hemavathi, K.G. A comparative clinical study of hypolipidemic efficacy of Amla (*Emblica officinalis*) with 3-hydroxy-3-methylglutaryl-coenzyme-a reductase inhibitor simvastatin. *Indian J. Pharmacol.* **2012**, *44*, 238.
74. Variya, B.C.; Bakrania, A.K.; Chen, Y.; Han, J.; Patel, S.S. Suppression of abdominal fat and anti-hyperlipidemic potential of Emblica officinalis: Upregulation of PPARs and identification of active moiety. *Biomed. Pharmacother.* **2018**, *108*, 1274–1281. [CrossRef]
75. Husain, I.; Zameer, S.; Madaan, T.; Minhaj, A.; Ahmad, W.; Iqubaal, A.; Ali, A.; Najmi, A.K. Exploring the multifaceted neuroprotective actions of *Emblica officinalis* (Amla): A review. *Metab. Brain Dis.* **2019**, *34*, 957–965. [CrossRef] [PubMed]
76. Shubhi, M.; Rohitash, J.; Radhey, S.; Dharmendra, K.M.; Kshipra, M.; Rajashree, P.; De, R.; Mukhopadhyay, A.; Srivastava, A.K.; Shoma, P.N. Anti-Helicobacter pylori and antioxidant properties of *Emblica officinalis* pulp extract: A potential source for therapeutic use against gastric ulcer. *J. Med. Plants Res.* **2011**, *5*, 2577–2583.
77. Al-Rehaily, A.J.; Al-Howiriny, T.S.; Al-Sohaibani, M.O.; Rafatullah, S. Gastroprotective effects of 'Amla'*Emblica officinalis* on in vivo test models in rats. *Phytomedicine* **2002**, *9*, 515–522. [CrossRef] [PubMed]
78. Santoshkumar, J.; Manjunath, S.; Sakhare, P.M. A study of anti-hyperlipedemia, hypolipedemic and anti-atherogenic activity of fruit of *Emblica officinalis* (amla) in high fat fed Albino rats. *Int. J. Med. Res. Health Sci.* **2013**, *2*, 70–77.
79. Kapoor, M.P.; Suzuki, K.; Derek, T.; Ozeki, M.; Okubo, T. Clinical evaluation of *Emblica officinalis* Gatertn (Amla) in healthy human subjects: Health benefits and safety results from a randomized, double-blind, crossover placebo-controlled study. *Contemp. Clin. Trials Commun.* **2020**, *17*, 100499. [CrossRef] [PubMed]

80. Gul, M.; Liu, Z.-W.; Haq, I.U.; Rabail, R.; Faheem, F.; Walayat, N.; Nawaz, A.; Shabbir, M.A.; Munekata, P.E.S.; Lorenzo, J.M.; et al. Functional and Nutraceutical Significance of Amla (*Phyllanthus emblica* L.): A Review. *Antioxidants* **2022**, *11*, 816. [CrossRef] [PubMed]

81. Kim, H.Y.; Okubo, T.; Juneja, L.R.; Yokozawa, T. The protective role of amla (*Emblica officinalis* Gaertn.) against fructose-induced metabolic syndrome in a rat model. *Br. J. Nutr.* **2010**, *103*, 502–512. [CrossRef]

82. Nisar, M.F.; He, J.; Ahmed, A.; Yang, Y.; Li, M.; Wan, C. Chemical components and biological activities of the genus *Phyllanthus*: A review of the recent literature. *Molecule* **2018**, *23*, 2567. [CrossRef]

83. Kaur, J.; Kaur, D.; Singh, H.; Khan, M.U. *Emblica officinalis*: A meritocratic drug for treating various disorders. *Indo Am. J. Pharm. Res.* **2013**, *3*, 2231–6876.

84. Zahid, S.; Anjum, K.M.; Mughal, M.S.; Yaqub, A.; Yameen, M. Evaluation of antioxidant and antihyperlipidemic activity of Indian gooseberry (*Emblica officinalis*) fruit in high fat-fed rabbits. *JAPS J. Animal Plant Sci.* **2018**, *28*, 1007–1013.

85. Iyer, U.; Shah, A.; Venugopal, S. Amla (*Emblica officinalis*) and Guava (*Psidium guajava*) supplementation: Impact of Low Carbon footprint local seasonal fruits on Lipemic Status of Morning Walkers. *Eco. Environ. Cons.* **2021**, *27*, S233–S238.

86. Sandeep, B.S.; Panda, N.; Sethy, K.; Nath, S. Effect of dietary supplementation of amla (*Emblica officinalis*) powder and equivalent synthetic vitamin C on growth performance in black rock broiler chicken. *Pharma Innov. J.* **2022**, *11*, 2002–2007.

87. Dabulici, C.M.; Sârbu, I.; Vamanu, E. The Bioactive Potential of Functional Products and Bioavailability of Phenolic Compounds. *Foods* **2020**, *9*, 953. [CrossRef]

88. Taleuzzaman, M.; Mahapatra, D.K.; Gupta, D.K. Emblicanin-A and Emblicanin-B: Pharmacological and Nano-Pharmacotherapeutic Perspective for Healthcare Applications. In *Applied Pharmaceutical Practice and Nutraceuticals*; Apple Academic Press: Waretown, NJ, USA, 2021; pp. 13–27.

89. Li, T.; Liang, W.; Xiao, X.; Qian, Y. Nanotechnology, an alternative with promising prospects and advantages for the treatment of cardiovascular diseases. *Int. J. Nanomed.* **2018**, *13*, 7349–7362. [CrossRef] [PubMed]

90. Han, H.S.; Koo, S.Y.; Choi, K.Y. Emerging nanoformulation strategies for phytocompounds and applications from drug delivery to phototherapy to imaging. *Bioact. Mater.* **2022**, *14*, 182–205. [CrossRef] [PubMed]

91. Moradi, S.Z.; Momtaz, S.; Bayrami, Z.; Farzaei, M.H.; Abdollahi, M. Nanoformulations of Herbal Extracts in Treatment of Neurodegenerative Disorders. *Front. Bioeng. Biotechnol.* **2020**, *8*, 238. [CrossRef]

92. Mitchell, M.J.; Billingsley, M.M.; Haley, R.M.; Wechsler, M.E.; Peppas, N.A.; Langer, R. Engineering precision nanoparticles for drug delivery. *Nat. Rev. Drug Discov.* **2021**, *20*, 101–124. [CrossRef]

93. Kapoor-Narula, U.; Lenka, N. Phytochemicals and their nanoformulation in sustained drug delivery and therapy. In *Innovations in Fermentation and Phytopharmaceutical Technologies*; Elsevier: Amsterdam, The Netherlands, 2022; pp. 181–220.

94. Enrico, C. Nanotechnology-Based Drug Delivery of Natural Compounds and Phytochemicals for the Treatment of Cancer and Other Diseases. *Stud. Nat. Prod. Chem.* **2019**, *62*, 91–123. [CrossRef]

95. Omran, Z.; Bader, A.; Porta, A.; Vandamme, T.; Anton, N.; Alehaideb, Z.; El-Said, H.; Faidah, H.; Essa, A.; Vassallo, A.; et al. Evaluation of Antimicrobial Activity of Triphala Constituents and Nanoformulation. *Evid.-Based Complement. Altern. Med.* **2020**, *2020*, 6976973. [CrossRef]

96. Rosarin, S.; Fathima, S.; Vadivel, A.; Samuthira, N.; Sankaran, M. Antiproliferative effect of silver nanoparticles synthesized using amla on Hep2 cell line. *Asian Pac. J. Trop Med.* **2013**, *6*, 1–10. [CrossRef]

97. Abitha, S.T.; Rajeshkumar, S.; Lakshmi, T.; Roy, A. Cytotoxic effects of silver nanoparticles synthesized using amla fruit seed. *Drug Invent. Today* **2019**, *1*, 11.

98. Soundarajan, S.; Sankari, M.; Rajeshkumar, S. Antibacterial activity and cytotoxicity of amla seed mediated graphene oxide, silver nanoparticle and go-ag nanoparticle-an in vitro study. *Plant Cell Biotechnol. Mol. Biol.* **2020**, *21*, 55–67.

99. Ranjani, S.; Hemalatha, S. Triphala decorated multipotent green nanoparticles and its applications. *Mater. Lett.* **2021**, *308*, 131184. [CrossRef]

100. Naik, S.; Jarnain, T.; David, N. Phytofabrication of silver and zinc oxide nanoparticles using the fruit extract of *Phyllanthus emblica* and its potential anti-diabetic and anti-cancer activity. *Part Sci. Technol.* **2022**, 1–13. [CrossRef]

101. Harakeh, S.; Mohammed, A.; Al Soad, J.; Saad, A.; Saber, H.; Al Turki, A.; Najiah; Azhar, E. Antidiabetic effects of novel ellagic acid nanoformulation: Insulin-secreting and anti-apoptosis effects. *Saudi J. Biol. Sci.* **2020**, *27*, 3474–3480. [CrossRef]

102. Hosny, K.M.; Rizg, W.Y.; Khallaf, R.A. Preparation and Optimization of In Situ Gel Loaded with Rosuvastatin-Ellagic Acid Nanotransfersomes to Enhance the Anti-Proliferative Activity. *Pharmaceutics* **2020**, *12*, 263. [CrossRef] [PubMed]

103. Patil, P.; Suresh, K. Chitosan and glyceryl monooleate nanostructures containing gallic acid isolated from amla fruit: Targeted delivery system. *Heliyon* **2021**, *7*, e06526. [CrossRef]

104. Alfei, S.; Barbara, M.; Guendalina, Z.; Federica, T.; Cinzia, D. Dendrimer nanodevices and gallic acid as novel strategies to fight chemoresistance in neuroblastoma cells. *Nanomaterials* **2020**, *10*, 1243. [CrossRef]

105. Khan, B.A.; Mahmood, T.; Menaa, F.; Shahzad, Y.; Yousaf, A.M.; Hussain, T.; Ray, S.D. New perspectives on the efficacy of gallic acid in cosmetics & nanocosmeceuticals. *Curr. Pharm. Des.* **2018**, *24*, 5181–5187.

106. Alfei, S.; Turrini, F.; Catena, S.; Zunin, P.; Parodi, B.; Zuccari, G.; Pittaluga, A.M.; Boggia, R. Preparation of ellagic acid micro and nano formulations with amazingly increased water solubility by its entrapment in pectin or non-PAMAM dendrimers suitable for clinical applications. *New J. Chem.* **2019**, *43*, 2438–2448. [CrossRef]

of food ingredients. The modern medicine system includes expensive disease treatment methods. People are unsatisfied with moving toward alternative substitutes (beneficial products or prebiotics and probiotics) in maintaining a healthy diet to protect against diseases [2]. Nutraceuticals are beneficial compounds generally attained from nutrients, herbal products, dietary supplements, and diets for genetically engineered "designed foods". Food products include soups, cereals, and beverages [3,4].

Physico-chemical stability of these bioactive compounds and nutraceuticals are significantly affected by the harsh environmental condition during production, processing, and storage. In the meantime, due to the physiological conditions generated within the body system during digestion and absorption, the organic functions of bioactive compounds are altered. With an ever-rising demand for healthy foodstuff, substantial interest has been drawn to renovating food processing technologies to build up novel, purposeful, or functional food products with enhanced health benefits. In this perception, nanotechnology or nanofabricated bioactive compounds/nutraceuticals with superior delivery strategies have been launched in the field of the food sector to improve the biological efficacy and physicochemical stability of naturally available bioactive compounds/Nutraceuticals in food and also to shield the food quality [1,3]. Nanofabrication is a promising and exceptional tool in food biotechnology to augment the efficiency of biomolecules in the past, present, and future.

Table 1. Biochemical profiling of principle bioactive compounds and their functions.

Bioactive Compounds	Health-Promoting Property	Occurrence	Molecular Weight (g/mol)	Structure	Reference
Quercetin	Promoting cardiovascular health properties and helping in blood flow.	Fruits and vegetables, especially in onions, grapes, lemon tea, citrus, etc.,	302.236		[5]
Luteolin	Anticarcinogenic activity	Green pepper carrots, Broccoli, oregano	286.24		[6]
Kaempferol	Antioxidant activity and Anticarcinogenic activity	Tomatoes, apples, grapes, green tea, broccoli, lettuce, peaches	286.23		[7]
Curcumin	Antibacterial activity, antioxidant activity, and anti-inflammatory activity	Turmeric	368.38		[8]
Berberine	Treatment of breast cancer, colon cancer, pancreatic cancer, gastric cancer, liver cancer, oral cancer, etc.,	Widely present in barks, leaves, twigs, rhizomes, roots, and stems of several medicinal plant species.	336.3612 g/mol		[9]

Table 1. *Cont.*

Bioactive Compounds	Health-Promoting Property	Occurrence	Molecular Weight (g/mol)	Structure	Reference
Rutin& Quercetin	Protection against cancer and some other diseases. Lowers cholesterol, mainly used in skin aging.	invasive plant species, *Carpobrotus edulis*	610.517 g/mol		[10]
Astaxanthin	Protection from UV skin damage, Reduction in inflammation, Supports Immune system.	algae, yeast, salmon, trout, krill, shrimp, and crayfish	596.841 g/mol		[11]
Vitamin E	helps maintain healthy skin and eyes, and strengthen the body's natural defense against illness and infection (the immune system).	Plant-based oils, nuts, seeds, fruits, and vegetables.	430.71 g/mol		[12]

2. Nano Formulation of Bioactive Compounds

Bioactive compounds such as vitamins, flavonoids, phenolics, proteins, etc., are essential for proper cell growth and differentiation in living organisms; these also play a crucial role in combating diseases by scheming biological mechanisms (Figure 1 and Table 1). Recent studies on coupling the bioactive compounds with existing drugs such as cyclophosphamide and paclitaxel have enhanced breast cancer's healing effect [13]. Consequently, combining medications with bioactive compounds opens a new window in the medical field to treat various diseases. In essence, the formulation of bioactive molecules and drugs also complements the negative aspect of the overdose of drug molecules [14]. On the other hand, the desired amount of bioactive compounds is not consumed in some cases. As a result, unregulated uptake also leads to deterioration in health. Henceforth functional food or food supplements are formulated to tackle these issues. Out of this biological importance, other fields such as biomedical, optoelectronics, and health care with light modification and functionalization were also the beneficiaries. They offered an excellent demand for bioactive compounds. Hence the development of nano-formulation bioactive compounds with an improved adaptation of nanofabrication tools of delivery strategies is in great need or challenge in scientific research [14,15]. Advancement in nano-formulation procedures certainly reduces the limitation of bioactive molecules such as bioavailability, specificity, and solubility. That paves new insights into discoveries of novel materials principally apt for diverse biomedical applications (Table 2).

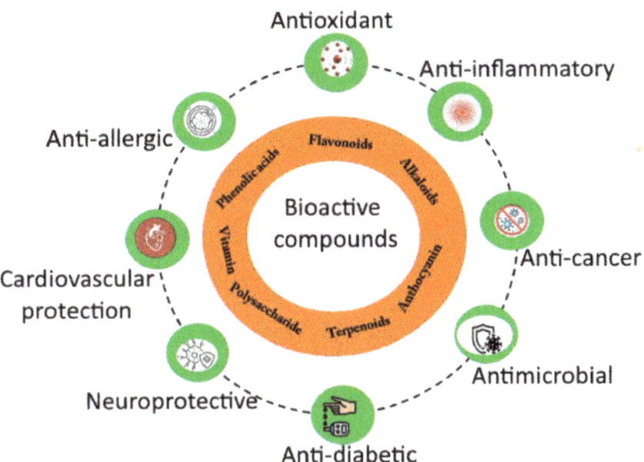

Figure 1. Bioactive compounds for different applications.

2.4. Functional Foods

Functional foods offer health benefits beyond their nutritional value, including whole foods and fortified [47,48]. Proponents of functional foods also enriched with dietary components say that they promote optimal health and may help reduce the risk of chronic diseases. It gives health benefits compared to its traditional nutrients [49]. According to the Canadian Health Organization (CHO), functional foods are "regular recipes that seem to have elements introduced in order to give it a proper scientific or physiological advantage, aside from a merely dietary effect". According to Japan, all pragmatic ingredients have to be available and arising in their herbal version [43] instead of a tablet, capsule, or powder [50] consumed in the nutrition plan on even a regular or daily basis [51] and should revise a biological system in order to avoid disease or disorder-like instances [43].

2.5. Medicinal Nutraceuticals and Pharmaceuticals

Medicinal foods are specially formulated and intended for the dietary management of a disease that has acquired distinctive nutritional needs and cannot be met by the regular diet itself [52]. These medicinal foods are absorbed internally under the consultation of a qualified physician. It acts as precise dietary management of a particular disease under which the nutritional requirements of individuals are confirmed by medical evaluation and based on recognized scientific principles [53]. Pharmaceuticals are used in diagnosing, preventing, or treating the disease and for modifying, restoring, and correcting organic functions [54]. These are enormously helpful medical products obtained from animals or crops. The term is derived from the combination of the words 'farm' and 'pharmaceuticals' [55]. Consumers' well-being is increasing, with a focus on scientific research and asserts reassurance of bioactive molecules, along with novel therapeutic dose and distribution forms [56,57]. The ability of innovative medical and drug technologies to diagnose disease and eventually expand the average lifespan has also provided us with a unique outlook. All food products could have nutritious, sensory characteristics, and preventative measures features. Countless biologically active components have already been commercialized as pharmaceuticals (pills, capsules, solutions, gels, liquors, powders, granules, etc.). These are known as "nutraceuticals", and they help to improve public health. These provide a physiological reward or grant immunity against serious illness, support healthy, and slow the aging process [56,57].

2.6. Role of Nutraceuticals in Human Health

Nutraceuticals are concentrated compounds from food sources that can provide health benefits, including disease prevention and health promotion [58]. Examples of nutraceuticals include substances such as antioxidants, probiotics, and omega 3. It combines nutrition and pharmaceuticals [59]. Nutraceuticals play a preeminent, significant, and valuable role in maintaining normal physiological function, which helps to keep healthy human beings; nutraceuticals are part of food [60]. There has been a fast growth in the nutraceuticals market worldwide that are present in the population and the health trends. In nutraceuticals, the food products used can be classified as fiber, polyunsaturated fatty acids, and other diverse types of herbal/natural foods [61]. Nutraceuticals are the source that is crucially utilized to regulate some of the century's major problems that involve cancer, diabetes, cardiovascular diseases, osteoporosis, cholesterol, arthritis, etc. [62,63]. Furthermore, nutraceuticals have led to a new era of health and medicine in the food industry that has become a research-oriented sector. In this decade, people are more conscious of the correlation between life-threatening diseases and eating habits differing and increasing daily [29]. For this reason, people have become more attentive to the quality consumed and show interest in the foods that have beneficial effects on one's health and avoid the onset of life-threatening diseases [64,65]. The above features help the researchers study the potential effects of beneficial nutraceuticals on their mechanism of action and human health. On an identical note, industries, mainly the food industry, were much more prominent and were interested in developing food products that attract consumers or people to purchase them for their health [66,67].

It is a comprehensive technology that has allowed researchers to improve enormous limitations related to the use of nutraceuticals to retinue their encapsulation into these structures, which involves their stability, poor bioavailability, and low solubility [68,69]. It is mandatory to know about the fact of bioavailability on a notice that it's a fraction of taken compound, which is absorbed and available for physiological functions and is the critical characteristic for nutraceuticals compounds, because of its effectiveness which is strictly associated to their bioavailability [70,71]. Few other important factors deal with the bioavailability of nutraceuticals, and the factors are various endogenous and exogenous, which are involved in the biochemical transformation that undergoes into food storage, Epithelial cells, physio-chemical features, and so on [72–74]. Due to this, more advanced technologies have been established to exert their beneficial effects when introduced into the organism. Different nano-formulations have been developed between them to increase the needful effects of nutraceuticals [75,76]. In the presence of the highest quantity of bioavailability, there promotes a controlled release and targeted delivery of encapsulated bioactive compounds, which increases the biological efficacy of food. Nanotechnology develops the framework of the agriculture and food industry by offering various advantages [77,78]. To design the nutraceutical delivery system, multiple criteria should be enrolled into an account which involves the formulation it must have good physical and chemical properties, sustainable production costs, and food-grade materials used [79,80].

2.7. Favorable Nutraceuticals Properties on Human Health

Nutraceuticals are those substances that have a physiological benefit or protect against chronic diseases. Most therapeutics are involved in it and have had various beneficial effects on human health in recent years. These includes diabetes, arthritis, inflammation, cholesterol, and many other disease conditions [81,82]. One of the best nutraceutical products is natural honey which can be used to feed infants, as it is known to be a food with great benefits and beneficial nutritional properties for human health [83]. Natural honey is made up of antioxidants and enzymes that are essential to the digestive process. Using raw honey allows for cloudiness. Since the extract inhibits the angiotensin-I converting enzyme (ACE1), it is also used to treat high blood pressure. Finally, it is important to emphasize that the skin of the allis shad, which is a nutraceutical, has a reduced benefit compared to some traditional medicines that have been used in the treatment of type 2 diabetes as conditions such as cardiovascular disease and eye disease Side effects on the patient provides problems and ulcers [84]. The honey obtained has antimicrobial activity and can be used to disinfect the infected wound [85]. Thanks to the increase in experimental evidence of the efficacy of nutraceuticals in the prevention or treatment of various pathological conditions, numerous efforts have been made to apply nanotechnology to encapsulate these natural products [86]. A combination of nutraceuticals (bergamot extract, chlorogenic acid, phytosterols, and vitamin C) was administered to overweight subjects with dyslipidemia, increased glucose, and lipid metabolism compared to subjects treated with overnight placebo [87]. Numerous scientific studies have conclusively shown that controlling blood cholesterol levels can reduce the likelihood of developing cardiovascular diseases [88–90]. It is unacceptable that the food industry is responsible for producing a significant part of its waste. It has a major impact on the environment and can be used as an essential source of nutraceuticals. The example concerns a citrus peel, fruit waste rich in phenolic compounds with strong antioxidant activity [87,91]. The antioxidant activity of natural compounds has multiple beneficial effects on human health [92].

Antioxidants act as a barrier to protect the organism and as free radical scavengers, preventing oxidative damage [93,94]. In addition, their effectiveness helps prevent cancer and cardiovascular disease, and their anti-inflammatory effects and ability to establish and prevent Alzheimer's disease are better known [95]. It has been scientifically proven

3.2. Nano Prebiotics

Nano-prebiotics are foods that promote the growth of beneficial bacteria in the gut. These are a type of fire that the human body cannot digest; they serve as food for probiotics and are tiny living organisms that involve bacteria and yeast (Table 4).

Table 4. List of prebiotics and their applications.

Type	Active Compounds	Activity	Study	Reference
Prebiotic	Chondroitin Sulfate Disaccharide	Anticancer effect	The growth of HT-29, human colon cancer cell line, was controlled by Chondroitin sulfate (CS)-Keel disaccharide (CSD) generated by chondroitin AC lyase, estimated at 80% antiproliferative activity.	[121]
Prebiotic	Blueberry anthocyanins	Antioxidant effect	The density and composition of intestinal microbiota in human models were increased by consumption of high purity blueberry anthocyanins through the increase in the modulatory and prebiotic activities.	[122]
Prebiotic	Short-chain fatty acids	Antiproliferative effects	The administration of short-chain fatty acids (SCFAs) prevented the expression of genes involved in the human colorectal cancer cell.	[123]
Prebiotic	Oligosaccharides	Antioxidant effect	The water-soluble oligosaccharide of EMOS-1a showed a 1420% proliferation level	[124]

Prebiotics are found in various sources, including non-digestible carbohydrates, non-digestible oligosaccharides, unrefined wheat, unrefined barley, and soybeans [125]. Some compounds found in prebiotics are xylooligosaccharides, lactulose, and non-starch polysaccharides (NSP) [126].

3.3. Food Grade-Nanofabricated Delivery Systems

Incorporating functional food components for the better functionality of food products to meet consumer needs for a healthy lifestyle is in great demand. It has become a delightful trend in the market. These programmed food or functional food components is not limited only to their inhibitory action against unsolicited microbial growth. Still, it enhances the product's taste, flavor, and color and attributes beneficial health effects. Active food constituents possess some properties such as low water solubility, unappetizing sensory character, constrained oral bioavailability, and intolerance to chemical degradation, so most of these functional food constituents are antagonistic to the food matrix. Nanoencapsulation technology deals with encapsulating bioactive compounds using different nano-fabricated delivery systems [127]. Even though these delivery systems can be used as biocompatible and biodegradable to make them reachable for food applications, these delivery systems are generally classified into carbohydrates, lipids, or protein-based [128].

3.4. Carbohydrate/Polysaccharides Delivery Systems

Carbohydrate delivery systems can interact with a wide range of bioactive components because these carbohydrate-based delivery systems possess exceptional properties like biodegradability, abundance, and a high potential for adaptation to enviable functionality so that these delivery systems are more appropriate for high-temperature stability over lipid and protein-based carriers which undergo degradation and denaturation. Cyclodextrin (α, β, γ) is a carbohydrate-based delivery system. It is a prominent condensed cone-shaped oligosaccharide with a hydrophobic central cavity and hydrophilic end on the other. So, in carbohydrate-based delivery systems, the present trend is to study these cyclodextrins. These are useful for the confinement of less soluble, temperature susceptible, or chemically liable food bioactive. On the other hand, a mixture or combination of two or more carbohydrate-based delivery systems gives inventive frontiers for increased efficiency. Nanofabrication techniques like spray drying, electro-spraying, coacervation, and electro-spinning are extensively used to fabricate carbohydrate-based delivery systems [129,130] (Table 5) (Figure 3).

Table 5. Nutraceutical encapsulation by the carbohydrate-based delivery system.

Nutraceuticals	Encapsulating Materials	Encapsulation Techniques	Target	Reference
Catechin	Azivash (*Corchorus olitorius* L.) gum-polyvinyl alcohol	Electrospinning process	Simulated gastric fluid and simulated intestinal fluid, EE%.	[131]
Hesperetin (HSP)	Nanofibers: Basil seed mucilage/polyvinylalcohol	Electrospinning	Characterization of nanofibers. Release models, EE%	[132]
Rutin	Quinoa and maize starch NPs	Ultrasonication	Characterization, EE%, simulated in vitro digestion.	[133]
Saffron Bioactive components	Nanoparticles: chitosan (CS) and gum arabic (GA)	Ionic gelation (IG)	Characterization NPs, EE%, Release of Saffron in acidic and natural media	[134]

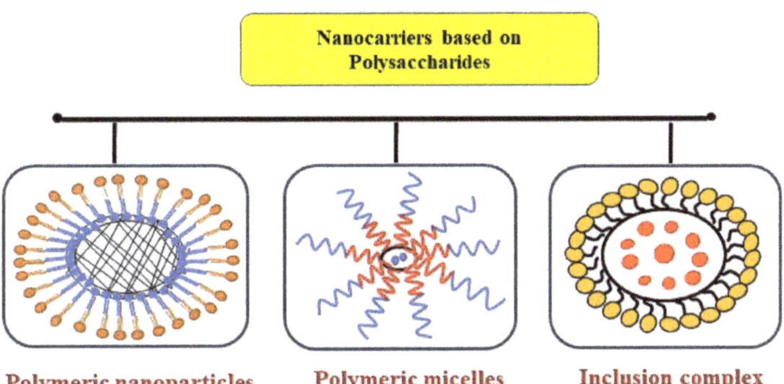

Figure 3. Polysaccharides-based nanofabrication systems.

3.5. Protein-Based Delivery Systems

Protein-based delivery systems also have captivated great recognition. The exterior surface of protein-based delivery systems possesses many functional groups that facilitate their ability to encapsulate both hydrophilic and lipophilic bioactive constituents. They are mainly obtained from plant, animal, and microbial sources. Coacervation, gelation, spray drying, and electro-hydrodynamic process are commonly used methods for fabricating protein-based delivery systems. [135]. In recent years plant-protein-based delivery systems have also emerged in the field of nanotechnology to shield and manage lipophilic bioactive compounds (polyphenols, polyunsaturated fatty acids, vitamins, carotenoids), has increased attention in nutraceuticals, food, and pharmaceutical fields.

The promising role of plant proteins from legumes, nuts, tuber, oil/edible seeds, and cereals is aggravated by their sustainable, eco-friendly, and vigorous nature collating with other sources. Before using raw materials carriers, particular challenges need to be addressed to overcome the adverse consequences by modifying approaches to improve their techno-functionality and limitations, aiming to fabricate innovative production of plant-based carriers (emulsions, hydrogels, structures, and self-assembled films). This plant protein delivery system is a limited application that only acts as a carrier for lipophilic compounds. Nevertheless, bottomless insights into new purification or extraction technologies and protein sources are necessary to validate their functional properties for the evolution of an effective delivery platform [136] (Figure 4).

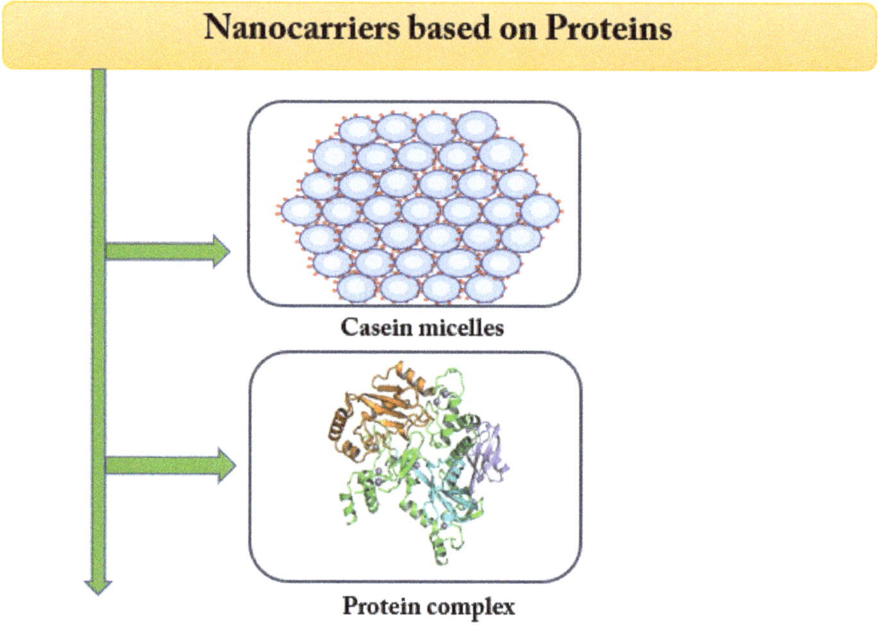

Figure 4. Protein-based nanofabrication system. Adapted with permission from Ref. Singh et al. [136]. Copyright © 2021, Controlled Release Society).

The spatial shape, size, colloidal stability, water dispersibility, etc., of protein-based nanocarriers, were determined by the physicochemical molecular principles of protein. Some of the available preparative methods could be used to fabricate several protein-based nanocarriers, including antisolvent precipitation, pH–driven, gelation, and electrospray methods. Recent studies have evidenced that protein-based nanocarriers provide numerous advantages in aid of the application of bioactive compounds to the food, medical, and cosmetic sectors [137].

3.6. Lipid-Based Nano Delivery Systems

Compared to protein and carbohydrate-based delivery systems, the lipid-based nano delivery system is the most auspicious encapsulation system that offers peerless advantages over the other two systems due to its ability to entrap materials with different solubilities enhanced targeted delivery and ability to guard a constituent against free radical's pH and enzymes [138]. Lipid-based nanocarriers have been fabricated for food applications such as nanoemulsions, nanoliposomes, nanostructured lipid carriers (NLC), and solid lipid nanoparticles [139] (Figure 5).

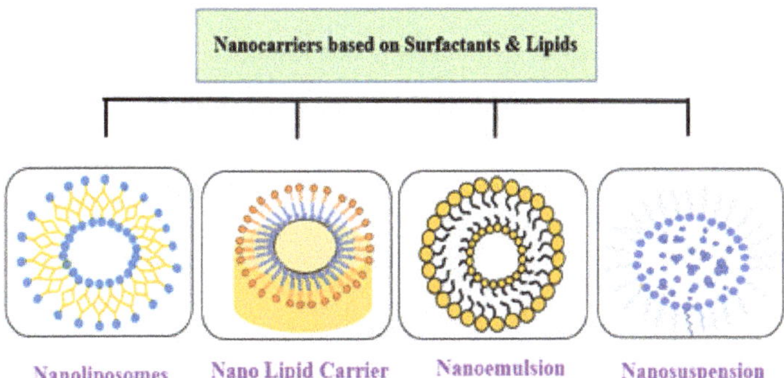

Figure 5. Lipid-based nanofabrication system.

3.7. Nano-Emulsions

Nano-emulsions are liquid droplets with colloidal dispersion sizes less than 100 nm. They can be either oil-in-water or water-in-oil (w/o) [140]. They are made by entrapping bioactive constituents within the distribute phase, which is self-stabilized by surfactants/biopolymers in the continuous phase [141]. They are made by entrapping bioactive constituents within the distribute phase, which is self-stabilized by surfactants/biopolymers in the continuous phase [141]. These can be manufactured using high-energy emulsification techniques such as homogenization, ultrasonication, and microfluidization.

3.8. Nanoliposomes

Liposomes are nano-sized spherical vesicles composed of a phospholipid, a lipid bilayer with an aqueous core. Nanoliposomes have a superior surface area and tolerable stability profile to safeguard their size within nanometric scales such as small liposomes (20–100 nm) and large liposomes (>100 nm) [105]. These carries mainly contain lipids and phospholipids. Conversely, some include molecules such as antioxidants, carbohydrates, protein, or sterols in their structure. Their amphiphilic nature makes them potent nanocarriers since these have the potential to encapsulate and massive release hydrophobic and hydrophilic compounds concurrently, providing mutual benefit. Distinguishing phospholipids/bilayer structure is highly attuned to the skin surface, allowing them to act as penetration enhancers of nutraceuticals and bioactive compounds towards targeted sites [142]. Some techniques such as microfluidization, extrusion, and ultrasonication are usually used for nanoliposomes production [143]. The main restrictions lie with poor solubility towards low pH and enzymes after intake, low loading ability, and higher cost of food-grade fabrication wall materials. The present trend is to tackle these barriers for the expansion of nanoliposomes optimistic nanocarriers on a large scale [144].

Examples of drugs in liposomal carriers are doxorubicin, amphotericin B, nystatin, and vincristine. In addition, ZawnVillines 2018 [145] has reported that carotenoids encapsulated in lipid bilayers can impair carotenoid stability. This, in turn, affects the properties of liposomes. This is because types of carotenoids and different concentrations could modulate the structure, hydrophobicity, and dynamics of the liposomal membrane [146]. In turn, the structural integrity of the liposome bilayers can protect the encapsulated compounds from degradation, thereby enhancing their physicochemical stability. In recent years, some liposomes have been manufactured for the simultaneous delivery of bioactive compounds. For example, pectin-coated phosphatidylcholine (PC) liposomes have been engineered to encapsulate two important lysozyme and nisin compounds [147].

Furthermore, liposomes containing both the compounds lysozyme and nisin showed higher stability and antimicrobial activity compared to liposomes containing lysozyme, representing a promising co-delivery system for bioactive substances [148]. Water-soluble vitamin E derivative of PEG, d–tocopheryl polyethylene glycol 1000 succinate (TPGS). Such fabricated liposomes are more stable in biological environments and have proven to be a valuable, safe tool for incorporation into liposome membranes [149]. Some of the beneficial activities of liposomes include increased bioavailability and absorption rate compared to oral supplements; The micronized encapsulation protects against the harsh environment of the gastrointestinal tract and improves absorption [150]. It also has the potential to hold both hydrophilic and hydrophobic molecules. A list of Nutraceuticals encapsulated by liposomes is shown in Figure 6.

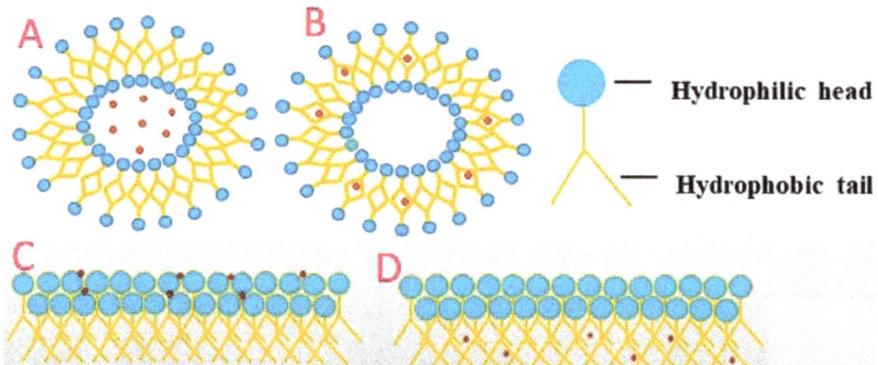

Figure 6. Schematic representation of bioactive compounds encapsulated in various nano-fabricated liposome systems. (**A**), bioactive compound entrapped in the interior of liposome region; (**B**), bioactive compound entrapped in the fatty acid region of liposome region; (**C**), bioactive compound entrapped in the hydrophilic head region of phospholipid bilayer; (**D**), bioactive compound entrapped in the hydrophobic head region of the phospholipid bilayer.

3.9. Solid Lipid Nanoparticles (SLNs) and Nanostructured Lipid Carrier (NLC)

(SLNs) and (NLC) are the other two important lipid-based nano delivery systems. SLNs are submicron emulsions established by homogenizing two immiscible phases in the presence of amphiphilic surfactant. These SLNs facilitate the maintenance of the liquid phase during the process. The lipid phase is melted at an elevated temperature to offer higher encapsulation efficiency, a slower degradation rate, and flexibility in the release profile [151]. The commonly used method for large-scale production of SLNs for food processing homogenization, the major problem associated with SLNs is the low encapsulation load. So that the manufacturing of nanostructured based lipid carrier as partially crystallized lipid nanocarriers provide an effective solution with enhanced physical stability, improved dispersibility, and higher encapsulation load as well as to move up a remarkable trap of hydrophilic and hydrophobic nutrients [152–154].

3.10. Nanogels

Nanogels are nanoparticles synthesized by combining hydrogel and a cross-linked chemically or physically with hydrophilic polymer chains [155]. It also signifies a biocompatible and biodegradable drug delivery system obtained from a nano-sized polymeric network; the swelling attains it under the action of a suitable solvent [156]. There are different ways to be administered these drug delivery systems. Nanogels can be expressed as they must be responsive to various stimuli involving magnetic field, temperature, and pH [157]. For example, the nanogels, which are made of dextran and ovalbumin, were produced by Maillard reaction and were effectively employed to improve curcumin bioaccessibility in the gastrointestinal tract, as was established in the form of an in vitro reproduced model [137].

4. Future Perspectives and Regulatory Outlook

The toxicological valuation of nano-fabricated material and their different delivery systems for bioactive compounds and nutraceuticals holds immense importance and vigilant assessment. The utilization of nano-fabricated materials may be allied with numerous safety reservations and necessitate implementing practices that perilously take safety and human health into consideration. The use of nano-fabricated materials in the food industry in food packaging is efficiently browbeaten with rapid growth but still coupled with end-user regulatory and safety issues that should be assessed and addressed critically [158] (Figure 6).

Health concerns, authoritarian policies, and safety must be well-thought-out during the production, processing, and wrapping of nano-based programmed food products. In terms of regulatory aspects of nanomaterials and nanostructure in the medical and food sector, there is no explicit legislation applied worldwide. Many countries still do not have specific authorization for nano-encapsulated products' safety and risk assessments. The lack of reliability in the functional switching information between the countries is an outlined risk to mankind and the environment. It may perhaps minimize the promotion of novel and valuable products globally [159]. Due to the lack of specified regulations for food nanotechnology, European Union regulations and U.S. Food and Drug Administration has established guidelines regarding the use of nanotechnology in food and to get safety assessment approval before its use. In one of the articles on the nature of nanotechnology, Steffen Foss Hansen put forth a framework named 'React Now,' which stands for (Registration, Evaluation, Authorization, Categorization, and Tools to Evaluate Nanomaterials Opportunities and Weaknesses). All users of nano-fabricated materials should be evaluated according to the Nano-Risk-Cat-methodology [159]. Therefore, organizations and companies functioning with nano-fabricated materials have to carefully think about all the above-mentioned aspects to contest food safety concerns and ensure successful promotions of nano-fabricated programmed food products or nutraceuticals in the marketplace.

5. Conclusions

In this review, we focused on the biological functionality of bioactive compounds and nutraceuticals and their beneficial activity for maintaining a healthy lifestyle. Nutraceutical compounds derived from natural origin can treat several pathologies, inflammation, cancer, and cardiovascular diseases. In contrast, we reviewed the physicochemical instability of these significant compounds and declined bioavailability due to environmental conditions. Furthermore, we addressed the promising contribution of food technology throughout the sustained release/targeted delivery of bioactive compounds/nutraceuticals and about their defense from harsh environments through the use of various nanofabricated delivery systems. Various delivery systems such as carbohydrates, proteins, lipids (nano-emulsion, liposome-mediated delivery systems), and many other techniques have drawn scientific interest and allowed researchers to eradicate many of the limitations in using nutraceuticals. Food nanotechnology could be an alternative method to generate future trends in nutraceuticals and uphold bioactive compounds' stability. However, certain regulatory aspects and safety issues need to be addressed by the European Union regulations and U.S. Food and Drug Administration to make the best use of nanofabricated materials as nanofabricated delivery systems to enhance the activity and retain the shelf-life of bioactive compounds and nutraceuticals to treat various disease and to protect or improve food constituents, flavor and taste for the betterment of mankind in maintaining a healthy standard of living.

Author Contributions: Conceptualization, V.B.R., M.S. and K.S.; methodology, R.P., R.L. and N.L.S.; software, V.K.G., B.S.I. and Z.U.; validation, V.K.G., V.B.R., B.S.I. and Z.U.; formal analysis, R.P., R.L., M.S. and N.L.S.; investigation, R.P., R.L. and N.L.S.; resources, V.B.R., R.P., K.S.; data curation, R.P., R.L. and N.L.S.; writing—original draft preparation, R.P., R.L. and N.L.S.; writing—review and editing, V.K.G., B.S.I., V.B.R., M.S., K.S., R.P. and Z.U; visualization, V.B.R., M.S., K.S. and Z.U; supervision, V.B.R., M.S. and K.S.; project administration, V.B.R., M.S., R.P. and K.S.; funding acquisition, V.B.R., M.S. and K.S. All authors have read and agreed to the published version of the manuscript.

Funding: This research received no external funding.

Institutional Review Board Statement: Not applicable.

Informed Consent Statement: Not applicable.

Data Availability Statement: Data that support the findings are available within the manuscript.

Acknowledgments: We are very thankful to Bhagya Lakshmi Neel Warne, for her timely support of the corrections of the manuscript.

Conflicts of Interest: The authors declare no conflict of interest.

References

1. Luo, Y.; Wang, Q.; Zhang, Y. Biopolymer-Based Nanotechnology Approaches to Deliver Bioactive Compounds for Food Applications: A Perspective on the Past, Present, and Future. *J. Agric. Food Chem.* **2020**, *68*, 12993–13000. [CrossRef] [PubMed]
2. Daliu, P.; Santini, A.; Novellino, E. From pharmaceuticals to nutraceuticals: Bridging disease prevention and management. *Expert Rev. Clin. Pharmacol.* **2019**, *12*, 1–7. [CrossRef] [PubMed]
3. Sarwal, A.; Rawat, N.; Singh, G.; Sinha, V.R.; Sharma, S.; Kumar, D. Nanonutraceuticals in Central Nervous System Disorders. In *NanoNutraceuticals*; CRC Press: Boca Raton, FL, USA, 2018; pp. 91–104. [CrossRef]
4. Sasi, S. Nutraceuticals—A review. *Int. J. Ind. Biotechnol. Biomater.* **2017**, *3*, 25–29.
5. Deepika; Maurya, P.K. Health Benefits of Quercetin in Age-Related Diseases. *Molecules* **2022**, *27*, 2498. [CrossRef] [PubMed]
6. Sagi, S.S. Quercetin: A Potential Flavanol with Multiple Health Benefits. *Arch. Food Sci. Nutr. Res.* **2021**, *2*, 1002.
7. Yang, L.; Gao, Y.; Bajpai, V.K.; El-Kammar, H.A.; Simal-Gandara, J.; Cao, H.; Cheng, K.W.; Wang, M.; Arroo, R.R.; Zou, L.; et al. Advance toward isolation, extraction, metabolism and health benefits of kaempferol, a major dietary flavonoid with future perspectives. *Crit. Rev. Food Sci. Nutr.* **2021**, 1–17. [CrossRef]
8. Guest, P.C.; Sahebkar, A. Research in the Middle East into the Health Benefits of Curcumin. In *Studies on Biomarkers and New Targets in Aging Research in Iran*; Springer: Cham, Switzerland, 2021; pp. 1–13. [CrossRef]
9. Xu, X.; Yi, H.; Wu, J.; Kuang, T.; Zhang, J.; Li, Q.; Du, H.; Xu, T.; Jiang, G.; Fan, G. Therapeutic effect of berberine on metabolic diseases: Both pharmacological data and clinical evidence. *Biomed. Pharmacother.* **2021**, *133*, 110984. [CrossRef]
10. Zhou, L.; Cai, L.; Ruan, H.; Zhang, L.; Wang, J.; Jiang, H.; Wu, Y.; Feng, S.; Chen, J. Electrospun chitosan oligosaccharide/polycaprolactone nanofibers loaded with wound-healing compounds of Rutin and Quercetin as antibacterial dressings. *Int. J. Biol. Macromol.* **2021**, *183*, 1145–1154. [CrossRef]
11. Raza, S.H.A.; Naqvi, S.R.Z.; Abdelnour, S.A.; Schreurs, N.; Mohammedsaleh, Z.M.; Khan, I.; Shater, A.F.; El-Hack, M.E.A.; Khafaga, A.F.; Quan, G.; et al. Beneficial effects and health benefits of Astaxanthin molecules on animal production: A review. *Res. Vet. Sci.* **2021**, *138*, 69–78. [CrossRef]
12. Zaaboul, F.; Liu, Y. Vitamin E in foodstuff: Nutritional, analytical, and food technology aspects. *Compr. Rev. Food Sci. Food Saf.* **2022**, *21*, 964–998. [CrossRef]
13. Mangla, B.; Kohli, K. Combination of natural agent with synthetic drug for the breast cancer therapy. *Int. J. Drug Dev. Res.* **2018**, *10*, 22–26.
14. Majumdar, M.; Shivalkar, S.; Pal, A.; Verma, M.L.; Sahoo, A.K.; Roy, D.N. Nanotechnology for enhanced bioactivity of bioactive compounds. In *Biotechnological Production of Bioactive Compounds*; Elsevier: Amsterdam, The Netherlands, 2019; pp. 433–466. [CrossRef]
15. McClements, D.J.; Decker, E.; Weiss, J. Emulsion-Based Delivery Systems for Lipophilic Bioactive Components. *J. Food Sci.* **2007**, *72*, R109–R124. [CrossRef] [PubMed]
16. Ghanbari-Movahed, M.; Mondal, A.; Farzaei, M.H.; Bishayee, A. Quercetin- and rutin-based nano-formulations for cancer treatment: A systematic review of improved efficacy and molecular mechanisms. *Phytomedicine* **2022**, *97*, 153909. [CrossRef] [PubMed]
17. Sharma, M.; Inbaraj, B.S.; Dikkala, P.K.; Sridhar, K.; Mude, A.N.; Narsaiah, K. Preparation of Curcumin Hydrogel Beads for the Development of Functional *Kulfi*: A Tailoring Delivery System. *Foods* **2022**, *11*, 182. [CrossRef]
18. Lv, S.; Zhang, Y.; Tan, H.; Zhang, R.; McClements, D.J. Vitamin E Encapsulation within Oil-in-Water Emulsions: Impact of Emulsifier Type on Physicochemical Stability and Bioaccessibility. *J. Agric. Food Chem.* **2019**, *67*, 1521–1529. [CrossRef]
19. Wang, P.-P.; Luo, Z.-G.; Peng, X.-C. Encapsulation of Vitamin E and Soy Isoflavone Using Spiral Dextrin: Comparative Structural Characterization, Release Kinetics, and Antioxidant Capacity during Simulated Gastrointestinal Tract. *J. Agric. Food Chem.* **2018**, *66*, 10598–10607. [CrossRef]
20. Fan, L.; Lu, Y.; Ouyang, X.-K.; Ling, J. Development and characterization of soybean protein isolate and fucoidan nanoparticles for curcumin encapsulation. *Int. J. Biol. Macromol.* **2021**, *169*, 194–205. [CrossRef]
21. Bolat, Z.B.; Islek, Z.; Demir, B.N.; Yilmaz, E.N.; Sahin, F.; Ucisik, M.H. Curcumin- and Piperine-Loaded Emulsomes as Combinational Treatment Approach Enhance the Anticancer Activity of Curcumin on HCT116 Colorectal Cancer Model. *Front. Bioeng. Biotechnol.* **2020**, *8*, 50. [CrossRef]

22. Oueslati, M.H.; Ben Tahar, L.; Harrath, A.H. Catalytic, antioxidant and anticancer activities of gold nanoparticles synthesized by kaempferol glucoside from Lotus leguminosae. *Arab. J. Chem.* **2020**, *13*, 3112–3122. [CrossRef]

23. Loo, Y.S.; Madheswaran, T.; Rajendran, R.; Bose, R.J. Encapsulation of berberine into liquid crystalline nanoparticles to enhance its solubility and anticancer activity in MCF7 human breast cancer cells. *J. Drug Deliv. Sci. Technol.* **2020**, *57*, 101756. [CrossRef]

24. Campos, E.; Proença, P.L.F.; Oliveira, J.L.; Pereira, A.; Ribeiro, L.N.D.M.; Fernandes, F.O.; Gonçalves, K.C.; Polanczyk, R.A.; Pasquoto-Stigliani, T.; Lima, R.; et al. Carvacrol and linalool co-loaded in β-cyclodextrin-grafted chitosan nanoparticles as sustainable biopesticide aiming pest control. *Sci. Rep.* **2018**, *8*, 7623. [CrossRef] [PubMed]

25. Burgos-Díaz, C.; Opazo-Navarrete, M.; Soto-Añual, M.; Leal-Calderon, F.; Bustamante, M. Food-grade Pickering emulsion as a novel astaxanthin encapsulation system for making powder-based products: Evaluation of astaxanthin stability during processing, storage, and its bioaccessibility. *Food Res. Int.* **2020**, *134*, 109244. [CrossRef] [PubMed]

26. Aronson, J.K. Defining 'nutraceuticals': Neither nutritious nor pharmaceutical. *Br. J. Clin. Pharmacol.* **2017**, *83*, 8–19. [CrossRef] [PubMed]

27. Prabu, S.L.; Suriyaprakash, T.; Dinesh, K.; Suresh, K.; Ragavendran, T. Nutraceuticals: A review. *Elixir Pharm.* **2012**, *46*, 8372–8377.

28. Crawford, C.; Boyd, C.; Paat, C.F.; Meissner, K.; Lentino, C.; Teo, L.; Berry, K.; Deuster, P. Dietary Ingredients as an Alternative Approach for Mitigating Chronic Musculoskeletal Pain: Evidence-Based Recommendations for Practice and Research in the Military. *Pain Med.* **2019**, *20*, 1236–1247. [CrossRef]

29. Zanella, M.; Ciappellano, S.G.; Venturini, M.; Tedesco, E.; Manodori, L.; Benetti, F. Nutraceuticals and nanotechnology. *Diet. Ingred. Suppl.* **2015**, *26*, 26–31.

30. Subramani, T.; Ganapathyswamy, H. An overview of liposomal nano-encapsulation techniques and its applications in food and nutraceutical. *J. Food Sci. Technol.* **2020**, *57*, 3545–3555. [CrossRef]

31. Singh, A.R.; Desu, P.K.; Nakkala, R.K.; Kondi, V.; Devi, S.; Alam, M.S.; Hamid, H.; Athawale, R.B.; Kesharwani, P. Nanotechnology-based approaches applied to nutraceuticals. *Drug Deliv. Transl. Res.* **2022**, *12*, 485–499. [CrossRef]

32. Sahani, S.; Sharma, Y.C. Advancements in applications of nanotechnology in global food industry. *Food Chem.* **2021**, *342*, 128318. [CrossRef]

33. Manocha, S.; Dhiman, S.; Grewal, A.S.; Guarve, K. Nanotechnology: An approach to overcome bioavailability challenges of nutraceuticals. *J. Drug Deliv. Sci. Technol.* **2022**, *72*, 103418. [CrossRef]

34. Bronzwaer, S.; Kass, G.; Robinson, T.; Tarazona, J.; Verhagen, H.; Verloo, D.; Vrbos, D.; Hugas, M. Food Safety Regulatory Research Needs 2030. *EFSA J.* **2019**, *17*, e170622. [CrossRef] [PubMed]

35. Whitman, M. Understanding the perceived need for complementary and alternative nutraceuticals: Lifestyle issues. *Clin. J. Oncol Nurs.* **2001**, *5*, 190–194.

36. Ghani, U.; Naeem, M.; Rafeeq, H.; Imtiaz, U.; Amjad, A.; Ullah, S.; Rehman, A.; Qasim, F. A Novel Approach towards Nutraceuticals and Biomedical Applications. *Sch. Int. J. Biochem.* **2019**, *2*, 245–252. [CrossRef]

37. Palmer, M.E.; Haller, C.; McKinney, P.E.; Klein-Schwartz, W.; Tschirgi, A.; Smolinske, S.C.; Woolf, A.; Sprague, B.M.; Ko, R.; Everson, G.; et al. Adverse events associated with dietary supplements: An observational study. *Lancet* **2003**, *361*, 101–106. [CrossRef]

38. Dable-Tupas, G.; Otero, M.C.B.; Bernolo, L. Functional foods and health benefits. In *Functional Foods and Nutraceuticals*; Springer: Cham, Switzerland, 2020; pp. 1–11.

39. Crowe, K.M.; Francis, C. Position of the Academy of Nutrition and Dietetics: Functional Foods. *J. Acad. Nutr. Diet.* **2013**, *113*, 1096–1103. [CrossRef] [PubMed]

40. Dwyer, J.T.; Coates, P.M.; Smith, M.J. Dietary Supplements: Regulatory Challenges and Research Resources. *Nutrients* **2018**, *10*, 41. [CrossRef] [PubMed]

41. Rawson, E.S.; Miles, M.P.; Larson-Meyer, D.E. Dietary supplements for health, adaptation, and recovery in athletes. *Int. J. Sport Nutr. Exerc. Metab.* **2018**, *28*, 188–199. [CrossRef]

42. Psota, T.L.; Gebauer, S.K.; Kris, E.P. Dietary omega-3 fatty acid intake and cardiovascular risk. *Am. J. Cardiol.* **2006**, *98*, 3–18. [CrossRef]

43. Chandra, S.; Saklani, S.; Kumar, P.; Kim, B.; Coutinho, H.D.M. Nutraceuticals: Pharmacologically Active Potent Dietary Supplements. *BioMed Res. Int.* **2022**, *2022*, 2051017. [CrossRef] [PubMed]

44. Abbadi, A.; Domergue, F. Biosynthesis of very long- chain polyunsaturated fatty acids in transgenic oilseeds: Constraints on their accumulation. *Plant Cell* **2004**, *16*, 2734–2748. [CrossRef]

45. Wu, G.; Truksa, M.; Datla, N. Stepwise engineering to produce high yields of very long-chain polyunsaturated fatty acids in plants. *Nat. Biotechnol.* **2005**, *23*, 1013–1017. [CrossRef] [PubMed]

46. Das, L.; Bhaumik, E.; Raychaudhuri, U.; Chakraborty, R. Role of nutraceuticals in human health. *J. Food Sci. Technol.* **2012**, *49*, 173–183. [CrossRef] [PubMed]

47. Siciliano, R.; Reale, A.; Mazzeo, M.; Morandi, S.; Silvetti, T.; Brasca, M. Paraprobiotics: A New Perspective for Functional Foods and Nutraceuticals. *Nutrients* **2021**, *13*, 1225. [CrossRef] [PubMed]

48. Adefegha, A. Functional Foods and Nutraceuticals as Dietary Intervention in Chronic Diseases; Novel Perspectives for Health Promotion and Disease Prevention. *J. Diet. Suppl.* **2018**, *15*, 977–1009. [CrossRef]

49. Gruss, S.M.; Nhim, K.; Gregg, E.; Bell, M.; Luman, E.; Albright, A. Public Health Approaches to Type 2 Diabetes Prevention: The US National Diabetes Prevention Program and Beyond. *Curr. Diabetes Rep.* **2019**, *19*, 78. [CrossRef]

50. De Felice, S. FIM rationale and proposed guidelines for the nutraceutical research & education act–NREA. In Proceedings of the FIM's 10th Nutraceutical Conference, The Waldorf-Astoria, New York, NY, USA, 10–11 November 2002; Available online: https://fimdefelice.org/fim-rationale-and-proposed-guidelines-for-the-nutraceutical-research-education-act-nrea/ (accessed on 12 August 2022).
51. Rajasekaran, A.; Sivagnanam, G.; Xavier, R. Nutraceuticals as therapeutic agents, a review research. *Res. J. Pharm. Technol.* **2008**, *1*, 328–340.
52. Ozdal, T.; Tomas, M.; Toydemir, G.; Kamiloglu, S.; Capanoglu, E. Introduction to nutraceuticals, medicinal foods, and herbs. In *Aromatic Herbs in Food*; Academic Press: Cambridge, MA, USA, 2021; pp. 1–34. [CrossRef]
53. Jagannathan, R.; Patel, S.A.; Ali, M.K.; Narayan, K.M.V. Global Updates on Cardiovascular Disease Mortality Trends and Attribution of Traditional Risk Factors. *Curr. Diabetes Rep.* **2019**, *19*, 44. [CrossRef]
54. Inoue, M.; Sumii, Y.; Shibata, N. Contribution of Organofluorine Compounds to Pharmaceuticals. *ACS Omega* **2020**, *5*, 10633–10640. [CrossRef]
55. Gonçalves, R.F.S.; Martins, J.T.; Duarte, C.M.M.; Vicente, A.A.; Pinheiro, A.C. Advances in nutraceutical delivery systems: From formulation design for bioavailability enhancement to efficacy and safety evaluation. *Trends Food Sci. Technol.* **2018**, *78*, 270–291. [CrossRef]
56. Gopi, S.; Balakrishnan, P. *Handbook of Nutraceuticals and Natural Products*; John Wiley & Sons, Inc.: Hoboken, NJ, USA, 2022.
57. Shahgholian, N. Introduction to Nutraceuticals and Natural Products. In *Handbook of Nutraceuticals and Natural Products: Biological, Medicinal, and Nutritional Properties and Applications*; John Wiley & Sons, Inc.: Hoboken, NJ, USA, 2002; pp. 1–14.
58. Durazzo, A.; Lucarini, M.; Santini, A. Nutraceuticals in Human Health. *Foods* **2020**, *9*, 370. [CrossRef]
59. Natarajan, T.D.; Ramasamy, J.R.; Palanisamy, K. Nutraceutical potentials of synergic foods: A systematic review. *J. Ethn. Foods* **2019**, *6*, 1–7. [CrossRef]
60. Ghaffari, S.; Roshanravan, N. The role of nutraceuticals in prevention and treatment of hypertension: An updated review of the literature. *Food Res. Int.* **2020**, *128*, 108749. [CrossRef] [PubMed]
61. Roshanravan, N.; Mohammadi, H. The Role of Nutraceuticals in Gestational Diabetes Mellitus. In *Nutraceuticals for Prenatal, Maternal and Offspring's Nutritional Health*; CRC Press: Boca Raton, FL, USA, 2019; pp. 153–167. [CrossRef]
62. Allahveran, A.; Farokhzad, A.; Asghari, M.; Sarkhosh, A. Foliar application of ascorbic and citric acids enhanced 'Red Spur' apple fruit quality, bioactive compounds and antioxidant activity. *Physiol. Mol. Biol. Plants* **2018**, *24*, 433–440. [CrossRef]
63. Langer, R.; Peppas, N.A. Advances in biomaterials, drug delivery, and bionanotechnology. *AIChE J.* **2003**, *49*, 2990–3006. [CrossRef]
64. Estevinho, L.M. Special issue "Nutraceuticals in human health and disease". *Int. J. Mol. Sci.* **2018**, *19*, 1213. [CrossRef] [PubMed]
65. Ting, Y.; Jiang, Y.; Ho, C.-T.; Huang, Q. Common delivery systems for enhancing in vivo bioavailability and biological efficacy of nutraceuticals. *J. Funct. Foods* **2014**, *7*, 112–128. [CrossRef]
66. Meléndez-Martínez, A.J.; Böhm, V.; Borge, G.I.A.; Cano, M.P.; Fikselová, M.; Gruskiene, R.; Lavelli, V.; Loizzo, M.R.; Mandić, A.I.; Brahm, P.M.; et al. Carotenoids: Considerations for Their Use in Functional Foods, Nutraceuticals, Nutricosmetics, Supplements, Botanicals, and Novel Foods in the Context of Sustainability, Circular Economy, and Climate Change. *Annu. Rev. Food Sci. Technol.* **2021**, *12*, 433–460. [CrossRef] [PubMed]
67. Leonard, N.B. Stability testing of nutraceuticals and functional foods. In *Handbook of Nutraceuticals and Functional Foods*; CRC Press: Boca Raton, FL, USA, 2000.
68. Zheng, B.; McClements, D.J. Formulation of More Efficacious Curcumin Delivery Systems Using Colloid Science: Enhanced Solubility, Stability, and Bioavailability. *Molecules* **2020**, *25*, 2791. [CrossRef]
69. Jones, D.; Caballero, S.; Davidov-Pardo, G. Chapter Six—Bioavailability of nanotechnology-based bioactives and nutraceuticals. In *Advances in Food and Nutrition Research*; Academic Press: Cambridge, MA, USA, 2019; Volume 88, pp. 235–273.
70. Punia, S.; Sandhu, K.S.; Kaur, M.; Siroha, A.K. Nanotechnology: A successful approach to improve nutraceutical bioavailability. In *Nanobiotechnology in Bioformulations*; Springer: Cham, Switzerland, 2019; pp. 119–133.
71. Helal, N.A.; Eassa, H.A.; Amer, A.M.; Eltokhy, M.A.; Edafiogho, I.; Nounou, M.I. Nutraceuticals' novel formulations: The good, the bad, the unknown and patents involved. *Recent Pat. Drug Deliv. Formul.* **2019**, *13*, 105–156. [CrossRef]
72. Prasad, R.; Kumar, V.; Kumar, M.; Choudhary, D.K. (Eds.) *Nanobiotechnology in Bioformulations*; Springer International Publishing: Cham, Switzerland, 2019.
73. Huang, Q.; Yu, H.; Ru, Q. Bioavailability and Delivery of Nutraceuticals Using Nanotechnology. *J. Food Sci.* **2010**, *75*, R50–R57. [CrossRef]
74. Pandey, M.; Verma, R.K.; Saraf, S.A. Nutraceuticals: New era of medicine and health. *Asian J. Pharm. Clin. Res.* **2010**, *3*, 11–15.
75. Rachmale, M.; Rajput, N.; Jadav, T.; Sahu, A.K.; Tekade, R.K.; Sengupta, P. Implication of metabolomics and transporter modulation based strategies to minimize multidrug resistance and enhance site-specific bioavailability: A needful consideration toward modern anticancer drug discovery. *Drug Metab. Rev.* **2022**, *54*, 101–119. [CrossRef] [PubMed]
76. Balkrishna, A.; Sakat, S.S.; Joshi, K.; Joshi, K.; Sharma, V.; Ranjan, R.; Bhattacharya, K.; Varshney, A. Cytokines driven anti-Inflammatory and anti-psoriasis like efficacies of nutraceutical sea buckthorn (hippophaerhamnoides) oil. *Front. Pharm.* **2019**, *10*, 1186. [CrossRef] [PubMed]

77. Younis, S.A.; Kim, K.-H.; Shaheen, S.M.; Antoniadis, V.; Tsang, Y.F.; Rinklebe, J.; Deep, A.; Brown, R.J. Advancements of nanotechnologies in crop promotion and soil fertility: Benefits, life cycle assessment, and legislation policies. *Renew. Sustain. Energy Rev.* **2021**, *152*, 111686. [CrossRef]

78. Bortolotti, M.; Mercatelli, D.; Polito, L. Momordica charantia, a Nutraceutical Approach for Inflammatory Related Diseases. *Front. Pharmacol.* **2019**, *10*, 486. [CrossRef]

79. Gonçalves, A.; Estevinho, B.N.; Rocha, F. Methodologies for simulation of gastrointestinal digestion of different controlled delivery systems and further uptake of encapsulated bioactive compounds. *Trends Food Sci. Technol.* **2021**, *114*, 510–520. [CrossRef]

80. DeRosa, G.; Limas, C.P.; Macías, P.C.; Estrella, A.; Maffioli, P. Dietary and nutraceutical approach to type 2 diabetes. *Arch. Med Sci.* **2014**, *10*, 336–344. [CrossRef]

81. Sachdeva, V.; Roy, A.; Bharadvaja, N. Current Prospects of Nutraceuticals: A Review. *Curr. Pharm. Biotechnol.* **2020**, *21*, 884–896. [CrossRef]

82. Castrogiovanni, P.; Trovato, F.M.; Loreto, C.; Nsir, H.; Szychlinska, M.A.; Musumeci, G. Nutraceutical Supplements in the Management and Prevention of Osteoarthritis. *Int. J. Mol. Sci.* **2016**, *17*, 2042. [CrossRef]

83. Pashte, V.V.; Pashte, S.V.; Said, P.P. Nutraceutical properties of natural honey to fight health issues: A comprehensive review. *J. Pharmacogn. Phytochem.* **2020**, *9*, 234–242.

84. Abbas, M.; Shah, I.; Nawaz, M.A.; Pervez, S.; Hussain, Y.; Niaz, K.; Khan, F. Honey Against Cancer. In *Nutraceuticals and Cancer Signaling*; Springer: Cham, Switzerland, 2021; pp. 401–418.

85. Ajibola, A.; Chamunorwa, J.P.; Erlwanger, K.H. Nutraceutical values of natural honey and its contribution to human health and wealth. *Nutr. Metab.* **2012**, *9*, 61. [CrossRef] [PubMed]

86. Cianciosi, D.; Forbes-Hernández, T.Y.; Ansary, J.; Gil, E.; Amici, A.; Bompadre, S.; Simal-Gandara, J.; Giampieri, F.; Battino, M. Phenolic compounds from Mediterranean foods as nutraceutical tools for the prevention of cancer: The effect of honey polyphenols on colorectal cancer stem-like cells from spheroids. *Food Chem.* **2020**, *325*, 126881. [CrossRef] [PubMed]

87. Cicero, A.F.G.; Fogacci, F.; Bove, M.; Giovannini, M.; Borghi, C. Three-arm, placebo-controlled, randomized clinical trial evaluating the metabolic effect of a combined nutraceutical containing a bergamot standardized flavonoid extract in dyslipidemic overweight subjects. *Phytother. Res.* **2019**, *33*, 2094–2101. [CrossRef] [PubMed]

88. Martínez-Puc, J.F.; Cetzal-Ix, W.; Basu, S.K.; Enríquez-Nolasco, J.R.; Magaña-Magaña, M.A. Nutraceutical and medicinal properties of native stingless bees honey and their contribution to human health. In Functional Foods and Nutraceuticals in Metabolic and Non-Communicable Diseases, Academic Press: Cambridge, MA, USA, 2022; pp. 481–489.

89. Poli, A.; Barbagallo, C.M.; Cicero, A.F.; Corsini, A.; Manzato, E.; Trimarco, B.; Bernini, F.; Visioli, F.; Bianchi, A.; Canzone, G.; et al. Nutraceuticals and functional foods for the control of plasma cholesterol levels. An intersociety position paper. *Pharmacol. Res.* **2018**, *134*, 51–60. [CrossRef]

90. Cicero, A.F.G.; Fogacci, F.; Stoian, A.P.; Vrablik, M.; Al Rasadi, K.; Banach, M.; Toth, P.P.; Rizzo, M. Nutraceuticals in the Management of Dyslipidemia: Which, When, and for Whom? Could Nutraceuticals Help Low-Risk Individuals with Non-optimal Lipid Levels? *Curr. Atheroscler. Rep.* **2021**, *23*, 57. [CrossRef]

91. Chen, Y.; Pan, H.; Hao, S.; Pan, D.; Wang, G.; Yu, W. Evaluation of phenolic composition and antioxidant properties of different varieties of Chinese citrus. *Food Chem.* **2021**, *364*, 130413. [CrossRef]

92. Alcaide-Hidalgo, J.M.; Romero, M.; Duarte, J.; López-Huertas, E. Antihypertensive Effects of Virgin Olive Oil (Unfiltered) Low Molecular Weight Peptides with ACE Inhibitory Activity in Spontaneously Hypertensive Rats. *Nutrients* **2020**, *12*, 271. [CrossRef]

93. Akbari, B.; Baghaei-Yazdi, N.; Bahmaie, M.; Abhari, F.M. The role of plant-derived natural antioxidants in reduction of oxidative stress. *BioFactors* **2022**, *48*, 611–633. [CrossRef]

94. Sahebkar, A.; Serban, M.-C.; Gluba-Brzózka, A.; Mikhailidis, D.P.; Cicero, A.F.; Rysz, J.; Banach, M. Lipid-modifying effects of nutraceuticals: An evidence-based approach. *Nutrition* **2016**, *32*, 1179–1192. [CrossRef]

95. Bianconi, V.; Mannarino, M.R.; Sahebkar, A.; Cosentino, T.; Pirro, M. Cholesterol-lowering nutraceuticals affecting vascular function and cardiovascular disease risk. *Curr. Cardiol. Rep.* **2018**, *20*, 1–20. [CrossRef]

96. Paolino, D.; Mancuso, A.; Cristiano, M.; Froiio, F.; Lammari, N.; Celia, C.; Fresta, M. Nanonutraceuticals: The New Frontier of Supplementary Food. *Nanomaterials* **2021**, *11*, 792. [CrossRef] [PubMed]

97. Liu, X.; Raghuvanshi, R.; Ceylan, F.D.; Bolling, B.W. Quercetin and Its Metabolites Inhibit Recombinant Human Angiotensin-Converting Enzyme 2 (ACE2) Activity. *J. Agric. Food Chem.* **2020**, *68*, 13982–13989. [CrossRef] [PubMed]

98. Cimaglia, P.; Sega, F.V.D.; Vitali, F.; Lodolini, V.; Bernucci, D.; Passarini, G.; Fortini, F.; Marracino, L.; Aquila, G.; Rizzo, P.; et al. Effectiveness of a Novel Nutraceutical Compound Containing Red Yeast Rice, Polymethoxyflavones and Antioxidants in the Modulation of Cholesterol Levels in Subjects with Hypercholesterolemia and Low-Moderate Cardiovascular Risk: The NIRVANA Study. *Front. Physiol.* **2019**, *10*, 217. [CrossRef] [PubMed]

99. Barreca, D.; Laganà, G.; Leuzzi, U.; Smeriglio, A.; Trombetta, D.; Bellocco, E. Evaluation of the nutraceutical, antioxidant and cytoprotective properties of ripe pistachio (*Pistacia vera* L., variety Bronte) hulls. *Food Chem.* **2016**, *196*, 493–502. [CrossRef]

100. Mandalari, G.; Barreca, D.; Gervasi, T.; Roussell, M.A.; Klein, B.; Feeney, M.J.; Carughi, A. Pistachio Nuts (*Pistacia vera* L.): Production, Nutrients, Bioactives and Novel Health Effects. *Plants* **2021**, *11*, 18. [CrossRef]

101. Kumar, K.; Yadav, A.N.; Kumar, V.; Vyas, P.; Dhaliwal, H.S. Food waste: A potential bioresource for extraction of Nutraceuticals And bioactive compounds. *Bioresour. Bioprocess.* **2017**, *4*, 18. [CrossRef]

102. Tenore, G.C.; Caruso, D.; D'Avino, M.; Buonomo, G.; Caruso, G.; Ciampaglia, R.; Schiano, E.; Maisto, M.; Annunziata, G.; Novellino, E. A Pilot Screening of Agro-Food Waste Products as Sources of Nutraceutical Formulations to Improve Simulated Postprandial Glycaemia and Insulinaemia in Healthy Subjects. *Nutrients* **2020**, *12*, 1292. [CrossRef]

103. Varzakas, T.; Zakynthinos, G.; Verpoort, F. Plant Food Residues as a Source of Nutraceuticals and Functional Foods. *Foods* **2016**, *5*, 88. [CrossRef]

104. Enrico, C. Nanotechnology-Based Drug Delivery of Natural Compounds and Phytochemicals for the Treatment of Cancer and Other Diseases. In *Studies in Natural Products Chemistry*; Elsevier: Amsterdam, The Netherlands, 2019; Volume 62, pp. 91–123. [CrossRef]

105. Khorasani, S.; Danaei, M.; Mozafari, M. Nanoliposome technology for the food and nutraceutical industries. *Trends Food Sci. Technol.* **2018**, *79*, 106–115. [CrossRef]

106. Salehi, B.; Venditti, A.; Sharifi-Rad, M.; Kręgiel, D.; Sharifi-Rad, J.; Durazzo, A.; Martins, N. The therapeutic potential of apigenin. *Int. J. Mol. Sci.* **2019**, *20*, 1305. [CrossRef]

107. Tahir, A.; Ahmad, R.S.; Imran, M.; Ahmad, M.H.; Khan, M.K.; Muhammad, N.; Nisa, M.U.; Nadeem, M.T.; Yasmin, A.; Tahir, H.S.; et al. Recent approaches for utilization of food components as nano-encapsulation: A review. *Int. J. Food Prop.* **2021**, *24*, 1074–1096. [CrossRef]

108. Patra, J.K.; Das, G.; Fraceto, L.F.; Campos, E.V.R.; del Pilar Rodriguez-Torres, M.; Acosta-Torres, L.S.; Diaz-Torres, L.A.; Grillo, R.; Swamy, M.K.; Sharma, S.; et al. Nano based drug delivery systems: Recent developments and future prospects. *J. Nanobiotechnol.* **2018**, *16*, 71. [CrossRef] [PubMed]

109. Martínez-Ballesta, M.; Gil-Izquierdo, Á.; García-Viguera, C.; Domínguez-Perles, R. Nanoparticles and controlled delivery for bioactive compounds: Outlining challenges for new "smart-foods" for health. *Foods* **2018**, *7*, 72. [CrossRef]

110. Persano, F.; Gigli, G.; Leporatti, S. Lipid-polymer hybrid nanoparticles in cancer therapy: Current overview and future directions. *Nano Express* **2021**, *2*, 012006. [CrossRef]

111. Mitchell, M.J.; Billingsley, M.M.; Haley, R.M.; Wechsler, M.E.; Peppas, N.A.; Langer, R. Engineering precision nanoparticles for drug delivery. *Nat. Rev. Drug Discov.* **2021**, *20*, 101–124. [CrossRef]

112. Sosa-Hernández, J.E.; Villalba-Rodríguez, A.M.; Romero-Castillo, K.D.; Zavala-Yoe, R.; Bilal, M.; Ramirez-Mendoza, R.A.; Parra-Saldivar, R.; Iqbal, H.M. Poly-3-hydroxybutyrate-based constructs with novel characteristics for drug delivery and tissue engineering applications—A review. *Polym. Eng. Sci.* **2020**, *60*, 1760–1772. [CrossRef]

113. Sonawane, P.G.; Doshi, Y.S.; Shah, M.U.; Bajaj, M.; Kevadia, V. Probiotics: The Nano Soldiers for Periodontium. *J. PEARLDENT* **2017**, *8*, 1. [CrossRef]

114. Wang, X.; Cheng, F.; Wang, X.; Feng, T.; Xia, S.; Zhang, X. Chitosan decoration improves the rapid and long-term antibacterial activities of cinnamaldehyde-loaded liposomes. *Int. J. Biol. Macromol.* **2021**, *168*, 59–66. [CrossRef]

115. Hills, S.; Walker, M.; Barry, A.E. Sport as a vehicle for health promotion: A shared value example of corporate social responsibility. *Sport Manag. Rev.* **2019**, *22*, 126–141. [CrossRef]

116. Khangwal, I.; Yadav, M.; Mandeep; Shukla, P. Probiotics and Prebiotics: Techniques Used and Its Relevance. In *Microbial Enzymes and Biotechniques*; Springer: Singapore, 2020; pp. 193–206. [CrossRef]

117. Ayyash, M.; Abu-Jdayil, B.; Itsaranuwat, P.; Galiwango, E.; Tamiello-Rosa, C.; Abdullah, H.; Esposito, G.; Hunashal, Y.; Obaid, R.S.; Hamed, F. Characterization, bioactivities, and rheological properties of exopolysaccharide produced by novel probiotic Lactobacillus plantarum C70 isolated from camel milk. *Int. J. Biol. Macromol.* **2020**, *144*, 938–946. [CrossRef]

118. Ghanavati, R.; Asadollahi, P.; Shapourabadi, M.B.; Razavi, S.; Talebi, M.; Rohani, M. Inhibitory effects of Lactobacilli cocktail on HT-29 colon carcinoma cells growth and modulation of the Notch and Wnt/β-catenin signaling pathways. *Microb. Pathog.* **2020**, *139*, 103829. [CrossRef] [PubMed]

119. Rupasinghe, H.P.V.; Parmar, I.; Neir, S.V. Biotransformation of Cranberry Proanthocyanidins to Probiotic Metabolites by *Lactobacillus rhamnosus* Enhances Their Anticancer Activity in HepG2 Cells In Vitro. *Oxidative Med. Cell. Longev.* **2019**, *2019*, 4750795. [CrossRef] [PubMed]

120. Kang, D.; Su, M.; Duan, Y.; Huang, Y. *Eurotium cristatum*, a potential probiotic fungus from Fuzhuan brick tea, alleviated obesity in mice by modulating gut microbiota. *Food Funct.* **2019**, *10*, 5032–5045. [CrossRef] [PubMed]

121. Rani, A.; Baruah, R.; Goyal, A. Prebiotic Chondroitin Sulfate Disaccharide Isolated from Chicken Keel Bone Exhibiting Anticancer Potential Against Human Colon Cancer Cells. *Nutr. Cancer* **2019**, *71*, 825–839. [CrossRef]

122. Zhou, L.; Xie, M.; Yang, F.; Liu, J. Antioxidant activity of high purity blueberry anthocyanins and the effects on human intestinal microbiota. *LWT* **2020**, *117*, 108621. [CrossRef]

123. Ohara, T.; Mori, T. Antiproliferative Effects of Short-chain Fatty Acids on Human Colorectal Cancer Cells *via* Gene Expression Inhibition. *Anticancer Res.* **2019**, *39*, 4659–4666. [CrossRef]

124. Li, E.; Yang, S.; Zou, Y.; Cheng, W.; Li, B.; Hu, T.; Li, Q.; Wang, W.; Liao, S.; Pang, D. Purification, Characterization, Prebiotic Preparations and Antioxidant Activity of Oligosaccharides from Mulberries. *Molecules* **2019**, *24*, 2329. [CrossRef]

125. Davani-Davari, D.; Negahdaripour, M.; Karimzadeh, I.; Seifan, M.; Mohkam, M.; Masoumi, S.J.; Ghasemi, Y. Prebiotics: Definition, types, sources, mechanisms, and clinical applications. *Foods* **2019**, *8*, 92. [CrossRef]

126. Pandey, K.; Naik, S.; Vakil, B. Probiotics, prebiotics and synbiotics—a review. *J. Food Sci. Technol.* **2015**, *52*, 7577–7587. [CrossRef]

127. Nuruzzaman, M.; Rahman, M.M.; Liu, Y.; Naidu, R. Nanoencapsulation, Nano-guard for Pesticides: A New Window for Safe Application. *J. Agric. Food Chem.* **2016**, *64*, 1447–1483. [CrossRef]

128. Xiao, J.; Cao, Y.; Huang, Q. Edible Nanoencapsulation Vehicles for Oral Delivery of Phytochemicals: A Perspective Paper. *J. Agric. Food Chem.* **2017**, *65*, 6727–6735. [CrossRef] [PubMed]

129. Fathi, M.; Martín, A.; McClements, D.J. Nanoencapsulation of food ingredients using carbohydrate based delivery systems. *Trends Food Sci. Technol.* **2014**, *39*, 18–39. [CrossRef]

130. Samaranayaka, A.G.; Li-Chan, E.C. Food-derived peptidic antioxidants: A review of their production, assessment, and potential applications. *J. Funct. Foods* **2011**, *3*, 229–254. [CrossRef]

131. Hoseyni, S.Z.; Jafari, S.M.; Tabarestani, H.S.; Ghorbani, M.; Assadpour, E.; Sabaghi, M. Release of catechin from Azivash gum-polyvinyl alcohol electrospun nanofibers in simulated food and digestion media. *Food Hydrocoll.* **2021**, *112*, 106366. [CrossRef]

132. Kurd, F.; Fathi, M.; Shekarchizadeh, H. Nanoencapsulation of hesperetin using basil seed mucilage nanofibers: Characterization and release modeling. *Food Biosci.* **2019**, *32*, 100475. [CrossRef]

133. Remanan, M.K.; Zhu, F. Encapsulation of rutin using quinoa and maize starch nanoparticles. *Food Chem.* **2021**, *353*, 128534. [CrossRef]

134. Rajabi, H.; Jafari, S.M.; Rajabzadeh, G.; Sarfarazi, M.; Sedaghati, S. Chitosan-gum Arabic complex nanocarriers for encapsulation of saffron bioactive components. *Colloids Surf. A Physicochem. Eng. Asp.* **2020**, *578*, 123644. [CrossRef]

135. Fathi, M.; Donsi, F.; McClements, D.J. Protein-Based Delivery Systems for the Nanoencapsulation of Food Ingredients. *Compr. Rev. Food Sci. Food Saf.* **2018**, *17*, 920–936. [CrossRef]

136. Ahmad, U.; Ali, A.; Khan, M.M.; Siddiqui, M.A.; Akhtar, J.; Ahmad, F.J. Nanotechnology-Based Strategies for Nutraceuticals: A Review of Current Research Development. *Nanosci. Technol. Int. J.* **2019**, *10*, 133–155. [CrossRef]

137. Zhang, R.; Han, Y.; Xie, W.; Liu, F.; Chen, S. Advances in Protein-Based Nanocarriers of Bioactive Compounds: From Microscopic Molecular Principles to Macroscopical Structural and Functional Attributes. *J. Agric. Food Chem.* **2022**, *70*, 6354–6367. [CrossRef]

138. Mozafari, M.R.; Flanagan, J.; Matia-Merino, L.; Awati, A.; Omri, A.; Suntres, Z.E.; Singh, H. Recent trends in the lipid-based nanoencapsulation of antioxidants and their role in foods. *J. Sci. Food Agric.* **2006**, *86*, 2038–2045. [CrossRef]

139. Akhavan, S.; Assadpour, E.; Katouzian, I.; Jafari, S.M. Lipid Nano Scale Cargos for The Protection and Delivery of Food Bioactive Ingredients and Nutraceuticals. *Trends Food Sci. Technol.* **2018**, *74*, 132–146. [CrossRef]

140. Livney, Y.D. Nanostructured delivery systems in food: Latest developments and potential future directions. *Curr. Opin. Food Sci.* **2015**, *3*, 125–135. [CrossRef]

141. Assadpour, E.; Jafari, S.M. An overview of specialized equipment for nanoencapsulation of food ingredients. In *Nanoencapsulation of Food Ingredients by Specialized Equipment*; Jafari, S., Ed.; Academic Press: Cambridge, MA, USA, 2019; pp. 1–30. [CrossRef]

142. Rashidi, L. Different nano-delivery systems for delivery of nutraceuticals. *Food Biosci.* **2021**, *43*, 101258. [CrossRef]

143. Ajeeshkumar, K.K.; Aneesh, P.A.; Raju, N.; Suseela, M.; Ravishankar, C.N.; Benjakul, S. Advancements in liposome technology: Preparation techniques and applications in food, functional foods, and bioactive delivery: A review. *Compr. Rev. Food Sci. Food Saf.* **2021**, *20*, 1280–1306. [CrossRef]

144. Rostamabadi, H.; Falsafi, S.R.; Jafari, S.M. Nanoencapsulation of carotenoids within lipid-based nanocarriers. *J. Control. Release* **2019**, *298*, 38–67. [CrossRef]

145. Xia, S.; Tan, C.; Zhang, Y.; Abbas, S.; Feng, B.; Zhang, X.; Qin, F. Modulating effect of lipid bilayer–carotenoid interactions on the property of liposome encapsulation. *Colloids Surf. B Biointerfaces* **2015**, *128*, 172–180. [CrossRef]

146. Sercombe, L.; Veerati, T.; Moheimani, F.; Wu, S.Y.; Sood, A.K.; Hua, S. Advances and Challenges of Liposome Assisted Drug Delivery. *Front. Pharmacol.* **2015**, *6*, 286. [CrossRef]

147. Lopes, N.A.; Pinilla, C.M.B.; Brandelli, A. Antimicrobial activity of lysozyme-nisin co-encapsulated in liposomes coated with polysaccharides. *Food Hydrocoll.* **2019**, *93*, 1–9. [CrossRef]

148. Zhou, W.; Cheng, C.; Ma, L.; Zou, L.; Liu, W.; Li, R.; Cao, Y.; Liu, Y.; Ruan, R.; Li, J. The Formation of Chitosan-Coated Rhamnolipid Liposomes Containing Curcumin: Stability and In Vitro Digestion. *Molecules* **2021**, *26*, 560. [CrossRef]

149. Dalmoro, A.; Bochicchio, S.; Lamberti, G.; Bertoncin, P.; Janssens, B.; Barba, A.A. Micronutrients encapsulation in enhanced nanoliposomal carriers by a novel preparative technology. *RSC Adv.* **2019**, *9*, 19800–19812. [CrossRef] [PubMed]

150. Peng, Y.; Meng, Q.; Zhou, J.; Chen, B.; Xi, J.; Long, P.; Zhang, L.; Hou, R. Nanoemulsion delivery system of tea polyphenols enhanced the bioavailability of catechins in rats. *Food Chem.* **2019**, *242*, 527–532. [CrossRef]

151. Katouzian, I.; Esfanjani, A.F.; Jafari, S.M.; Akhavan, S. Formulation and application of a new generation of lipid nano-carriers for the food bioactive ingredients. *Trends Food Sci. Technol.* **2017**, *68*, 14–25. [CrossRef]

152. Fathi, M.; Mozafari, M.R.; Mohebbi, M. Nanoencapsulation of food ingredients using lipid based delivery systems. *Trends Food Sci. Technol.* **2012**, *23*, 13–27. [CrossRef]

153. Dhiman, N.; Awasthi, R.; Sharma, B.; Kharkwal, H.; Kulkarni, G.T. Lipid nanoparticles as carriers for bioactive delivery. *Front. Chem.* **2021**, *9*, 580118. [CrossRef]

154. Zhang, Z.; Li, X.; Sang, S.; McClements, D.J.; Chen, L.; Long, J.; Jiao, A.; Wang, J.; Jin, Z.; Qiu, C. A review of nanostructured delivery systems for the encapsulation, protection, and delivery of silymarin: An emerging nutraceutical. *Food Res. Int.* **2022**, *156*, 111314. [CrossRef]

155. Peres, L.B.; dos Anjos, R.S.; Tappertzhofen, L.C.; Feuser, P.E.; de Araújo, P.H.; Landfester, K.; Muñoz-Espí, R. pH-responsive physically and chemically cross-linked glutamic-acid-based hydrogels and nanogels. *Eur. Polym. J.* **2018**, *101*, 341–349. [CrossRef]

156. Sahu, P.; Kashaw, S.K.; Sau, S.; Kushwah, V.; Jain, S.; Agrawal, R.K.; Iyer, A.K. pH Responsive 5-Fluorouracil Loaded Biocompatible Nanogels for Topical Chemotherapy of Aggressive Melanoma. *Colloids Surf. B Biointerfaces* **2019**, *174*, 232–245. [CrossRef]

157. Wang, J.; Xu, W.; Zhang, N.; Yang, C.; Xu, H.; Wang, Z.; Li, B.; Ding, J.; Chen, X. X-ray-responsive polypeptide nanogel for concurrent chemoradiotherapy. *J. Control. Release* **2021**, *332*, 1–9. [CrossRef]
158. Mullen, E.; Morris, M. Green Nanofabrication Opportunities in the Semiconductor Industry: A Life Cycle Perspective. *Nanomaterials* **2021**, *11*, 1085. [CrossRef]
159. Bazana, M.T.; Codevilla, C.F.; de Menezes, C.R. Nanoencapsulation of bioactive compounds: Challenges and perspectives. *Curr. Opin. Food Sci.* **2019**, *26*, 47–56. [CrossRef]

bioengineering

Review

Transformation of Agro-Waste into Value-Added Bioproducts and Bioactive Compounds: Micro/Nano Formulations and Application in the Agri-Food-Pharma Sector

Saroj Bala [1,†], Diksha Garg [1,†], Kandi Sridhar [2], Baskaran Stephen Inbaraj [3], Ranjan Singh [4], Srinivasulu Kamma [2], Manikant Tripathi [5,*] and Minaxi Sharma [6,*]

[1] Department of Microbiology, Punjab Agricultural University, Ludhiana 141004, India
[2] Department of Food Technology, Koneru Lakshmaiah Education Foundation Deemed to be University, Vaddeswaram 522502, India
[3] Department of Food Science, Fu Jen Catholic University, New Taipei City 242062, Taiwan
[4] Department of Microbiology, Dr. Rammanohar Lohia Avadh University, Ayodhya 224001, India
[5] Biotechnology Program, Dr. Rammanohar Lohia Avadh University, Ayodhya 224001, India
[6] Haute Ecole Provinciale de Hainaut-Condorcet, 7800 Ath, Belgium
* Correspondence: manikant.microbio@gmail.com (M.T.); minaxi86sharma@gmail.com (M.S.)
† These authors equally contributed to this work.

Abstract: The agricultural sector generates a significant amount of waste, the majority of which is not productively used and is becoming a danger to both world health and the environment. Because of the promising relevance of agro-residues in the agri-food-pharma sectors, various bioproducts and novel biologically active molecules are produced through valorization techniques. Valorization of agro-wastes involves physical, chemical, and biological, including green, pretreatment methods. Bioactives and bioproducts development from agro-wastes has been widely researched in recent years. Nanocapsules are now used to increase the efficacy of bioactive molecules in food applications. This review addresses various agri-waste valorization methods, value-added bioproducts, the recovery of bioactive compounds, and their uses. Moreover, it also covers the present status of bioactive micro- and nanoencapsulation strategies and their applications.

Keywords: agro-waste; valorization; value-added bioproducts; bioactive compounds; micro/nano encapsulation; agri-food-pharma applications

Citation: Bala, S.; Garg, D.; Sridhar, K.; Inbaraj, B.S.; Singh, R.; Kamma, S.; Tripathi, M.; Sharma, M. Transformation of Agro-Waste into Value-Added Bioproducts and Bioactive Compounds: Micro/Nano Formulations and Application in the Agri-Food-Pharma Sector. *Bioengineering* **2023**, *10*, 152. https://doi.org/10.3390/bioengineering10020152

Academic Editor: Giovanni Esposito

Received: 27 December 2022
Revised: 19 January 2023
Accepted: 20 January 2023
Published: 23 January 2023

1. Introduction

The disposal of agricultural waste is one of the major environmental issues, which poses adverse effects on ecosystems due to the careless dumping of agricultural waste into the environment. Most reports say that untreated and underutilized agro-industrial waste is disposed of by burning, dumping, or placing it in a landfill [1,2]. Untreated garbage contributes to several greenhouse gas emissions, which in turn exacerbate climate change in a variety of ways. In addition to negative effects on the climate, this also results in the release of additional, undesirable gaseous byproducts [3]. Therefore, significant interventions are required for the sustainable use of agro-waste. This can take the form of the development of sustainable energy technologies and the creation of value-added bioproducts. A larger strain on the environment has resulted from increased global output of wastes, with detrimental effects on soil, air, and water resources [4], which in turn threaten the health of populations and the long-term viability of ecosystems. About 21–37% of greenhouse gases are produced by the agricultural sector [5]. This new reality has prompted a model of sustainable development in recent years that calls for substantial shifts in conventional agricultural production methods and waste utilization.

High concentrations of complex carbohydrates, proteins, fibers, polyphenolic components, bioactive compounds, etc., are found in agro-wastes [6]. Despite the fact that organic compound wastes pose a threat to the environment, they could be used as a raw material in a wide range of agricultural, food, and pharmaceutical goods [7]. The agro-residues are not considered a waste because of their high nutrient content; rather, they are used as a source material for new products. Effective pretreatment technologies for agro-waste biomass, its biochemical characteristics, and advanced conversion processes can improve the cost-effectiveness of conversion processes for bioproducts development [8]. Microbial biotechnology and nanotechnology play a pivotal role in the bioconversion of agricultural waste into enzymes and other bioactive compounds in the pharmaceutical sector, vermicompost, organic fertilizers, and biofuels in the agriculture sector, and nutraceuticals and food products [9–13]. Due to its many potential applications, nanotechnology is increasingly being explored in the food and healthcare industries. Due to the aforementioned enhancements in bioavailability and levels of bioactive compounds, the capability for guided administration of bioactive compounds to specific tissues or organs is also boosted. Nanostructured materials have the potential to revolutionize the food business because of their unique features and large surface area. While the potential benefits of nanotechnology to society as a whole have been widely recognized, recently, the field of food science has begun to explore its applications [14]. Using a matrix or inert material, nanoencapsulation is a method for keeping coated substances (food or flavor molecules/ingredients) that are in a liquid, solid, or gaseous form. Stabilizing bioactive compounds via nanoencapsulation allows for more precise control over their release at physiologically active locations [15]. Therefore, this review addresses the valorization of agro-waste into value-added bioproducts, bioactive compounds, and their applications in the agri-energy-pharma sector, along with the present status of bioactive micro/nanoencapsulation strategies and their applications, with the ultimate goal of reducing waste for a sustainable and green environment. In this review, we retrieved 93 articles on agriculture waste, valorization, pretreatment methods, value-added products, and micro- and nanoencapsulation from the Web of Science™ Core Collection.

2. Availability of Agro-Waste

There are various sorts of agro-waste in the environment, depending on their origin and availability [16]. Agro-waste consists primarily of cellulose, hemicellulose, and lignin. Due to their structural complexity, lignocelluloses are difficult to break [17]. Various types of wastes, including crop residue, aquaculture waste, and other wastes of agricultural origin, have been reported [18,19]. The two largest available categories of waste are crop residue and agro-industrial waste, both of which are produced in huge quantities every day in the agricultural and food processing industries [20]. On the other hand, the absence of appropriate management techniques for these wastes, which may be traced back to a lack of or limited access to adequate information, has steadily become a significant obstacle that cannot be minimized due to its magnitude.

Because the manure from cattle is a major source of noxious gases, harmful microorganisms, and odor, there is cause for concern regarding both public health and the environment [21]. Unconsumed feed and undigested substances, both of which are expelled from fish as waste in the form of feces, are the two main contributors to aquaculture solid waste [22]. The possibility for contamination of both surface water and groundwater exists if there is an excessive application of a mixture, an incorrect disposal of agro-residue, undiluted chemicals, or even pesticide containers. Utilizing farm waste in a managed manner can improve irrigation and erosion management [23]. The distinction between agricultural residue and other solid fuels, such as charcoal, wood, and charcoal briquettes, is based on their availability and distinguishable qualities.

3. Valorization Technologies for Agro-Waste

Agro-waste cannot be utilized without a proper pretreatment since modified agro-waste is more effective and useful [24]. Agricultural waste undergoes multiple treatment procedures prior to being utilized. Residues of agro-waste are made of lignocellulosic materials that require chemical, physical, and biological treatments to break the complexity [25]. The transformation of agro-waste's complex molecular structures into simpler monomers is typically regarded as a necessary pretreatment step [26]. Table 1 details the pros and cons of various types of pretreatment procedures. For better understanding of agro-waste utilization, the graphical mapping of the keyword co-occurrence and co-authorship was constructed by the noncommercial visualization of similarities (VOS) viewer VOS viewer 1.6.18 (https://www.vosviewer.com/, accessed on 20 November 2022). The volume of published work is a strong predictor of future research directions. A keyword list can be used to categorize various fields of study. Figure 1 depicts network visualization, density visualization, and overlay visualization maps of the most-cited keywords from 93 articles. The majority of current research is focused on the development of new value-added products, and the valorization techniques suggest that agricultural waste is among one of the most explored domains that have applications in the agri-food-pharma sector.

Table 1. Application based pretreatment methods and related pros as well as cons.

Application		Pretreatment Methods	Pros	Cons	Refs.
Agriculture sector Biofuels and manure Enzymatic digestibility Ethanol production Bio-oil and biochar formation **Food sector** Bioactive compounds Nutraceuticals Ethanol and enzyme production	Physical	Grinding	From biomass, a fine powder with a crystallinity of up to 0.2 mm is produced.	Lack of long-term viability in technique calls for a lot of energy.	[27–32]
		Ultrasonic	Easing the process of breaking down a variety of lignocellulosic materials.	Collisions between particles during prolonged sonication could result in an antagonistic effect.	
		Steaming explosion	Minimal need for energy.	Incomplete lignin-carbohydrate matrix cleavage, xylan fraction destruction, creation of hydrolysis, and fermentation inhibitors.	
		Microwave	Easily functional, and with efficiency in handling large agro-waste with fewer inhibitors being formed.	This causes both a rise in temperature and an increase in the amount of electricity used.	
		Pyrolysis	The highest possible rate of cellulose sugar conversion.	High-cost technique.	
		Irradiations	The surface area was increased, crystallinity was reduced, hemicelluloses were hydrolyzed, and the structure of lignin was altered.	Expensive method.	
Pharma sectorSugars (glucose, xylose, mannose, and galactose) and organic acids (formic, acetic acid) production **Agriculture sector** Enzymes production, organic acids, and hydrolysis of agro-waste to increase glucose yield Biorefinery Biomass saccharification, bioethanol and biogas production **Food sector** Extraction of phenolic compounds and acids productions	Chemical	Acid hydrolysis (HCl, CH$_3$COOH, H$_2$SO$_4$)	Change the structure of lignin, and hydrolyze hemicellulose to xylose and other sugars.	Corrosion of expensive equipment and the production of harmful byproducts are additional costs.	[2,29,33–35]
		Alkaline hydrolysis (KOH, NaOH, NH$_4$OH, Mg(OH)$_2$, Ca(OH)$_2$	Pretreatment under milder conditions. Removing lignin and hemicelluloses raises the available surface area.	High alkalinity concentrations and lengthy residence durations are necessary.	
		Ozonolysis	Decreases lignin content. Does not indicate the production of hazardous substances.	Method that is both expensive and demanding of a substantial quantity of ozone.	
		Organosolv	Hydrolyzes lignin and hemicellulose; helpful for lignin extraction.	Due to their high volatility, costly solvents are unsuitable for industrial use.	
		Wet oxidation	Effectively eliminated lignin and low formation inhibitors.	Expensive because of the utilization of oxygen and acid catalyst.	

Table 1. *Cont.*

Application	Pretreatment Methods		Pros	Cons	Refs.
Agriculture sector Animal manure and biofertilizers Biorefinery and animal feed **Pharma sector** Antibiotics production **Food sector** Single cell protein	Biological	Enzyme	Moderate circumstances are present, and minimal effort is necessary.	Low hydrolysis rate and a large sterile space requirement.	[13,23,36,37]
		Bacteria	Economical and requiring only mild reaction conditions.		
		Fungi	Inexpensive, destroys lignin and hemicelluloses, minimal energy needs.		
Food and pharma sectors Antibiotic production Antioxidant properties Antibacterial and anticancer properties	Green solvents	Ionic liquids	Effective at dissolving copious amounts of cellulose and recovering usable cellulose from lignin.	Possible toxicity, prohibitively expensive method, and a lack of practicality for mass production.	[38–40]
		Deep eutectic solvents	Conditions are modest but environmentally friendly and safe.	Creates undesirable contaminants and higher viscosity on occasion.	
		Natural deep eutectic solvents	Low-cost, readily available, highly modifiable, and less hazardous.		

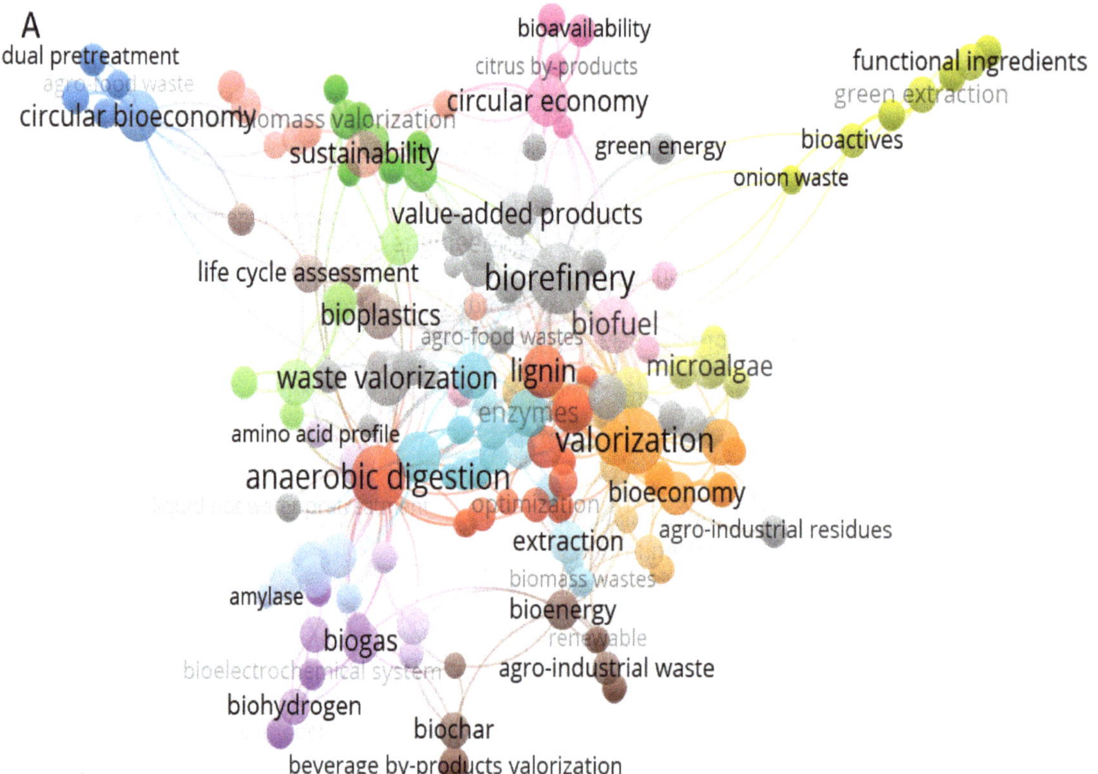

Figure 1. *Cont.*

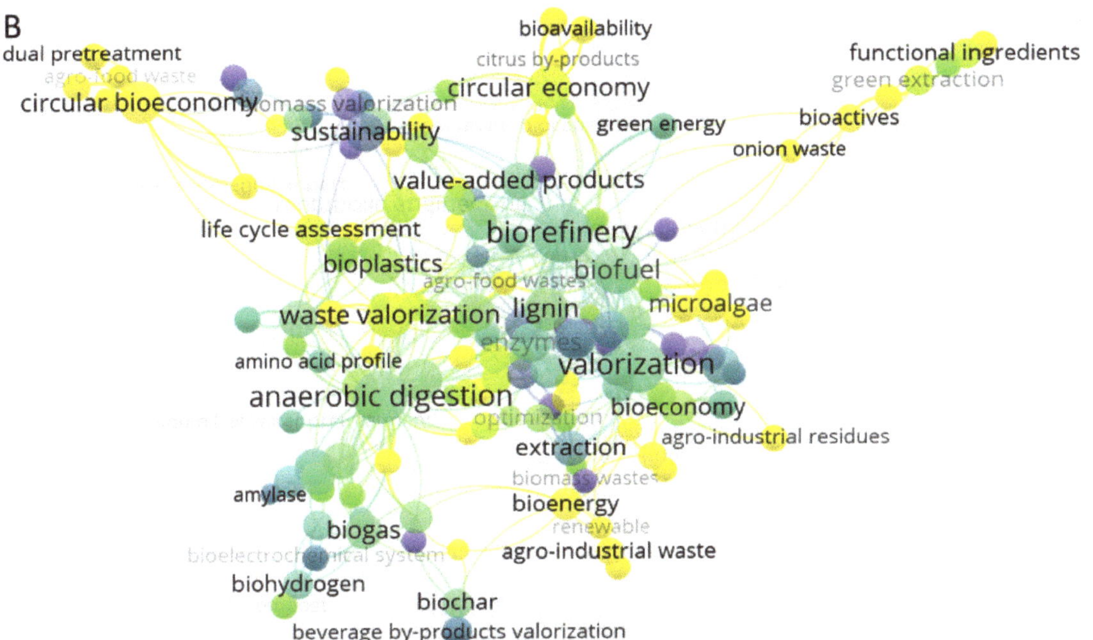

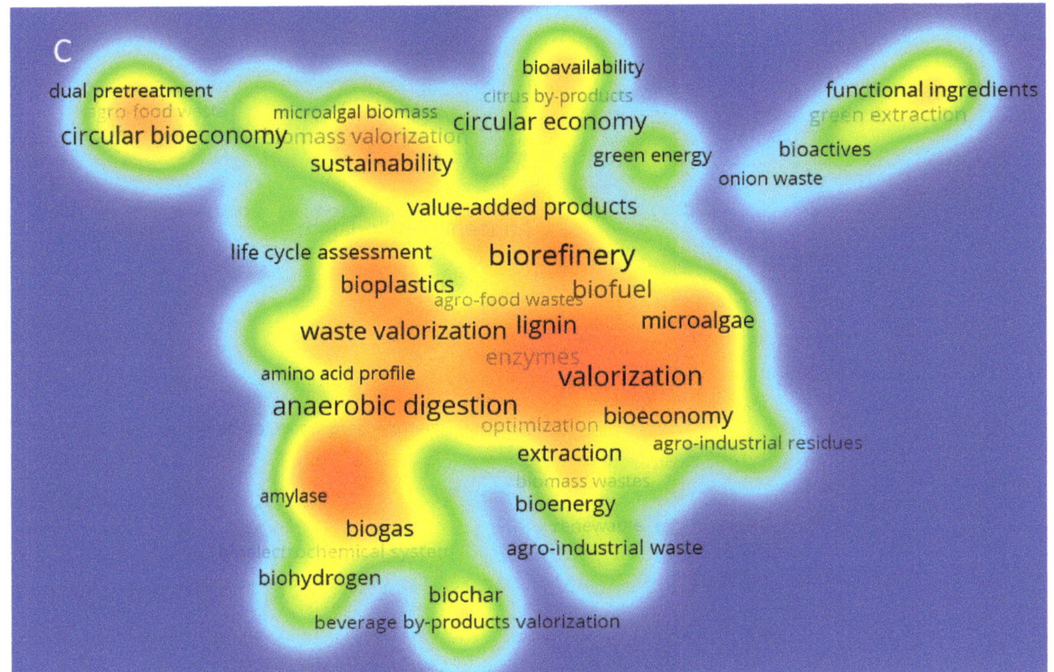

Figure 1. Most frequently used keywords over time, as well as a clustering of the keywords' citation networks. Network visualization (**A**), overlay visualization (**B**), and density visualization (**C**).

Wheat straw biodegradation kinetics are increased by mechanical, thermal, and sonic pretreatments but at the expense of high energy consumption [41]. This method decreases the size of biomass and increases its accessible surface area, which improves hydrolysis as a result of enhanced heat and mass transmission [42]. The internal linkages of lignin and hemicellulose are broken due to chemical pretreatment (e.g., the use of acids, alkalis, ozonolysis, organosolv, and wet oxidation) [43]. Pretreatment decreased cassava bagasse weight and boosted cellulose yield significantly compared to unpretreated bagasse. The maximum cellulose yield was achieved with the H_2SO_4 pretreatment [44]. The biological process (using microbes) includes hydrolysis and saccharification, which are mediated by exoenzymes. Fermentation using whole-cell systems or simply enzymatic digestion of a solid or liquid substrate is required. This procedure uses less energy, is eco-friendly, and produces no inhibitors [45]. Green pretreatment involves ionic liquids (ILs), deep eutectic solvents, and natural deep eutectic solvents. Pretreatment with ILs improved glucose production from enzymatic hydrolysis by increasing cellulose crystallinity and decreasing particle size and lignin level [46]. Enhanced lignin and cellulose breakdown occurred by breaking intramolecular and intermolecular hydrogen bonds [47]. These pretreatment approaches facilitate enzyme accessibility for hydrolysis, increasing surface area while decreasing operational expenses. Pathak et al. [12] discussed the strategies for the valorization of fruit wastes into value-added compounds. The possible valorization approaches for agro-waste conversion into useful bioproducts and biochemicals are shown in Figure 2.

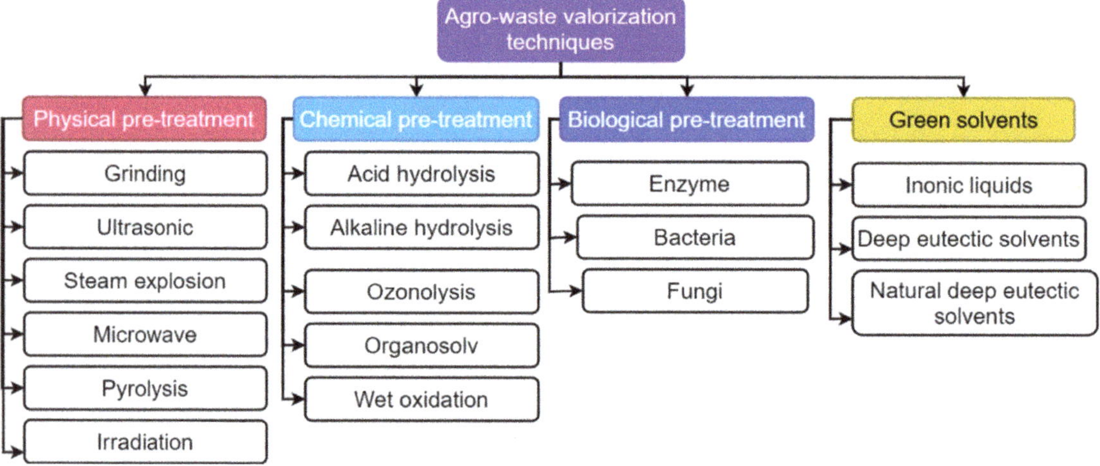

Figure 2. Agro-waste valorization methods for the development of value-added bioproducts.

4. Bioactive Compounds from Agro-Waste

In the scientific literature, bioactive compounds (BCs) are described as a natural compound capable of interacting with one or more components of living tissues and exerting a variety of effects [48,49]. In addition, the distinctions between the definitions of dietary supplements, nutraceuticals, and functional foods are sometimes misinterpreted. In contrast to dietary supplements and food additives, nutraceuticals and functional foods are also important [50]. In a nutshell, functional foods contain bioactives that may be helpful to health at higher levels compared to regular foods. These substances may contribute for the better health benefits, as they are deemed superior to conventional foods. Food additives are substances added during food processing to improve food quality and its shelf life [51]. Due to consumers' increased awareness of the health-promoting effects of nutraceuticals and dietary supplements, interest in adding functional and natural food additives has increased dramatically in recent years. Similarly, consumer interest in health and wellbeing has propelled the expansion of the dietary supplement and nutraceutical markets. Nutraceuticals and dietary supplements can refer to a variety of products with health benefits [52]. However, nutraceuticals are a distinct subset of dietary supplements because they contain pharmaceutical-grade substances but are exempt from the same testing requirements as medicines.

The interest in BCs for various culinary applications continues to expand, fueled by ongoing research efforts to determine the health qualities and prospective applications of these chemicals, which are primarily taken from natural sources, as well as public and consumer interest. Due to their pharmacological properties, they were commercialized as medications and derived mostly from plants, vegetables, and microorganisms [53].

5. Application of Agro-Waste in Agri-Food-Pharma

The biobased concept is the primary focus of current research because of its great potential to improve efficiency, cost, and yield, which support and protect environmental sustainability [54]. There are several valuable bioproducts that can be withdrawn from agricultural waste biomass using potential bioconversion pretreatment approaches, such as in the agriculture sector (vermicomposting, biofertilizers, biochar, wood vinegar), the bioenergy sector (biofuels), and the pharma sector (antimicrobials, antioxidants, and antibiotic production). There are various types of bioactives and bioproducts that can be developed from agro-waste, as shown in Figure 3. Agricultural wastes, as mentioned below, can be biotransformed into high-value-added products.

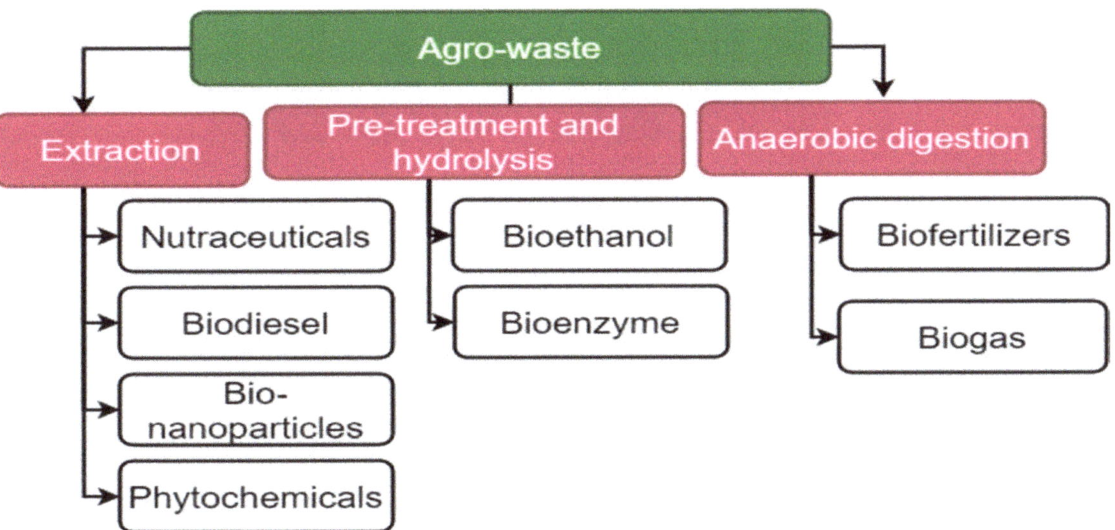

Figure 3. Agro-waste-derived bioproducts and bioactive compounds.

5.1. Agriculture Sector

5.1.1. Vermicomposting

Vermicomposting is the decomposition of organic household waste, and can be used in conjunction with inexpensive, space-saving models to improve soil fertility. Waste items from the agricultural industry that have not been reused or repurposed include both agricultural and processing byproducts [55]. These agri-horticultural wastes are highly biodegradable, making them a major issue in municipal landfills. These are the untapped raw materials with industrial applications [56]. Straw, leaves, twigs, stubbles, and vast quantities of grasses and weeds have all been produced during production of crops and agriculture. By working together, earthworms and aerobic microorganisms stabilize organic waste during the vermicomposting process. Vermicomposting is a low-cost and environmentally friendly method for dealing with agricultural waste. One of the most prevalent earthworm species utilized in vermicomposting is *Eisenia foetida* [57]. Results from physicochemical analyses have indicated that, in comparison to compost and other agricultural wastes, vermicomposting reduces total organic carbon (TOC) and the carbon–nitrogen (C/N) ratio while increasing nitrogen–phosphorus–potassium (NPK) content [58]. Additionally, it can be used to increase agricultural yields by eliminating harmful organisms in the environment (biocontrol), enhancing soil water retention, and manufacturing plant growth regulators. In an effort to revitalize depleted soils, Roohi et al. [59] fertilized maize with either bacteria or compost, or with both. Compost was found to increase both microbial activity and diversity in degraded irrigated soils. Das and Deka [60] demonstrated that nitrogen, phosphorus, and potassium concentrations in compost made from cow dung using *E. fetida* were significantly higher than in compost made from other types of waste. In addition, studies have found that certain species of earthworms secrete the decomposition-promoting phosphatase enzyme [61]. Therefore, vermicomposting can function as a tool for environmental conservation.

5.1.2. Biofertilizers

Soil biofertilizers, which can be made from microorganisms such as bacteria, cyanobacteria, fungi, and algae, are defined as those that improve soil quality by adding nutrients and carbon substrates. Green manures such as cyanobacterial supplements and bioformulations of bacteria such as *Azotobacter* sp., *Azospirillum* sp., *Trichoderma* sp., and arbuscular mycorrhizal fungus (AMF) are the most prevalent and widely used biofertilizers. In addition to microbial "biofertilizers", farmers frequently employ organic-based fertilizers, including vermicomposting residues, crop residues, and farmyard manure [62,63]. Efforts have been made to develop biofertilizers from agri-waste in the form of discarded melons, pineapples, oranges, bananas, and papayas. Biofertilizer was made using a solid-state fermentation technique and then employed in a vegetable garden. The physical properties of plant samples treated with biofertilizer made from watermelon, papaya, and banana wastes were positively represented in the experimental results. The maximum potassium concentration was found in the banana biofertilizer [64]. Composting and anaerobic digestion (AD) are the two major processes that make use of the metabolic capabilities of the thermophilic and decomposer microbial populations [65]. Enzymes found in native microbial communities aid in bioprocesses for the production of biofertilizers from agro-wastes.

5.1.3. Bioenergy (Biofuels)

Fuels made from renewable organic biomass, or biofuels, are considered a way to lessen reliance on fossil fuels, lower greenhouse gas (GHG) emissions, particularly from the transportation sector, and increase fuel supply security. Biofuel production from lignocellulosic materials and other wastes is reported by several researchers [10–13,66]. Enzymes are used to separate monosaccharides from polysaccharides after physicochemical preparation. Enzymes that break down cellulose, hemicellulose, and lignin are required for maximizing sugar monomer release. It has been noted that thermochemical pretreatment releases oligosaccharides that are not digestible by most of the relevant fermentation organisms [13,67]. The use of lignocellulose-derived sugars in ethanol synthesis has received the greatest attention.

Ethanol can be efficiently produced from glucose and other hexoses by using the yeast Saccharomyces cerevisiae, which is the most widely used organism for ethanol production. Brettanomyces bruxellensis and Zymomonas mobilis are yeasts and bacteria, respectively, that can produce ethanol in industrial settings. For example, B. bruxellensis has been proven to ferment oat straw hydrolysate to ethanol [68]. When using straw for biofuels, methane (biogas) production is an additional choice. Energy efficiency, greenhouse gas emissions, and biomass conversion are all improved when switching from ethanol production to biogas production via anaerobic digestion [69]. Compared to ethanol, butanol (both n-butanol and iso-butanol) offers superior fuel properties due to its higher energy density, lower corrosiveness, and greater compatibility with currently used engines. The use of solventogenic Clostridium has enabled the large-scale manufacture of acetone, butanol, and ethanol from starchy materials [70]. The potential of straw or agro-waste as a raw material can be used to manufacture biofuels for environmental sustainability.

5.2. Food Sector

Agro-waste is a good source of bioactive compounds for food applications. Bioactive compounds isolated from agricultural wastes or byproducts mainly include polyphenolic compounds, vitamins, minerals, fatty acids, volatiles, anthocyanins, and pigments that have valuable health benefits. For example, extracted bioactive compounds can be used in food fortification, such as the development of novel functional and health foods (Figure 4). The great diversity of different bioactive compounds from agro-waste can be a source of natural antioxidants and antimicrobials for wider application in the food and drug industries. Bioactive compounds are widely used in the meat processing industry to prevent lipid oxidation in meat products. In addition, a wider variety of micro- and nanoencapsulation technologies are used to encapsulate these bioactive compounds to increase their applications in the food sector. A study by Kaur et al. [71] extracted the betalains from red beetroot (*Beta vulgaris* L.) pomace and developed a functional Kulfi fortified with encapsulated betalains. This study improved the bioavailability, stability, and solubility of bioactive compounds (betalains) for wider application in the development of dairy-based food products. Likewise, different technologies are being used to improve the applicability of agro-waste-based bioactive compounds in the food sector.

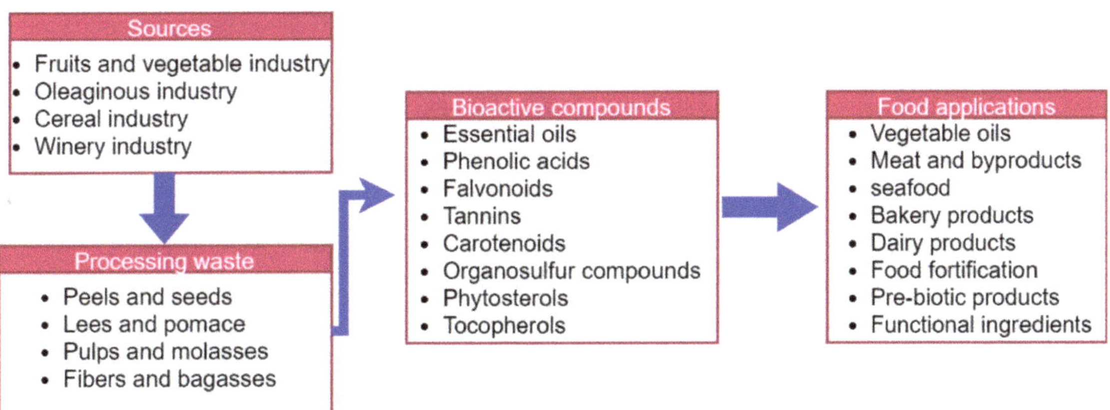

Figure 4. Agro-waste as natural sources of bioactive compounds for food applications. Reprinted from Shirahigue et al. [72] and is an open-access article (copyright © 2023 by authors) distributed under the terms and conditions of the Creative Commons Attribution (CC BY) license.

5.3. Pharma Sector

5.3.1. Antibiotic Production

Antibiotics are chemicals produced by certain microbes that inhibit or kill the growth of other germs at extremely low doses. Various agricultural byproducts are utilized in the synthesis of various antibiotics [1]. Several research projects employing agro-industrial waste to manufacture antibiotics were conducted. Asagbra et al. [73] investigated antibiotics out of a variety of agricultural byproducts and waste products. When it came to the generation of tetracycline, peanut shells proved to be the most productive substrate at 4.36 mg/g, followed by corncobs. The production of antibiotics was made far more affordable by the utilization of low-cost carbon sources derived from a wide variety of agricultural waste. These residues present an excellent opportunity for innovation in the manufacturing of neomycin and various other antibiotics. In the study conducted by Vastrad and Neelagund [74], the synthesis of extracellular rifamycin B was studied through the use of solid-state fermentation with oil-pressed cake, which is regarded as an agro-industrial waste. An investigation was made into the process of solid-state fermentation used by *Streptomyces speibonae* OXS1 to create oxytetracyline from cocoyam peels, which are considered to be household kitchen wastes of agricultural output [75].

5.3.2. Antioxidant Properties

Antioxidant and anticancer drugs were also developed using agro-waste as a substrate. The remains of various fruits and vegetables, such as fruit and vegetable peels, are usually referred to as waste or useless. Duda-Chodak and Tarko [76] studied the antioxidant capabilities of fruit seeds and peels. In their investigation, they discovered that the peels of selected fruits have the highest antioxidant activity and polyphenol concentration. Pomegranate byproducts from farming produce copious amounts of garbage. It contains ellagitannins, punicalagin, and punicalin, which are extremely powerful antioxidants [77]. Based on their broad range of bioactivities, natural products are often viewed as promising candidates for further value-added research. Phenolic chemicals play a particularly important role among them because of the widespread knowledge of their positive effects on human health, including their function in cancer and cardiovascular disease prevention [78]. Their ability to operate as powerful antioxidants, which are produced in oxidative stress situations and are responsible for the start of a number of inflammatory and degenerative disorders, has been linked to these effects.

5.3.3. Antibacterial and Anticancer Properties

Waste from the agriculture and food industries has the potential to boost the body's absorption of a wide variety of pharmaceuticals. These are a fantastic resource for a variety of essential nutrients as well as phytochemical substances that can help contribute to a balanced diet. They are a good source of organic and inorganic compounds, sugars, and many phenolic compounds. These phenolic compounds have therapeutic applications due to their anti-inflammatory, antibacterial, immunomodulatory, and antifungal properties. Additionally, they have antibacterial potential due to the production of many antibiotics by microbes produced through agricultural waste. Numerous studies have been conducted on the bioactive compounds found in agricultural residues [49,53]. From the results of orange and lemon processing, millions of tons of waste are generated annually from the industrial manufacture of orange and lemon juice, and this waste is a rich source of hydroxycinnamic acids and flavonoids, primarily flavanone glycosides, flavanones, and flavone aglycons [79]. As byproducts of potatoes, there is little doubt that potato peels are one of the most widely available vegetable waste products. There are a number of potential uses for their extracts, including in the food industry. Chlorogenic acid is the most abundant phenolic acid in peels [80]. Byproducts of lignocellulosic agriculture, such as wheat straw residues, wheat and rice bran, spent residues of coffee ground nuts, and sawdust, have been widely described as a clean source of phenolic compounds and have the potential to be utilized for application in a variety of industries due to the antioxidant and antimicrobial properties that they possess [79].

6. Bioactives from Agro Waste: Micro/Nano Formulation and Food Application

The main difference between microencapsulation and nanoencapsulation is the particle size, which normally ranges from 1 micrometer (m) to 1 mm (mm), and the length of time microencapsulation has been deployed [81]. Microencapsulation has been implemented in nearly all industrial sectors, from agriculture and the environment to household and personal care products [82]. Many nanoencapsulated and microencapsulated technologies rely on particle size and distribution homogeneity to be effective. The use of supercritical carbon dioxide (CO_2) has demonstrated promising advantages for ensuring the proper design of particle size in both microencapsulated and nanoencapsulated pharmaceuticals, as well as the capacity to manage drug-loading procedures across a variety of temperature, pressure, and flow ratios [83]. As noted previously, one of the most alluring features of nanoencapsulation is its capacity to prevent the degradation of active pharmaceutical ingredients. Nanoencapsulation has also enhanced the accuracy of medication delivery targets by coating or conjugating the surface to enable proper cell entrance [84]. In addition, nanoencapsulated pharmaceuticals can be labeled with fluorescent probes for imaging, which is particularly important for assessing therapeutic efficacy during preclinical and clinical research. In the agriculture sector, nanofertilizers are advantageous for nutrition management due to their high potential for enhancing nutrient utilization efficiency. Alone or in combination, nutrients are bonded to nanodimensional adsorbents, which release nutrients much more slowly than conventional fertilizers.

There are a number of conventional techniques for microencapsulating food components; however, no single technique is compatible with the core materials or product applications. As previously indicated, the process of microencapsulation is governed by the application and factors such as the target particle size, physicochemical properties of the external and interior phases, release mechanisms, total cost, etc. [85]. As the encapsulated chemicals are often in liquid form, drying techniques are commonly used to convert the liquid phase into a stable powder. To encapsulate active agents of food ingredients and nutraceuticals, various techniques are available [86]. Numerous BACs and other hydrophobic or poorly water-soluble nutrients are crucial for human nutrition and the maintenance of consumer health and wellbeing. Incorporating these compounds into the pharmaceutical and food sectors necessitated overcoming their poor solubility and bioavailability [15]. Nanoencapsulation is a promising strategy for protecting various food elements from certain physiological conditions or deterioration while also disguising some disagreeable aromas and flavors. Reducing the particle size of BACs could improve their stability, bioavailability, solubility, and delivery, hence enhancing their functional activity [87]. In addition, nanoencapsulation can regulate the release of active compounds and improve the final food product's qualities. Nanocapsulated bioactives and their advantages are presented in Figure 5.

Advances in nanotechnology have been made in the food science and industry recently. When it comes to protecting the public's health and enhancing the functional qualities of food formulations, nanoencapsulation is still one of the most promising options. Thus, new methods for encasing bioactive compounds are constantly being introduced, each with their own set of benefits and drawbacks [15]. By using nanoencapsulation, food scientists and engineers can create protective shields that can withstand harsh environments, hide off-putting flavors and odors, and improve nutrient absorption, release, and delivery while also increasing the solubility of lipophilic bioactive substances in water [88]. There is a high probability of overlap between the many uses of nanoencapsulation methods. Therefore, features such as release patterns, safety and toxicity concerns, and economic considerations are taken into account when selecting a nanocarrier for the bioactive molecule.

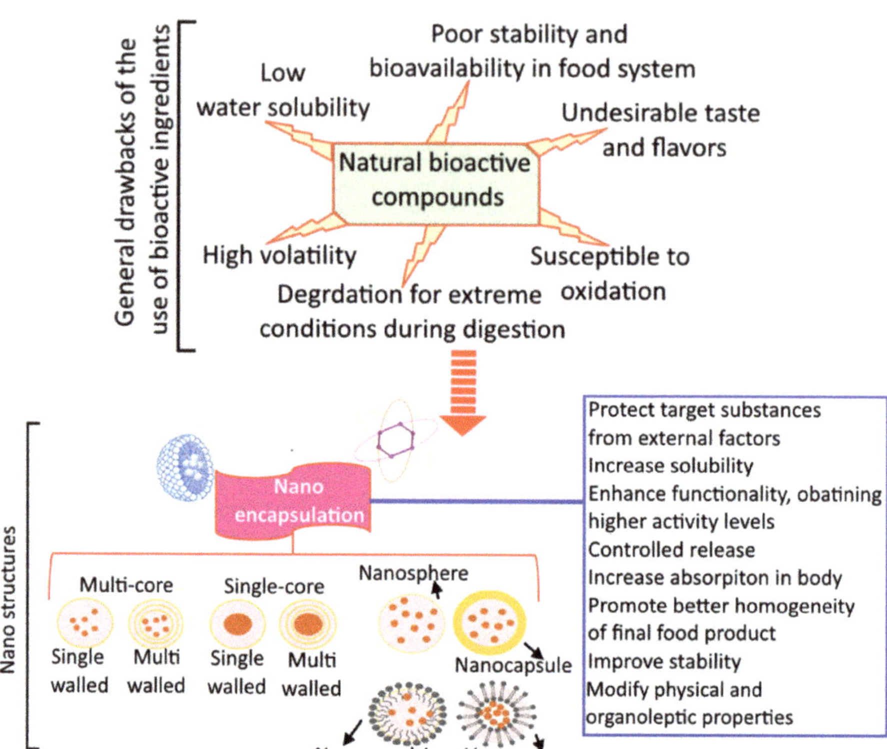

Figure 5. Limitations of bioactive compounds and application of nanoencapsulation to increase bioactive compounds applicability in food industry. Reprinted from Pateiro et al. [15] and is an open-access article (copyright © 2023 by authors) distributed under the terms and conditions of the Creative Commons Attribution (CC BY) license.

7. Future Prospective and Limitations

Agricultural residues are a potential resource for the synthesis of high-value compounds. In the future, biotechnological methods will be utilized to extract high-value bioproducts from agricultural waste. Because of their low energy requirements and low cost, integrated nano- and biotechnological techniques are preferred for industrial waste valorization. Furthermore, biotechnological methods for agricultural waste valorization are viable solutions for developing unique bioproducts for a variety of industries [54]. Therefore, it is necessary to design eco-friendly and cost-effective cascade conversion techniques. Agricultural waste is seen as a nutritional and functional raw resource that can be utilized in numerous applications, making it a viable solution to economic and environmental issues [89,90]. They can be employed directly as protein, fat, vitamin, fiber, carbohydrate, mineral, and antioxidant dietary components. Other biomolecules from agricultural waste can be recovered physically or chemically and employed as nutritional and functional components. In order to preserve the physicochemical and microbiological stability of bioproducts during the valorization of these agricultural wastes, the unitary drying process is required to prevent microbial hazards [91]. Therefore, governments should promote the installation of infrastructure and technology that enables the use of agricultural leftovers and trash in production and storage regions. The elimination of other dangers, such as poisonous substances and antinutritional factors, must also be considered [92]. Thus, such unique and worthwhile technical implementation may be helpful in addressing such problems across all sectors of agro-waste usage. A well-managed supply chain, supported by analysis, can also help in the disposal of agricultural waste. Agro-industrial food waste can be a useful low-cost feedstock for sustainable production of industrial products [93]. Thus, an appropriate green technology is required for bioproducts development. Moreover, the difficulties of economics and market resistance are reduced over time through effective policies and assessments of their execution.

8. Conclusions

Together, environmental policy and the growing demand for a wide variety of biobased products have led to a shift toward a greater focus on finding new uses for agricultural byproducts. Wastes from the food and agriculture industries are rich sources of both nutrients and bioactive compounds. It is more accurate to refer to these materials as "feedstock" rather than "waste" when addressing their possible use in the development of bioproducts due to the fact that their composition might vary, including the presence of sugars, minerals, and proteins. The analysis presented here focuses mostly on agricultural waste, which was valorized in a number of ways to produce value-added products. Valorization through biotechnological and green approaches are helpful in bioproducts development in a cost-effective and comprehensive way. In order to establish a safe and green environment, it is beneficial to recycle wastes from the agricultural and agro-based sectors and use them as feedstock for development of various value-added molecules. Besides these, nanomaterials-based technology can also be used to increase bioactive compounds applicability in various industrial sectors.

Author Contributions: Conceptualization, M.T. and M.S.; methodology, S.B. and D.G.; software, S.K. and R.S.; validation, K.S., B.S.I., M.T. and M.S.; formal analysis, S.B. and D.G.; investigation, S.B. and D.G.; resources, M.T. and M.S.; data curation, S.B. and D.G.; writing—original draft preparation, S.B., R.S. and D.G.; writing—review and editing, K.S., R.S., B.S.I., S.K., M.T. and M.S.; visualization, R.S., K.S., B.S.I., M.T. and M.S.; supervision, M.T. and M.S.; project administration, M.T. and M.S.; funding acquisition, M.T. and M.S. All authors have read and agreed to the published version of the manuscript.

Funding: This research received no external funding.

Institutional Review Board Statement: Not applicable.

Informed Consent Statement: Not applicable.

Data Availability Statement: Data are available within the article.

Acknowledgments: We are grateful to Biotechnology Program, Rammanohar Lohia Avadh University, Ayodhya, India.

Conflicts of Interest: The authors declare no conflict of interest.

References

1. Chilakamarry, C.R.; Sakinah, A.M.; Zularisam, A.; Sirohi, R.; Khilji, I.A.; Ahmad, N.; Pandey, A. Advances in solid-state fermentation for bioconversion of agricultural wastes to value-added products: Opportunities and challenges. *Bioresour. Technol.* **2022**, *343*, 126065. [CrossRef] [PubMed]
2. Awogbemi, O.; Von Kallon, D.V. Pretreatment techniques for agricultural waste. *Case Stud. Chem. Environ. Eng.* **2022**, *6*, 100229. [CrossRef]
3. Koul, B.; Yakoob, M.; Shah, M.P. Agricultural waste management strategies for environmental sustainability. *Environ. Res.* **2022**, *206*, 112285. [CrossRef] [PubMed]
4. Food and Agriculture Organization of the United Nations. *The State of Food and Agriculture 2016 (SOFA): Climate Change, Agriculture and Food Security*; Food and Agriculture Organization of the United Nations: Rome, Italy, 2016; Available online: https://www.fao.org/3/i6030e/i6030e.pdf (accessed on 5 December 2022).
5. Lynch, J.; Cain, M.; Frame, D.; Pierrehumbert, R. Agriculture's contribution to climate change and role in mitigation is distinct from predominantly fossil CO_2-emitting sectors. *Front. Sustain. Food Syst.* **2021**, *4*, 300. [CrossRef] [PubMed]
6. Dwivedi, S.; Tanveer, A.; Yadav, S.; Anand, G.; Yadav, D. Agro-Wastes for Cost Effective Production of Industrially Important Microbial Enzymes: An Overview. In *Microbial Biotechnology: Role in Ecological Sustainability and Research*; Chow-Dhary, P., Mani, S., Chaturvedi, P., Eds.; Wiley: Hoboken, NJ, USA, 2022; pp. 435–460.

7. Chojnacka, K.; Gorazda, K.; Witek-Krowiak, A.; Moustakas, K. Recovery of fertilizer nutrients from materials-Contradictions, mistakes and future trends. *Renew. Sustain. Energy Rev.* **2019**, *110*, 485–498. [CrossRef]

8. Reshmy, R.; Philip, E.; Madhavan, A.; Sirohi, R.; Pugazhendhi, A.; Binod, P.; Awasthi, M.K.; Vivek, N.; Kumar, V.; Sindhu, R. Lignocellulose in future biorefineries: Strategies for cost-effective production of biomaterials and bioenergy. *Bioresour. Technol.* **2022**, *344*, 126241. [CrossRef]

9. Balan, A.; Murthy, V.V.; Kadeppagari, R.K. Immobilized enzymes for bioconversion of waste to wealth. In *Biotechnology for Zero Waste: Emerging Waste Management Techniques*; Hussain, C.M., Kadeppagari, R.K., Eds.; Wiley: Hoboken, NJ, USA, 2022; pp. 33–46.

10. Bala, S.; Sharma, M.; Dashora, K.; Siddiqui, S.; Diwan, D.; Tripathi, M. Nanomaterials based sustainable bioenergy production systems: Current trends and future prospects. *Nanofabrication* **2022**, *7*, 314–324. [CrossRef]

11. Gupta, V.K.; Nguyen, Q.D.; Liu, S.; Taherzadeh, M.J.; Sirohi, R. Microbes in valorisation of biomass to value-added products (MVBVAP). *Bioresour. Technol.* **2022**, *347*, 126738. [CrossRef]

12. Pathak, N.; Singh, S.; Singh, P.; Singh, P.K.; Singh, R.; Bala, S.; Thirumalesh, B.V.; Gaur, R.; Tripathi, M. Valorization of jackfruit waste into value added products and their potential applications. *Front. Nutr.* **2022**, *9*, 1061098. [CrossRef]

13. Usmani, Z.; Sharma, M.; Diwan, D.; Tripathi, M.; Whale, E.; Jayakody, L.N.; Moreau, B.; Thakur, V.K.; Tuohy, M.; Gupta, V.K. Valorization of sugar beet pulp to value-added products: A review. *Bioresour. Technol.* **2022**, *346*, 126580. [CrossRef]

14. Kim, D.-Y.; Kadam, A.; Shinde, S.; Saratale, R.G.; Patra, J.K.; Ghodake, G. Recent developments in nanotechnology transforming the agricultural sector: A transition replete with opportunities. *J. Sci. Food Agric.* **2018**, *98*, 849–864. [CrossRef]

15. Pateiro, M.; Gómez, B.; Munekata, P.; Barba, F.; Putnik, P.; Kovačević, D.; Lorenzo, J. Nanoencapsulation of Promising Bioactive Compounds to Improve Their Absorption, Stability, Functionality and the Appearance of the Final Food Products. *Molecules* **2021**, *26*, 1547. [CrossRef]

16. Donner, M.; Gohier, R.; de Vries, H. A new circular business model typology for creating value from agro-waste. *Sci. Total. Environ.* **2020**, *716*, 137065. [CrossRef]

17. Verma, A.; Kumar, S.; Jain, P.K. Key pretreatment technologies on cellulosic ethanol production. *J. Sci. Res.* **2011**, *55*, 57–63.

18. Muhammad, S.; Khalil, H.P.S.A.; Hamid, S.A.; Albadn, Y.M.; Suriani, A.B.; Kamaruzzaman, S.; Mohamed, A.; Allaq, A.A.; Yahya, E.B. Insights into Agricultural-Waste-Based Nano-Activated Carbon Fabrication and Modifications for Wastewater Treatment Application. *Agriculture* **2022**, *12*, 1737. [CrossRef]

19. Popa, V.I. Biomass and Sustainability. In *Sustainability of Biomass through Bio-Based Chemistry*; CRC Press: Boca Raton, FL, USA, 2021; pp. 1–33.

20. Aravani, V.P.; Sun, H.; Yang, Z.; Liu, G.; Wang, W.; Anagnostopoulos, G.; Syriopoulos, G.; Charisiou, N.D.; Goula, M.A.; Kornaros, M.; et al. Agricultural and livestock sector's residues in Greece & China: Comparative qualitative and quantitative characterization for assessing their potential for biogas production. *Renew. Sustain. Energy Rev.* **2022**, *154*, 111821.

21. Sorathiya, L.M.; Fulsoundar, A.B.; Tyagi, K.K.; Patel, M.D.; Singh, R. Eco-friendly and modern methods of livestock waste recycling for enhancing farm profitability. *Int. J. Recycl. Org. Waste Agric.* **2014**, *3*, 1–7. [CrossRef]

22. Dauda, A.B.; Ajadi, A.; Tola-Fabunmi, A.S.; Akinwole, A.O. Waste production in aquaculture: Sources, components and managements in different culture systems. *Aquac. Fish.* **2019**, *4*, 81–88. [CrossRef]

23. Wang, R.; Wang, Q.; Dong, L.; Zhang, J. Cleaner agricultural production in drinking-water source areas for the control of non-point source pollution in China. *J. Environ. Manag.* **2021**, *285*, 112096. [CrossRef]

24. Dey, T.; Bhattacharjee, T.; Nag, P.; Ghati, A.; Kuila, A. Valorization of agro-waste into value added products for sustainable development. *Bioresour. Technol. Rep.* **2021**, *16*, 100834. [CrossRef]

25. de Souza, L.; Shivakumar, S. Conversion of Agro-industrial Wastes for the Manufacture of Bio-based Plastics. In *Bioplastics for Sustainable Development*; Springer: Singapore, 2021; pp. 177–204.

26. Saravanan, A.; Kumar, P.S.; Jeevanantham, S.; Karishma, S.; Vo, D.-V.N. Recent advances and sustainable development of biofuels production from lignocellulosic biomass. *Bioresour. Technol.* **2022**, *344*, 126203. [CrossRef] [PubMed]

27. Ndayisenga, F.; Yu, Z.; Zheng, J.; Wang, B.; Liang, H.; Phulpoto, I.A.; Habiyakare, T.; Zhou, D. Microbial electrohydrogenesis cell and dark fermentation integrated system enhances biohydrogen production from lignocellulosic agricultural wastes: Sub-strate pretreatment towards optimization. *Renew. Sustain. Energy Rev.* **2021**, *145*, 111078. [CrossRef]

28. Zhang, D.; Jiang, B.; Luo, Y.; Fu, X.; Kong, H.; Shan, Y.; Ding, S. Effects of ultrasonic and ozone pretreatment on the structural and functional properties of soluble dietary fiber from lemon peel. *J. Food Process. Eng.* **2022**, *45*, e13916. [CrossRef]

29. Periyasamy, S.; Karthik, V.; Kumar, P.S.; Isabel, J.B.; Temesgen, T.; Hunegnaw, B.M.; Melese, B.B.; Mohamed, B.A.; Vo, D.-V.N. Chemical, physical and biological methods to convert lignocellulosic waste into value-added products. A review. *Environ. Chem. Lett.* **2022**, *20*, 1129–1152. [CrossRef]

30. Indriani, D.W.; Susilo, B. The effect of microwave power on lignocellulose content, physical and chemical characteristics of biomass: A review. *IOP Conf. Ser. Earth Environ. Sci.* **2021**, *924*, 012069. [CrossRef]

31. Usmani, Z.; Sharma, M.; Tripathi, M.; Nizami, A.S.; Gong, L.; Nguyen, Q.D.; Reddy, M.S.; Thakur, V.K.; Gupta, V.K. Converting biowaste streams into energy–leveraging microwave assisted valorization technologies for enhanced conversion. *J. Energy Inst.* **2023**, *107*, 101161. [CrossRef]

32. Alcazar-Ruiz, A.; Ortiz, M.L.; Sanchez-Silva, L.; Dorado, F. Catalytic effect of alkali and alkaline earth metals on fast pyrolysis pre-treatment of agricultural waste. *Biofuels Bioprod. Biorefin.* **2021**, *15*, 1473–1484. [CrossRef]

33. Rasid, N.S.A.; Shamjuddin, A.; Amin, N.A.S. Chemical and Structural Changes of Ozonated Empty Fruit Bunch (EFB) in a Ribbon-Mixer Reactor. *Bull. Chem. React. Eng. Catal.* **2021**, *16*, 383–395. [CrossRef]

34. Kanrar, B.B.; Singh, S.; Pal, S.K.; Panda, D. Mild-temperature organosolv treatment of rice-straw: Extracting ability of dimethylformamide and material applications. *Int. J. Environ. Sci. Technol.* **2022**, 1–12. [CrossRef]

35. Khan, M.U.; Usman, M.; Ashraf, M.A.; Dutta, N.; Luo, G.; Zhang, S. A review of recent advancements in pretreatment tech-niques of lignocellulosic materials for biogas production: Opportunities and Limitations. *Chem. Eng. J. Adv.* **2022**, *10*, 100263. [CrossRef]

36. Zheng, Y.; Zhang, Q.; Zhang, Z.; Jing, Y.; Hu, J.; He, C.; Lu, C. A review on biological recycling in agricultural waste-based biohydrogen production: Recent developments. *Bioresour. Technol.* **2021**, *347*, 126595. [CrossRef]

37. Naik, G.P.; Poonia, A.K.; Chaudhari, P.K. Pretreatment of lignocellulosic agricultural waste for delignification, rapid hydrolysis, and enhanced biogas production: A review. *J. Indian Chem. Soc.* **2021**, *98*, 100147. [CrossRef]

38. Yin, X.; Wei, L.; Pan, X.; Liu, C.; Jiang, J.; Wang, K. The pretreatment of lignocelluloses with green solvent as biorefinery pre-process: A minor review. *Front. Plant Sci.* **2021**, *12*, 670061. [CrossRef]

39. Das, L.; Achinivu, E.C.; Barcelos, C.A.; Sundstrom, E.; Amer, B.; Baidoo, E.E.; Simmons, B.A.; Sun, N.; Gladden, J.M. Deconstruction of woody biomass via protic and aprotic ionic liquid pretreatment for ethanol production. *ACS Sustain. Chem. Eng.* **2021**, *9*, 4422–4432. [CrossRef]

40. Kumar, N.; Gautam, R.; Stallings, J.D.; Coty, G.G.; Lynam, J.G. Secondary Agriculture Residues Pretreatment Using Deep Eutectic Solvents. *Waste Biomass-Valoriz.* **2021**, *12*, 2259–2269. [CrossRef]

41. Rahmani, A.M.; Gahlot, P.; Moustakas, K.; Kazmi, A.A.; Ojha, C.S.P.; Tyagi, V.K. Pretreatment methods to enhance solubili-zation and anaerobic biodegradability of lignocellulosic biomass (wheat straw): Progress and challenges. *Fuel* **2022**, *319*, 123726. [CrossRef]

42. Prasad, B.R.; Padhi, R.K.; Ghosh, G. A review on key pretreatment approaches for lignocellulosic biomass to produce biofuel and value-added products. *Int. J. Environ. Sci. Technol.* **2022**, 1–16. [CrossRef]

43. Zhou, Z.; Ouyang, D.; Liu, D.; Zhao, X. Oxidative pretreatment of lignocellulosic biomass for enzymatic hydrolysis: Progress and challenges. *Bioresour. Technol.* **2022**, *367*, 128208. [CrossRef]

44. Igwo-Ezikpe, M.N.; Ayanshina, A.O.; Babalola, M. Optimization of Chemical Pre-Treatment of Cassava Bagasse for Reducing Sugar and Bioethanol Production. *Univ. Lagos J. Basic Med. Sci.* **2022**, *3*, 14–23.

45. Cheng, H.-H.; Whang, L.-M. Resource recovery from lignocellulosic wastes via biological technologies: Advancements and prospects. *Bioresour. Technol.* **2022**, *343*, 126097. [CrossRef]

46. Mudzakir, A.; Rizky, K.M.; Munawaroh, H.S.H.; Puspitasari, D. Oil palm empty fruit bunch waste pretreatment with benzotriazolium-based ionic liquids for cellulose conversion to glucose: Experiments with computational bibliometric analysis. *Indones. J. Sci. Technol.* **2022**, *7*, 291–310. [CrossRef]

47. Malolan, R.; Gopinath, K.P.; Vo, D.-V.N.; Jayaraman, R.S.; Adithya, S.; Ajay, P.S.; Arun, J. Green ionic liquids and deep eutectic solvents for desulphurization, denitrification, biomass, biodiesel, bioethanol and hydrogen fuels: A review. *Environ. Chem. Lett.* **2021**, *19*, 1001–1023. [CrossRef]

48. Zhang, Q.; Zhou, Y.; Yue, W.; Qin, W.; Dong, H.; Vasanthan, T. Nanostructures of protein-polysaccharide complexes or conjugates for encapsulation of bioactive compounds. *Trends Food Sci. Technol.* **2021**, *109*, 169–196. [CrossRef]

49. Guía-García, J.L.; Charles-Rodríguez, A.V.; Reyes-Valdés, M.H.; Ramírez-Godina, F.; Robledo-Olivo, A.; García-Osuna, H.T.; Cerqueira, M.A.; Flores-López, M.L. Micro and nanoencapsulation of bioactive compounds for agrifood applications: A re-view. *Ind. Crop. Prod.* **2022**, *186*, 115198. [CrossRef]

50. Santini, A.; Cammarata, S.M.; Capone, G.; Ianaro, A.; Tenore, G.C.; Pani, L.; Novellino, E. Nutraceuticals: Opening the debate for a regulatory framework. *Br. J. Clin. Pharmacol.* **2018**, *84*, 659–672. [CrossRef]

51. Banwo, K.; Olojede, A.O.; Adesulu-Dahunsi, A.T.; Verma, D.K.; Thakur, M.; Tripathy, S.; Singh, S.; Patel, A.R.; Gupta, A.K.; Aguilar, C.N.; et al. Functional importance of bioactive compounds of foods with Potential Health Benefits: A review on recent trends. *Food Biosci.* **2021**, *43*, 101320. [CrossRef]

52. Daliri, E.B.-M.; Lee, B.H. Current Trends and Future Perspectives on Functional Foods and Nutraceuticals. In *Beneficial Microorganisms in Food and Nutraceuticals*; Springer International Publishing: Cham, Switzerland, 2015; pp. 221–244.

53. Vilas-Boas, A.A.; Pintado, M.; Oliveira, A.L. Natural bioactive compounds from food waste: Toxity and safety concerns. *Foods* **2021**, *10*, 1564. [CrossRef]

54. Capanoglu, E.; Nemli, E.; Tomas-Barberan, F. Novel Approaches in the Valorization of Agricultural Wastes and Their Ap-plications. *J. Agric. Food Chem.* **2022**, *70*, 6787–6804. [CrossRef]

55. Singh, D.N.; Tripathi, M.; Singh, V.S.; Singh, R.; Gaur, R.; Pathak, N. Management of agriculture waste: Bioconversion of agro-waste into valued products. In *Bioremediation: Challenges and Advancements*; Tripathi, M., Singh, D.N., Eds.; Bentham Science Publisher: Singapore, 2022; pp. 225–253.

56. Siddiqua, A.; Hahladakis, J.N.; Al-Attiya, W.A.K.A. An overview of the environmental pollution and health effects associated with waste landfilling and open dumping. *Environ. Sci. Pollut. Res.* **2022**, *29*, 58514–58536. [CrossRef]

57. Ramnarain, Y.I.; Ansari, A.A.; Ori, L. Vermicomposting of different organic materials using the epigeic earthworm *Eisenia foetida*. *Int. J. Recycl. Org. Waste Agric.* **2022**, *8*, 23–36. [CrossRef]

58. Motamedi, A.; Jafarpour, M.; Oshaghi, M. Improving the vermicompost quality by using horticultural and agronomic residues. *J. Plant Nutr.* **2022**, *45*, 727–738. [CrossRef]

59. Roohi, M.; Arif, M.S.; Guillaume, T.; Yasmeen, T.; Riaz, M.; Shakoor, A.; Farooq, T.H.; Shahzad, S.M.; Bragazza, L. Role of fertilization regime on soil carbon sequestration and crop yield in a maize-cowpea intercropping system on low fertility soils. *Geoderma* **2022**, *428*, 116152. [CrossRef]

60. Das, D.; Deka, H. Vermicomposting of harvested waste biomass of potato crop employing *Eisenia fetida*: Changes in nutrient profile and assessment of the maturity of the end products. *Environ. Sci. Pollut. Res.* **2021**, *28*, 35717–35727. [CrossRef]

61. Giri, B.; Varma, A. (Eds.) *Soil Health*; Springer International Publishing: Cham, Switzerland, 2020. [CrossRef]

62. Sigurnjak, I.; Brienza, C.; Snauwaert, E.; De Dobbelaere, A.; De Mey, J.; Vaneeckhaute, C.; Michels, E.; Schoumans, O.; Adani, F.; Meers, E. Production and performance of bio-based mineral fertilizers from agricultural waste using ammonia (stripping-) scrubbing technology. *Waste Manag.* **2019**, *89*, 265–274. [CrossRef]

63. Diacono, M.; Persiani, A.; Testani, E.; Montemurro, F.; Ciaccia, C. Recycling Agricultural Wastes and By-products in Organic Farming: Biofertilizer Production, Yield Performance and Carbon Footprint Analysis. *Sustainability* **2019**, *11*, 3824. [CrossRef]

64. Lim, S.-F.; Matu, S.U. Utilization of agro-wastes to produce biofertilizer. *Int. J. Energy Environ. Eng.* **2015**, *6*, 31–35. [CrossRef]

65. Bian, B.; Hu, X.; Zhang, S.; Lv, C.; Yang, Z.; Yang, W.; Zhang, L. Pilot-scale composting of typical multiple agricultural wastes: Parameter optimization and mechanisms. *Bioresour. Technol.* **2019**, *287*, 121482. [CrossRef]

66. Passoth, V.; Sandgren, M. Biofuel production from straw hydrolysates: Current achievements and perspectives. *Appl. Microbiol. Biotechnol.* **2019**, *103*, 5105–5116. [CrossRef]

67. Almendro-Candel, M.B.; Gómez Lucas, I.; Navarro-Pedreño, J.; Zorpas, A.A. Physical Properties of Soils Affected by the Use of Agricultural Waste. *Agric. Waste Residues* **2018**, *2*, 9–27. [CrossRef]

68. Cheng, F.; Brewer, C. Conversion of protein-rich lignocellulosic wastes to bio-energy: Review and recommendations for hydrolysis + fermentation and anaerobic digestion. *Renew. Sustain. Energy Rev.* **2021**, *146*, 111167. [CrossRef]

69. Yong, K.J.; Wu, T.Y. Second-generation bioenergy from oilseed crop residues: Recent technologies, techno-economic assessments and policies. *Energy Convers. Manag.* **2022**, *267*, 115869. [CrossRef]

70. Karthick, C.; Nanthagopal, K. A comprehensive review on ecological approaches of waste to wealth strategies for production of sustainable biobutanol and its suitability in automotive applications. *Energy Convers. Manag.* **2021**, *239*, 114219. [CrossRef]

71. Kaur, N.; Kaur, A.; Sridhar, K.; Sharma, M.; Singh, T.P.; Kumar, S. Development and quality characteristics of functional *Kulfi* fortified with microencapsulated betalains. *Int. J. Food Sci. Technol.* **2021**, *56*, 5362–5370. [CrossRef]

72. Shirahigue, L.D.; Ceccato-Antonini, S.R. Agro-industrial wastes as sources of bioactive compounds for food and fermentation industries. *Ciênc. Rural. St. Maria* **2020**, *50*, 1–17. [CrossRef]

73. Asagbra, A.E.; Sanni, A.I.; Oyewole, O.B. Solid-state fermentation production of tetracycline by Streptomyces strains using some agricultural wastes as substrate. *World J. Microbiol. Biotechnol.* **2005**, *21*, 107–114. [CrossRef]

74. Vastrad, B.M.; Neelagund, S.E. Optimization and production of neomycin from different agro industrial wastes in solid state fermentation. *Int. J. Pharm. Sci. Drug Res.* **2011**, *3*, 104–111.

75. Ezejiofor, T.I.N.; Duru, C.I.; Asagbra, A.E.; Ezejiofor, A.N.; Orisakwe, O.E.; Afonne, J.O.; Obi, E. Waste to wealth: Production of oxytetracycline using Streptomyces species from household kitchen wastes of agricultural produce. *Afr. J. Biotechnol.* **2012**, *11*, 10115–10124.

76. Duda-Chodak, A.; Tarko, T. Antioxidant properties of different fruit seeds and peels. *Acta Sci. Pol. Technol. Aliment.* **2007**, *6*, 29–36.

77. Mourtzinos, I.; Goula, A. Polyphenols in agricultural byproducts and food waste. In *Polyphenols in Plants: Isolation, Purification and Extract Preparation*; Watson, R.R., Ed.; Academic Press: London, UK, 2019.

78. Jimenez-Lopez, C.; Fraga-Corral, M.; Carpena, M.; García-Oliveira, P.; Echave, J.; Pereira, A.G.; Lourenço-Lopes, C.; Prieto, M.A.; Simal-Gandara, J. Agriculture waste valorisation as a source of antioxidant phenolic compounds within a circular and sustainable bioeconomy. *Food Funct.* **2020**, *11*, 4853–4877. [CrossRef]

79. Fermoso, F.G.; Serrano, A.; Alonso-Fariñas, B.; Fernandez-Bolaños, J.; Borja, R.; Rodríguez-Gutiérrez, G. Valuable Compound Extraction, Anaerobic Digestion, and Composting: A Leading Biorefinery Approach for Agricultural Wastes. *J. Agric. Food Chem.* **2018**, *66*, 8451–8468. [CrossRef]

80. Joly, N.; Souidi, K.; Depraetere, D.; Wils, D.; Martin, P. Potato By-Products as a Source of Natural Chlorogenic Acids and Phenolic Compounds: Extraction, Characterization, and Antioxidant Capacity. *Molecules* **2021**, *26*, 177. [CrossRef]

81. Jiang, Y.; Lv, Y.; Wu, R.; Sui, Y.; Chen, C.; Xin, F.; Zhou, J.; Dong, W.; Jiang, M. Current status and perspectives on biobutanol production using lignocellulosic feedstocks. *Bioresour. Technol. Rep.* **2019**, *7*, 100245. [CrossRef]

82. Abdollahdokht, D.; Gao, Y.; Faramarz, S.; Poustforoosh, A.; Abbasi, M.; Asadikaram, G.; Nematollahi, M.H. Conventional agrochemicals towards nano-biopesticides: An overview on recent advances. *Chem. Biol. Technol. Agric.* **2022**, *9*, 1–19. [CrossRef]

83. Ghasemi, K.; Tasnim, S.; Mahmud, S. PCM, nano/microencapsulation and slurries: A review of fundamentals, categories, fabrication, numerical models and applications. *Sustain. Energy Technol. Assessments* **2022**, *52*, 102084. [CrossRef]

84. Das, C.A.; Kumar, V.G.; Dhas, T.S.; Karthick, V.; Kumar, C.V. Nanomaterials in anticancer applications and their mechanism of action—A review. *Nanomed. Nanotechnol. Biol. Med.* **2022**, *47*, 102613. [CrossRef]

85. Hosseini, H.; Jafari, S.M. Introducing nano/microencapsulated bioactive ingredients for extending the shelf-life of food prod-ucts. *Adv. Coll. Interface Sci.* **2020**, *282*, 102210. [CrossRef]

86. Alehosseini, E.; Jafari, S.M. Micro/nano-encapsulated phase change materials (PCMs) as emerging materials for the food in-dustry. *Trends Food Sci. Technol.* **2019**, *91*, 116–128. [CrossRef]

87. Chaudhari, A.K.; Singh, V.K.; Das, S.; Dubey, N.K. Nanoencapsulation of essential oils and their bioactive constituents: A novel strategy to control mycotoxin contamination in food system. *Food Chem. Toxicol.* **2021**, *149*, 112019. [CrossRef]

88. Comunian, T.; Babazadeh, A.; Rehman, A.; Shaddel, R.; Akbari-Alavijeh, S.; Boostani, S.; Jafari, S. Protection and controlled release of vitamin C by different micro/nanocarriers. *Crit. Rev. Food Sci. Nutr.* **2022**, *62*, 3301–3322. [CrossRef]

89. Puttasiddaiah, R.; Lakshminarayana, R.; Somashekar, N.L.; Gupta, V.K.; Inbaraj, B.S.; Usmani, Z.; Raghavendra, V.B.; Sridhar, K.; Sharma, M. Advances in Nanofabrication Technology for Nutraceuticals: New Insights and Future Trends. *Bioengineering* **2022**, *9*, 478. [CrossRef]

90. Das, A.; Nazni, P. Formulation and quality evaluation of (*Pennisetum glaucum* incorporated) value-added paneer by Response Surface Methodology. *Indian J. Dairy Sci.* **2021**, *74*, 131–137. [CrossRef]

91. Federici, F.; Fava, F.; Kalogerakis, N.; Mantzavinos, D. Valorisation of agro-industrial by-products, effluents and waste: Concept, opportunities and the case of olive mill wastewaters. *J. Chem. Technol. Biotechnol.* **2009**, *84*, 895–900. [CrossRef]

92. Gontard, N.; Sonesson, U.; Birkved, M.; Majone, M.; Bolzonella, D.; Celli, A.; Angellier-Coussy, H.; Jang, G.W.; Verniquet, A.; Broeze, J.; et al. A research challenge vision regarding management of agricultural waste in a circular bio-based econo-my. *Crit. Rev. Environ. Sci. Technol.* **2018**, *48*, 614–654. [CrossRef]

93. Sharma, V.; Tsai, M.-L.; Nargotra, P.; Chen, C.-W.; Kuo, C.-H.; Sun, P.-P.; Dong, C.-D. Agro-Industrial Food Waste as a Low-Cost Substrate for Sustainable Production of Industrial Enzymes: A Critical Review. *Catalysts* **2022**, *12*, 1373. [CrossRef]

MDPI

St. Alban-Anlage 66

4052 Basel

Switzerland

Tel. +41 61 683 77 34

Fax +41 61 302 89 18

www.mdpi.com

Bioengineering Editorial Office

E-mail: bioengineering@mdpi.com

www.mdpi.com/journal/bioengineering